Lecture Notes in Networks and Systems

Volume 264

The series "Lecture Notes in Networks and Systems" publishes the latest developments in Networks and Systems—quickly, informally and with high quality. Original research reported in proceedings and post-proceedings represents the core of LNNS.

Volumes published in LNNS embrace all aspects and subfields of, as well as new challenges in, Networks and Systems.

The series contains proceedings and edited volumes in systems and networks, spanning the areas of Cyber-Physical Systems, Autonomous Systems, Sensor Networks, Control Systems, Energy Systems, Automotive Systems, Biological Systems, Vehicular Networking and Connected Vehicles, Aerospace Systems, Automation, Manufacturing, Smart Grids, Nonlinear Systems, Power Systems, Robotics, Social Systems, Economic Systems and other. Of particular value to both the contributors and the readership are the short publication timeframe and the world-wide distribution and exposure which enable both a wide and rapid dissemination of research output.

The series covers the theory, applications, and perspectives on the state of the art and future developments relevant to systems and networks, decision making, control, complex processes and related areas, as embedded in the fields of interdisciplinary and applied sciences, engineering, computer science, physics, economics, social, and life sciences, as well as the paradigms and methodologies behind them.

Indexed by SCOPUS, INSPEC, WTI Frankfurt eG, zbMATH, SCImago.

All books published in the series are submitted for consideration in Web of Science.

More information about this series at http://www.springer.com/series/15179

Julia L. Wright · Daniel Barber ·
Sofia Scataglini · Sudhakar L. Rajulu
Editors

Advances in Simulation and Digital Human Modeling

Proceedings of the AHFE 2021 Virtual Conferences on Human Factors and Simulation, and Digital Human Modeling and Applied Optimization, July 25–29, 2021, USA

Springer

Editors
Julia L. Wright
U.S. Army Research Laboratory
Orlando, FL, USA

Daniel Barber
Soar Tech
Orlando, FL, USA

Sofia Scataglini
Department of Product Development
University of Antwerp
Antwerp, Belgium

Sudhakar L. Rajulu
Biomedical Research and Environmental
NASA Lyndon Johnson Space Center
Houston, TX, USA

ISSN 2367-3370 ISSN 2367-3389 (electronic)
Lecture Notes in Networks and Systems
ISBN 978-3-030-79762-1 ISBN 978-3-030-79763-8 (eBook)
https://doi.org/10.1007/978-3-030-79763-8

This Springer imprint is published by the registered company Springer Nature Switzerland AG
The registered company address is: Gewerbestrasse 11, 6330 Cham, Switzerland

Advances in Human Factors and Ergonomics 2021

AHFE 2021 Series Editors

Tareq Z. Ahram, Florida, USA
Waldemar Karwowski, Florida, USA

12th International Conference on Applied Human Factors and Ergonomics and the Affiliated Conferences (AHFE 2021)

Proceedings of the AHFE 2021 Virtual Conferences on Human Factors and Simulation, and Digital Human Modeling and Applied Optimization, July 25–29, 2021, USA.

Advances in Neuroergonomics and Cognitive Engineering	Hasan Ayaz, Umer Asgher and Lucas Paletta
Advances in Industrial Design	Cliff Sungsoo Shin, Giuseppe Di Bucchianico, Shuichi Fukuda, Yong-Gyun Ghim, Gianni Montagna and Cristina Carvalho
Advances in Ergonomics in Design	Francisco Rebelo
Advances in Safety Management and Human Performance	Pedro M. Arezes and Ronald L. Boring
Advances in Human Factors and Ergonomics in Healthcare and Medical Devices	Jay Kalra, Nancy J. Lightner and Redha Taiar
Advances in Simulation and Digital Human Modeling	Julia L. Wright, Daniel Barber, Sofia Scataglin and Sudhakar L. Rajulu
Advances in Human Factors and System Interactions	Isabel L. Nunes
Advances in the Human Side of Service Engineering	Christine Leitner, Walter Ganz, Debra Satterfield and Clara Bassano
Advances in Human Factors, Business Management and Leadership	Jussi Ilari Kantola, Salman Nazir and Vesa Salminen
Advances in Human Factors in Robots, Unmanned Systems and Cybersecurity	Matteo Zallio, Carlos Raymundo Ibañez and Jesus Hechavarria Hernandez
Advances in Human Factors in Training, Education, and Learning Sciences	Salman Nazir, Tareq Z. Ahram and Waldemar Karwowski

(continued)

(continued)

Advances in Human Aspects of Transportation	Neville Stanton
Advances in Artificial Intelligence, Software and Systems Engineering	Tareq Z. Ahram, Waldemar Karwowski and Jay Kalra
Advances in Human Factors in Architecture, Sustainable Urban Planning and Infrastructure	Jerzy Charytonowicz, Alicja Maciejko and Christianne S. Falcão
Advances in Physical, Social & Occupational Ergonomics	Ravindra S. Goonetilleke, Shuping Xiong, Henrijs Kalkis, Zenija Roja, Waldemar Karwowski and Atsuo Murata
Advances in Manufacturing, Production Management and Process Control	Stefan Trzcielinski, Beata Mrugalska, Waldemar Karwowski, Emilio Rossi and Massimo Di Nicolantonio
Advances in Usability, User Experience, Wearable and Assistive Technology	Tareq Z. Ahram and Christianne S. Falcão
Advances in Creativity, Innovation, Entrepreneurship and Communication of Design	Evangelos Markopoulos, Ravindra S. Goonetilleke, Amic G. Ho and Yan Luximon
Advances in Human Dynamics for the Development of Contemporary Societies	Daniel Raposo, Nuno Martins and Daniel Brandão

Preface

This volume is a compilation of cutting-edge research regarding how simulation and modeling supports human factors. The compilation of chapters is the result of efforts by the International Conference on Applied Human Factors and Ergonomics (AHFE), which provides the organization for several affiliated conferences. Specifically, the chapters herein represent the International Conference on Human Factors and Simulation and the International Conference on Digital Human Modeling and Applied Optimization.

Simulation is a technology that supports an approximation of real-world scenes and scenarios for a user. For example, a cockpit simulator represents the configuration of the inside of a cockpit and presents a sensory and motor experience to mimic flight. Simulations advance research by providing similar experiences to those scenarios that would otherwise be impractical to carry out in the real world for such reasons as monetary cost or safety concerns. Simulations can support numerous goals including training or practice on established skills.

Modeling is a somewhat different tool than simulation, though the two are often used interchangeably as they both imply estimation of real-world scenes or scenarios that bypass practical concerns. The difference in the context of this book is that modeling is not intended to provide a user with an experience, but rather to represent anything pertinent about the real world in computational algorithms, possibly including people and their psychological processing. Modeling may answer questions about large-scale scenarios that would be difficult to address otherwise, such as the effects of economic interventions or smaller-scale scenarios such as the cognitive processing required to perform a task that is otherwise undetectable by measurement devices.

The goal of the research herein is to bring awareness and attention to advances that human factors specialists may make in their field to address the design of programs of research, systems, policies and devices. This book provides a plethora of avenues for human factors research that may be helped by simulation and modeling.

The book is divided into the following nine sections:

Human Factors in Simulation

1. Modeling and Simulation for Human-Human and Human-Agent Teaming
2. Visualizing Team Trust and Cohesion
3. Computational Modeling Approaches in Politics and Economics
4. Virtual and Augmented Reality
5. Simulation and Human Factors for Nuclear Control Rooms
6. Data Essential for Human Factors Simulation Modeling
7. Human Sensing in Simulation

Digital Human Modeling

8. Digital Human Modeling and Applied Optimization
9. Digital Human Modeling by Women in Human Factors

Sections 1–7 cover topics in simulation and modeling while Sections 8–9 cover topics in Digital Human Modeling and Applied Optimization.

All papers in this book were either reviewed or contributed by the members of editorial board. For this, we would like to recognize the board members listed below:

H. Kim, USA
D. Regazzoni, Italy
A. Upmann, Germany
X. Wang, France
S. Wischniewski, Germany

This book covers diverse topics in simulation and modeling. I hope this book is informative and helpful for the researchers and practitioners in developing better products, services and systems.

July 2021

Julia Wright
Daniel Barber
Sofia Scataglini
Sudhakar Rajulu

Contents

Digital Human Modeling by Women in Human Factors

Modeling and Simulation for Human-Human and Human-Agent Teaming

Human-Autonomy Teaming with Learning Capable Agents: Performance and Workload Outcomes

Julia L. Wright[1](✉), Justin Lee[2], and Jacquelyn A. Schreck[2]

[1] US Army Research Laboratory, Orlando, FL, USA
Julia.l.wright8.civ@mail.mil
[2] University of Central Florida Institute for Simulation and Training, Orlando, FL, USA
{jlee,jschreck}@ist.ucf.edu

Abstract. Each of twenty participants teamed with a learning capable agent to conduct a threat classification task. The agent's reasoning and learning transparency varied across four scenarios. Access to agent reasoning transparency improved task performance as assessed by percent of correct classifications. Agent learning transparency of inferred knowledge improved task response time and reduced cognitive workload. However, when the human was burdened with directly teaching the agent, task completion time and perceived workload increased dramatically, while satisfaction in task performance decreased. These findings indicate that when teamed with learning capable agents, human performance and workload are best supported when the autonomy can derive its needed information with minimal human input.

Keywords: Human factors · Human-autonomy teaming · AI/ML · Human-robot interaction · Workload

1 Introduction

Soldiers collaborating with intelligent machine agents (e.g., robots, algorithmic planners) to conduct multiple tasks is quickly becoming a reality. Advances in artificial intelligence (AI) and machine learning (ML) continue to increase not only the abilities of these agents to conduct tasks autonomously, but also to conduct tasks of ever-increasing complexity and difficulty [1]. Learning capable agents (LCAs) are able to adapt to new information, updating their goals and reasoning without the knowledge of their human counterpart. The LCAs are not static; they dynamically learn from increasing amounts of data and experience. Thus, when human-agent teams (HATs) include LCAs, within-team communications become crucial to shared awareness and team effectiveness. The goal of this study is to examine how these advances in AI/ML task complexity and learning will affect the human teammate, specifically, how transparency of agent reasoning and agent learning affects human-agent team performance.

J. L. Wright et al. (Eds.): AHFE 2021, LNNS 264, pp. 3–10, 2021.
https://doi.org/10.1007/978-3-030-79763-8_1

2 Background

2.1 AI/ML

Advances in AI/ML have given rise to LCAs, agents that will adapt their behavior over time as their models of the world are updated by events encountered or additional data with which to update their algorithms. While the effects of agent transparency on operator performance have been studied with agents that are incapable of independently updating their goals and/or reasoning, little is known about outcomes related to human-agent teaming with a LCA; particularly how transparency of a LCA's reasoning and learning will affect the human operator.

Learning capable agents use their experiences and data to update the preprogrammed models that direct their actions and goals. One example of LCAs, image recognition agents, use camera images to understand their surroundings and inform their actions. These agents are trained for real-world tasks via presentation of a series of images prior to deployment (offline learning). During training, images are labelled and categorized, either *a priori* by the agent or post hoc with human assistance. However, it is not possible to generate labelled images of all potential threats and situations *a priori*, hence, the need for the agent to be able to continue learning after deployment [2].

There are several potential solutions for effective continual training for autonomous systems. One is that learning could occur after deployment by having the agent review the targets that the agent missed but the human identified, essentially repeating offline learning focusing on mistakes made in the real world. Another is real-time learning, where the agent infers meaning from human input and updates its programming, may create mis-interpretations initially that would reduce as more information is processed. Alternatively, on-the-job training, wherein the human teaches the agent, could be used to improve an agent's learning and performance while the team is conducting tasks in the field. There is utility to the agent being able to learn real-time after deployment, as this has the potential to create a more agile, responsive battle force in addition to reducing the time required for additional offline learning. In this study, the simulated agent will use information supplied by the human during missions to update its knowledge base offline (i.e., learn). Transparency of the agent's learning process will be manipulated by making it clear to the human how the information from human input is used (learning transparency: implicit, explicit, human-directed) in order to examine the effect on the human teammate and team performance.

2.2 Transparency

Knowledge of the agent's actions, its reasoning behind those actions, and the expected outcome of those actions can help the human teammate, improving task performance, SA, and reducing complacent behaviors, while not increasing their perceived workload [3, 4, 6]. Knowing what a machine agent is uncertain about can also improve human-agent team performance [5, 6]. Without knowledge of the agent's reasoning process, human operators may not know when the agent's planning is suboptimal for the mission's context. In this study, transparency of the agent's reasoning will be manipulated by the presence of icons displaying the five factors the agent evaluates when classifying a person as a potential threat or not (reasoning transparency: opaque, transparent) in order to examine the effect on the human teammate and team performance.

2.3 Workload

Mental workload has been described as "the relation between the function relating the mental resources demanded by a task and those resources available to be supplied by the human operator" [7]. Transparency of agent reasoning has been shown to improve task performance without increasing cognitive workload [4, 6], as such, its effect could be considered freeing resources so that more are available to accommodate task demands [8]. However, the influence of agent learning transparency on cognitive workload has yet to be examined. To that end, subjective workload measures will be used to inform as to the effect of learning transparency.

3 Current Study

In the current experiment, participants jointly conducted a threat classification task with a simulated learning-capable Aided Target Recognition (AiTR) agent while traversing a simulated urban environment. The agent identified persons that appeared along their route and classified each as either a potential threat or not. The agent either shared its reasoning for the classification in the robot dialogue window on the interface (transparent) or it did not (opaque). At times, the agent recognized a human or object, but was unable to classify it. When that occurred, the agent identified these potential threats and the participant would indicate yes/threat, no/nonthreat. The agent would share how the participant classification would be used to update its algorithm (explicit learning) or it would not (implicit learning), or the participant would specify what the missing information should be (human-directed learning). It is hypothesized that understanding how the agent used information supplied by the participant would result in improvements on the threat classification task and reduced cognitive workload [1, 9, 10], however when the participant had to direct learning it would increase workload.

Based on the review of the literature, the following outcomes are expected:

1. When agent reasoning is transparent, compared to opaque reasoning, there will be:

    a. Increased correct classifications on the threat classification task
    b. Increased time required to classify threats.
    c. Reduced participant perceived workload

2. When agent learning is explicit, compared to implicit learning, there will be:

    a. Increased correct classifications on the threat classification task
    b. Increased time required to classify threats
    c. Reducсd participant perceived workload

3. When the human assists in agent learning, compared to explicit learning, there will be:

a. Reduced correct classifications on the threat classification task
b. Increased time required to classify threats
c. Increased participant perceived workload

4 Methods

4.1 Participants

Twenty participants ($Min_{age} = 18$ years, $Max_{age} = 28$ years, $M_{age} = 21.7$ years) received $15/h compensation.

4.2 Equipment/Simulator

A custom software application was developed in the UnReal 4 engine. The simulation was delivered via a commercial desktop computer system, two 15″ monitors, standard keyboard, and three-button mouse.

4.3 Study Design

This 2 (reasoning transparency; opaque vs. transparent) ×3 (learning transparency; implicit, explicit, human-directed) fractional factorial within-subjects study assesses the effects of transparency of agent learning (TAL) on a threat classification task. Since reasoning transparency and learning transparency are conveyed using the same icons, it is not possible to have transparent learning with opaque reasoning (see Table 1). In the baseline condition, Opaque Reasoning/Implicit Learning (TAL-1), the agent conducts the task without sharing its reasoning for unclassified persons. In the Transparent Reasoning/Implicit Learning (TAL-2) condition, the agent shares its reasoning but not indicate how the participant's input is used in updating its underlying reasoning; in the Transparent Reasoning/Explicit Learning (TAL-3) condition, the agent shares its reasoning and backfills reasoning information based upon participant input to indicate what it inferred from the human's input. Finally, in the Human-Directed Learning (TAL-4) condition, the agent shares its reasoning, and the participant completes any missing reasoning information before identifying whether the target is a potential threat. Participants completed four scenarios, one in each transparency combination condition.

Table 1. Reasoning by Learning Transparency condition matrix.

		Learning		
		Implicit	Explicit	Human-Directed
Reasoning	Opaque	TAL-1	--	---
	Transparent	TAL-2	TAL-3	TAL-4

4.4 Dependent Measures

Dependent measures are task performance (evaluated using number of correct classifications and response time), and cognitive workload (assessed using the NASA-TLX raw scores).

4.5 Procedure

Upon arrival, participants signed the informed consent form, and then completed the demographics questionnaire and a brief Ishihara Color Vision Test [11]. Training lasted approximately 30 min, and consisted of self-paced PowerPoint slides, several assessments and a training scenario. Participants were required to achieve at least 80% proficiency on the assessments to continue.

The study session began immediately after the training session and lasted approximately 2.5 h. Participants completed four scenarios, one in each transparency condition assigned to one of four potential routes. The scenarios were presented in order of increasing transparency (i.e., TAL-1, TAL-2, TAL-3, and TAL-4); however, the route assignment was counterbalanced and randomized. After each scenario, participants completed a NASA-TLX questionnaire assessing their cognitive workload. Following completion of all scenarios, participants were debriefed and dismissed.

5 Results

Data were analyzed using repeated-measures ANOVAs ($\alpha = .05$) and planned comparisons (TAL-1 to TAL-2, TAL-2 to TAL-3, TAL-3 to TAL-4). Means and Standard Deviation (SD) are shown in Table 2.

Table 2. Means and Standard Deviation for Performance and NASA-TLX factors.

		TAL-1		TAL-2		TAL-3		TAL-4	
		M	SD	*M*	SD	*M*	SD	*M*	SD
Performance Measures	Correct Class %	91.4	4.7	93.8	3.8	94.4	3.1	93.7	3.8
	Response Time	11.5	4.2	9.9	2.6	8.3	1.4	21.2	19.0
NASA-TLX scores	MD	69.3	26.7	59.3	25.6	55.0	24.7	67.3	24.0
	TD	61.0	26.5	56.0	24.1	51.0	27.6	70.5	19.0
	PD	34.8	27.4	32.8	27.1	33.8	26.4	37.0	28.8
	Effort	65.8	15.2	54.0	24.3	51.8	21.4	66.8	21.0
	Frustration	38.5	29.1	32.3	25.1	29.8	23.6	38.3	26.7
	Perf. Satisfaction	32.5	21.4	33.5	22.0	26.8	17.9	41.0	23.8
	Overall	50.5	16.4	44.6	17.6	41.5	16.8	53.5	16.9

Note: MD - Mental Demand, TD - Temporal Demand, PD - Physical Demand

5.1 Correct Classifications

Agent transparency had a marginally significant effect on correct classifications, Wilkes' $\Lambda = 0.658$, $F(3,17) = 2.94$, $p = 0.063$. Examination using planned comparisons show that while agent reasoning transparency had a significant effect on correct classifications (Cohens $d_s = 0.56$), learning transparency did not.

5.2 Response Times

Both reasoning and learning transparency affected the participants response time, Wilkes' $\Lambda = 0.463$, $F(3,17) = 6.56$, $p = 0.004$. Time to respond decreased as transparency increased; it was longer in TAL-1 than in TAL-2 (Cohens $d_s = 0.44$), and longer in TAL-2 than in TAL-3 (Cohens $d_s = 0.78$). However, time to respond increased dramatically from TAL-3 to the Human-directed Learning condition TAL-4 (Cohens $d_s = 0.96$) See Fig. 1.

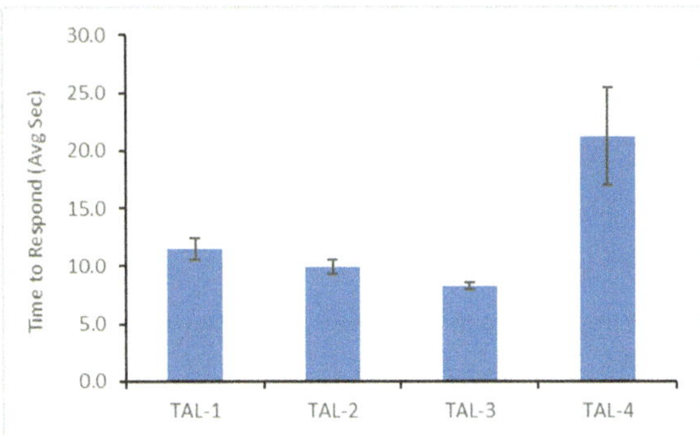

Fig. 1. Response Times in average seconds by transparency condition.

5.3 Workload

Overall, transparency level did have a significant effect on participant workload, Wilkes' $\Lambda = 0.451$, $F(3,17) = 6.91$, $p = 0.003$ (see Fig. 2). Access to agent reasoning (TAL-1 to TAL-2) reduced Mental Demand scores, (Cohens $d_s = 0.38$; $F(1, 19) = 8.54$, $p = 0.009$) and Effort scores (Cohens $d_s = 0.58$; $F(1, 19) = 7.08$, $p = 0.015$). Access to learning transparency (TAL-2 to TAL-3) did not affect most workload factors, except for an increase in performance satisfaction (as indicated by lower scores, Cohens $d_s = 0.34$; $F(1, 19) = 8.60$, $p = 0.009$).

Human-Directed Learning (TAL-4) increased workload and decreased performance satisfaction, as compared to Explicit Learning Transparency (TAL-3). Mental Demand (Cohens $d_s = 0.50$; $F(1, 19) = 11.47$, $p = 0.003$), Temporal Demand (Cohens $d_s = 0.82$; $F(1, 19) = 15.72$, $p = 0.001$), Effort (Cohens $d_s = 0.71$; $F(1, 19) = 29.48$, $p < 0.001$), and Performance Satisfaction (Cohens $d_s = 0.68$; $F(1, 19) = 19.47$, $p < 0.001$) scores all increased dramatically in TAL-4 as compared to TAL-3.

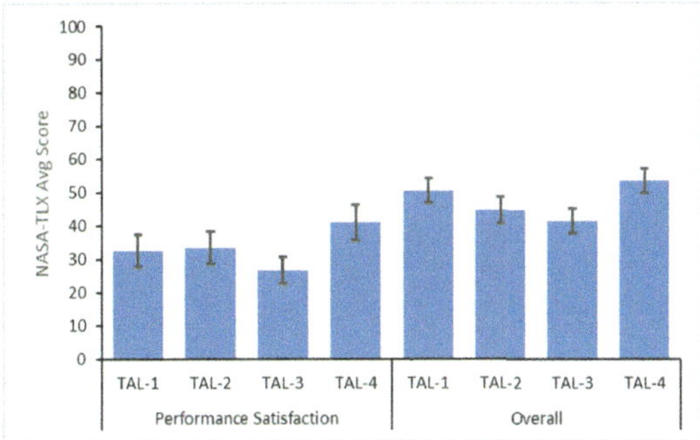

Fig. 2. Performance Satisfaction and Overall NASA-TLX scores.

6 Discussion

The goal of this study was to examine how transparency of agent reasoning and agent learning interact to influence the human's task performance. Transparency of agent reasoning is well known to support improved task performance in human-agent teaming without concomitant increases in perceived workload [5, 6, 10], and this study adds more evidence to support that position. What is still unknown is how transparency of agent learning affects the human-agent team's task performance.

This study examined a scenario that is very likely to become a reality in the near future – a human teaming with a learning capable agent that collects information from the human and uses that input to update its algorithms later. There was no difference in correct classifications due to learning transparency. There are several potential explanations for this result. First, the task itself was not difficult; each encountered person was visible and unobstructed for a minimum of 6 s and snapshots were provided for later review. Second, the agent classified most persons prior to the participant classifying, so it would be easy for the participant to agree with the agent's classification. However, correct classifications were expected to drop off in the human-directed learning condition because of the increase in task demand, so the consistency in performance is surprising. Reviewing the associated response times, it appears there was a speed-accuracy trade-off [13] in this condition.

Increased learning transparency reduced task completion time and perceived workload in the explicit learning condition, demonstrating that understanding how the agent is using the humans input can be beneficial. However, burdening the human with directly teaching the agent resulted in dramatic increases in task completion times and perceived workload, with a concomitant decrease in task performance satisfaction. An increase in response time was expected; when reasoning was missing, the participant had to supply the needed information. However, the amount of increase was surprising, as response times more than doubled those in the explicit learning condition. Clearly, there is more happening in this condition than an extra mouse click or two. These findings indicate that when teamed with learning capable agents, human performance and workload are best supported when the autonomy can derive its needed information with minimal human direction.

Acknowledgment. The authors wish to thank Jonathan Coole for his input and contributions to this project. The views and conclusions contained in this paper are those of the authors and should not be interpreted as presenting the official policies or position, either expressed or implied, of the U.S. Army Research Laboratory or the U.S. Government.

References

1. David, R.A., Nielsen, P.: Defense science board summer study on autonomy. Defense Science Board Washington United States (2016)
2. Kim, D., Oh, S.M., Rehg, J.M.: Traversability classification for UGV navigation: A comparison of patch and superpixel representations. In: IEEE/RSJ International Conference on Intelligent Robots and Systems 2007, pp. 3166–3173. IEEE (2007)
3. Kilgore, R., Voshell, M.: Increasing the transparency of unmanned systems: Applications of ecological interface design. In: International Conference on Virtual, Augmented and Mixed Reality 2014 pp. 378–389. Springer, Cham.
4. Wright, J.L., Chen, J.Y., Barnes, M.J., Hancock, .PA.: Agent reasoning transparency: the influence of information level on automation induced complacency. US Army Research Laboratory Aberdeen Proving Ground United States (2017)
5. Chen, J.Y., Procci, K., Boyce, M., Wright, J., Garcia A, Barnes M.: Situation awareness-based agent transparency (ARL-TR-6905). Aberdeen Proving Ground, MD: US. Army Research Laboratory (2014)
6. Mercado, J.E., Rupp, M.A., Chen, J.Y., Barnes, M.J., Barber, D., Procci, K.: Intelligent agent transparency in human–agent teaming for Multi-UxV management. Hum. Fact. **58**(3), 401–415 (2016)
7. Parasuraman, R., Sheridan, T.B., Wickens, C.D.: Situation awareness, mental workload, and trust in automation: Viable, empirically supported cognitive engineering constructs. J. Cog Eng and Decision Making. **2**(2), 140–160 (2008)
8. Wickens, C.D.: Multiple resources and mental workload. H factors. **50**(3), 449–455 (2008)
9. Morrow, P.B., Fiore, SM.: Supporting human-robot teams in social dynamicism: An overview of the metaphoric inference framework. In: Proceedings Human Factors and Ergonomics Society, vol. 56,1, pp. 1718–1722. SAGE Publications, Los Angeles (2012)
10. Chen, J.Y., Lakhmani, S.G., Stowers, K., Selkowitz, A.R., Wright, J.L., Barnes, M.: Situation awareness-based agent transparency and human-autonomy teaming effectiveness. Theo Iss Ergo Sc. **19**(3), 259–282 (2018)
11. Ishihara, S.: Tests for color-blindness. Handaya, Hongo Harukicho (1917)
12. Lyons, J.B., Havig, P.R.: Transparency in a human-machine context: approaches for fostering shared awareness/intent. In International Conference on Virtual, Augmented and Mixed Reality, pp. 181–190. Springer, Cham (2014)
13. Wickelgren, W.A.: Speed-accuracy tradeoff and information processing dynamics. Acta Physiol. (Oxf) **41**(1), 67–85 (1977)

Examining Vigilance in a Simulated Unmanned Aircraft System (UAS) Monitoring Task

Maria Chaparro(✉) and Meredith Carroll

Florida Institute of Technology, 150 W University Boulevard, Melbourne, FL, USA
mchaparro2016@my.fit.edu

Abstract. Tasks which require individuals to sustain attention to seldomly occurring stimuli for long periods of time can induce workload, stress, and often lead to the vigilance decrement [4, 10]. Research into potential methods to mitigate the effects of these tasks have proposed training, video game experience, and the potential of using music. This study examined the impact of listening to music during performance of a simulated UAS monitoring task on the vigilance decrement using twenty-eight participants from a southeastern private Florida university. Music was not found to relieve the vigilance decrement, however, interesting results were found related to gaming experience and workload. Findings were consistent with increased workload and mental demand for participants with gaming experience. These findings are discussed in the context of operator selection.

Keywords: Workload · Sustained attention · Mental demand · Operator selection

1 Introduction

Several modern operational tasks, such as Unmanned Aerial Systems (UAS) monitoring tasks, require sustained attention, as the operator must focus on a task for long periods of time [5, 10]. Tasks which require sustained attention, such as a UAS pilot's monitoring task, are a prime example of a typical vigilance task, as they require individuals to detect signals that are seldom and challenging to identify [10]. Studies have revealed that people are not able to sustain the same level of attention for long periods of time without a decrease in attention [8, 15, 22]. This decrease in an individual's capacity to sustain attention at an effective level typically surfaces after approximately thirty minutes and is referred to as the vigilance decrement [23]. High stress and workload are commonly cited as a byproduct of sustained attention tasks and can cause personnel retention issues [7, 9, 10, 28]. Increased workload has been found to be due to the pressure of staying attentive without a high task load [10, 31]. Given ongoing issues with recruitment and retention for the U.S. Air Force's UAS workforce [3] understanding how to ameliorate negative byproducts of vigilance is important.

Research has shown that there are individual difference factors that influence the ability to sustain attention. Specifically, video game players (VGP) have been found to have improved cognitive functions that are important to vigilance tasks, such as spatial ability, attention allocation skills, and rapid target identification [1, 9, 24]. However, it

J. L. Wright et al. (Eds.): AHFE 2021, LNNS 264, pp. 11–18, 2021.
https://doi.org/10.1007/978-3-030-79763-8_2

is important to note that the lack of excitement and interactivity during a vigilance task, may yield differences in workload and mental demand for VGPs as, compared to non-gaming populations [18, 26, 29]. Studies looking into methods to combat the vigilance decrement have proposed the inclusion of music during task performance to increase arousal [12, 19]. Positive valence music that participants like has been found to lead to positive performance in the form of decreased misses in a sustained attention to response task [2]. Further, several studies have found support for music increasing arousal [5, 20, 32]. Per and Zhang [24] examined the effects of music on a simulated strategy game and found that music had a significant benefit to proactive control while those who played the simulated strategy game without music had significant improvements to reactive control. A study by Navarro et al., [20] found that music led to increased arousal and better driving in a driving simulator. Studies have also found that music can lead to distraction and increased reaction times due to arousal or relaxation caused by the music [6, 19]. Given the rise in human performance tasks that require sustained attention (e.g., UAS inspections and surveillance), identifying factors which may overcome the vigilance decrement is important, including obtaining more conclusive evidence regarding the use of music to achieve this. This study examined the impact of listening to music during performance of a simulated sustained attention UAS task, on target detection performance, stress and workload. Additionally, the effect of individual differences related to prior UAS experience and gaming experience were examined. This paper describes the methods, results and implications of the experimental findings.

2 Methods

Participants. Twenty-eight participants from a southeastern private Florida university performed a simulated UAS monitoring task. Participants were recruited using a University list service. Inclusion criteria included normal color vision. Participation was voluntary and the study was approved by a university Institutional Review Board. Participants were entered into a raffle to win a $100 gift card; some received extra credit.

Experimental Design. The study was a between groups, randomized posttest only control group experimental design. The independent variable was presence of music during task performance (present vs. not). There were three dependent variables: performance, stress, and workload. Participants were assigned to groups using a latin squares method. It was hypothesized that those who received music during task performance would have (a) better performance, (b) lower workload levels, and (c) lower stress levels compared to those who performed the task with no music. Additionally, an exploratory analysis was conducted to examine the impact of UAS and gaming experience.

Music. A playlist of nine songs was created on Spotify and played for participants in the music group. The music selected followed criteria for positive-valence, slow-tempo songs, which have been found to promote vigilance and performance [2]. Valence is defined here as the emotion emitted or perceived by the song. Positive valence was measured using the Positive and Negative Affect scale (PANAS) [30]. A group of four students varying in age and gender rated 40 songs for their affect level. Songs which scored in the top quadrant for valence were assessed for slow tempo (i.e., in the range of 50 to 100 beats per minute (bpm). Songs not in the 50 to 100 bpm, based on the bpm calculation software, GetSongbpm, were removed from the playlist.

Measures. Target detection was measured as either a "hit" (i.e., correct detection) or a miss (i.e., target not detected). Each correct detection input was recorded in the software output log with a timestamp, allowing for the calculation of reaction time, which was calculated as the difference between the time in which the UAS became visible on the screen and when the participant hit the space bar. A post-experiment survey was administered after the study via Qualtrics and included demographic information such as gender, age, academic level, gaming experience and UAS experience. Workload and its subscales of mental demand, physical demand, temporal demand, performance, effort, and frustration level were also measured post-experiment using the NASA-TLX short, a 6-item, 20-point scale, from low to high.

Additionally, despite attempts to equalize target difficulty, a pilot study using four participant's revealed variability across target detection and reaction time as well as target difficulty ratings for all targets. Therefore, a difficulty scale was included in the study to allow difficulty to be taken into account during analyses. To rate difficulty, after task performance, participants re-watched the points in the simulation where the UAS came onto the screen then rated the difficulty of each target on a 5-point scale (1 = very easy and 5 = very difficult). Additionally, participants in the music group were administered the PANAS, a validated measure of evoked positive emotion, after testing to rate the song playlist for positive affect [11].

Procedure. When participants first arrived, they reviewed the consent form followed by a pre-survey. Once completed, they were given details about how to perform the simulated task. After the scenario and instructions were read aloud, participants were presented with a piece of paper showing an example of the Heads Up Display (HUD) information they would be seeing on the screen. The illustration showed where the information needed to perform the task was located, including the altitude of the UAS, and what the target UAS looked like. Participants in both the music and non-music conditions were then fitted with MDRZX110NC SONY headphones which had 13 dB of noise suppression. At this time, participant's in the music condition were told they would be listening to music during the task and set a comfortable music volume as a song played. After selecting the volume, they were then asked to remove the headphones until the simulation commenced. Both groups were given time to ask questions and instructions were repeated. The participant then performed a 31-min sustained attention task, in which they monitored a video presenting a landscape from a simulated UAS first-person view. Two inputs were required from the participant, to hit space bar when seeing the UAS and type the number 1 into a document located on a separate laptop, when the UAS rose above an altitude of 100 AGL. The simulated task followed parameters set by Daly et al., [7] as closely as possible, on how to develop a laboratory vigilance task for a UAS environment. Therefore, the target (i.e., a covert UAS) was presented seven times within the span of the 31-min simulation of a barren junkyard that could be circled by the UAS in approximately five minutes, ensuring that the changing environment did not appear novel. Knowledge of results feedback was incorporated through a window that appeared stating: "Target identified!" if space bar was selected and the UAS was on the screen.

If the UAS was not in sight and the space bar was hit a prompt displaying: "Target not in sight!" appeared. After the simulation ended, participants were asked to complete the post survey and re-watch the areas in the simulation where the target was present to rate each target's difficulty.

3 Results

Preliminary Analysis. Three participants were excluded from the analysis, as two were unable to complete the study due to time constraints, and the third participant had a personal issue which required the study to be stopped. This resulted in 25 participants' data being included in the analysis. Participant's target difficulty ratings were examined using a 2 x 7 (music x target) repeated-measures ANOVA. Results of repeated-measures revealed a significant main effect of target on difficulty ratings, (F (6, 18) = 21.633, p > .001, partial $\eta^2 = .878$. There was no significant main effect of music on difficulty ratings, (F (1, 23) = .041, p > .05, partial $\eta^2 = .054$. There was also not a significant interaction between target and music for difficulty ratings, F (6, 18) = 1.367, p > .05, partial $\eta^2 = .313$. These results reveal that, despite our efforts to ensure consistent difficulty, there was a significant variation in difficulty across targets, and this was consistent across both treatment conditions. As such, the variability in target difficulty resulted in a confounding variable. To account for this, the decision was made to focus the repeated measures analyses on the last four targets as they all had target difficulty greater than two and a half and all took place after approximately 15 min, when the vigilance decrement has the potential to commence.

Primary Analysis. A 2 x 4 (music presence x target) repeated-measures MANOVA was conducted on measures of target detection and reaction time, to examine the impact of music on performance. At the multivariate level, there was not a significant main effect of music presence, $F(2, 22) = 1.276$, p > .05, partial $\eta^2 = .104$. However, within subjects, there was a significant main effect of target, $F(2, 22) = 41.218$, p < .001, partial $\eta^2 = .932$. There was also a significant interaction between music presence and target, $F(2, 22) = .2.651$, p = .05, partial $\eta^2 = .469$. Univariate results revealed that there was a significant main effect of target on detection, F (1, 23) = 6.526, p = .001, $\eta^2 = \cdot 221$ and reaction time $F(1, 23) = 19.288$, p < 001, $\eta^2 = .456$. There was not a significant effect of music on detection $F(1, 23) = .270$, p > .05, $\eta^2 = .012$ or reaction time, $F(1, 23) = 2.064$, p > .05, $\eta^2 = .082$. There was also not a significant interaction between target and music on detection, $F(1, 23) = 1.543$, $\eta^2 = .063$ or reaction time, $F(1, 23) = .926$, $\eta^2 = .039$.

An ANCOVA examining the influence of music on workload with a covariate of gaming experience was conducted. There was not a significant interaction between music presence and gaming experience, F(3, 21) = 1.983, p = .1474. There was, however, a significant effect of gaming experience on workload, F(2, 2) = 4.80, p = .03. To examine whether music that evoked emotions could have had a relationship with workload, correlational analysis of affect scores, overall workload, and the workload subscales was conducted. A positive significant relationship was found for positive affect and mental demand, r = .54, p = .05.

To examine the potential impact of individual difference factors, a correlation analysis was conducted including: gaming experience, UAS experience, workload and each workload subscales. Gaming experience was significantly and positively correlated with the mental demand subscale of the NASA TLX, $r = .60$, $p < .001$. Gaming experience was significantly and positively correlated with the effort subscale of the NASA TLX used to measure stress, $r = .53$, $p < .01$. Gaming experience was significantly and positively correlated with overall workload measured by the NASA TLX, $r = .46$, $p < .05$. Results revealed that as gaming experience increased, mental demand, stress, and overall workload also increased. However, no significant correlation was found for UAS experience, physical demand, temporal demand, or frustration. In order to examine the effects of gaming, t-tests with unequal variances were used to compare the means with respect to the variables [13]. A two samples t-test assuming unequal variances was conducted comparing the mental demand of those with low gaming experience to those with high gaming experience. Mental demand was significantly different between these groups, $t(22) = 4.35$, $p = .0002$. A two samples t-test assuming unequal variances was conducted comparing the effort of those with low to those with high gaming experience. Effort was significantly different between these groups, $t(20) = 3.26$, $p = .004$. When examining the means, those with high gaming experience scored the task on average 4.1 points higher on the effort domain than those with low gaming experience. A two samples t-test assuming unequal variances was conducted comparing the workload of those with low to those with high gaming experience. Workload was significantly different between these groups, $t(16) = 2.390$, $p = .0294$.

4 Discussion

This study aimed to address whether music could be a viable mitigation technique to prevent the vigilance decrement. The vigilance decrement is characterized by an increase in reaction time, over time, and a decrease in detection [8, 14]. No significant difference in reaction time between music and non-music groups was found. Due to the similarity in the reaction time trends for both the music and non-music groups, it appears that the addition of music did not affect reaction time. However, the variability in difficulty across targets could be to blame. Target difficulty for the simulated vigilance task should have been equivalent across targets to facilitate the identification of the treatment effect [7]. The significant variation in difficulty may have washed out any treatment effects or the increase in difficulty in the last 15 min (i.e., target 4 and on) may have prevented a vigilance decrement altogether. Difficulty has been found to lead to stimulation [17, 25]. A study by Thomson et al., [27] found that varying task difficulty can alleviate performance degradation in the vigilance task. Furthermore, the simulation may not have been long enough to create a vigilance decrement [5, 8], as other simulated UAS tasks measuring the effects of performance have been longer [5, 7].

No significant difference in target detection between music and non-music groups was found in the analyses when attempting to account for difficulty by examining detection in the last four targets. Interestingly, the music group had a higher detection trend until the sixth target where there was a large decrease in detection, however, detection increased again after the sixth target. It is not clear why there is such a discrepancy.

A potential reason could be the song being played before target 6 may have been too arousing, to the point of distraction. Previous studies have found that music can lead to distraction in vigilance scenarios such as driving [6, 21]. A study by North and Hargreaves [21] found that arousing music can compete with cognitive processing and lead to performance decrements. Music may aide in detection as long as difficulty is not too high, at which point it becomes a distraction. Additionally, when examining the music affect scores, the higher the evoked positive emotion the higher the mental demand scores. This may have played a role. Some of the music may have led to increased mental demand rather than alleviating the mental demand, thus muddying the waters.

Both groups had perfect detection for the final target. The trend for the participants in the non-music group seemed to stay constant until the final target. The perfect detection ($M = 1.0$) of the final target may been due partially to a ceiling effect, which may have led to inaccuracy in the data. However, another reason that the average rating did not fall under "very easy", is due to knowledge of the study coming to an end, leading to an increase in arousal. Malhotra [16] found that when a challenge is nearing its end participant's arousal levels sometimes increase as they attempt to finish strong.

There were some interesting findings with respect to some of the individual difference variables. When examining the relationship between music and mental demand with a covariate of gaming, it appears that gaming did not make any difference in the amount of mental demand experienced between groups. However, when looking solely at gaming experience, it did have an effect on workload. Those who had high gaming experience self-reported higher workload and mental demand than those with low gaming experience. These findings could be helpful to narrowing down good candidates for UAS operators. Although performance was not significantly different, studies have found that high workload is negatively correlated with retention and job satisfaction [33]. Those with high gaming experience may potentially be more likely to leave the job than those who have low gaming experience. Additionally, given that retention is an issue for certain UAS domains [3], finding ways to select candidates which would potentially pose less retention issues would beneficial.

5 Conclusion

Although the findings with respect to music alleviating the vigilance decrement were inconclusive, there were several limitations of the study that could have prevented conclusive results emerging. First, the variability in difficulty could have led to arousal, preventing the vigilance decrement, the ceiling effect, and leading to the variability seen in reaction time for certain targets. Second, the novelty of the simulation may have prevented the vigilance decrement from appearing. Future research should attempt to ensure that difficulty is held constant across targets and extend the length of the study to counteract the novelty effect. Interesting results were found with respect to gaming experience, workload, effort, and mental demand. Results indicated that gaming experience is positively correlated to workload, mental demand and stress, therefore indicating those with high gaming experience may also experience higher workload, mental demand, and stress. This may not be ideal for tasks requiring sustained attention. These findings could have implications for job selection parameters for roles which require sustained attention, such as, UAS operations and air traffic control, leading to more optimal employee selection with greater retention. More research should be conducted with alternative populations such as with a larger sample and a UAS task that is more closely indicative of the respective population's task. As modern operational tasks increase, the domain's that utilize them and the components of each task may vary, yielding differing levels of workload and mental demand. Understanding what can be done to offset the negative effects these tasks produce can aide in a more seamless and safe transition.

References

1. Achtman, R.L., Green, C.S., Bavelier, D.: Video games as a tool to train visual skills. Restorative Neurol. Neurosci. **26**(4, 5), 435–446 (2008)
2. Baldwin, C.L., Lewis, B.A.: Positive valence music restores executive control over sustained attention. PloS One **12**(11), e0186231 (2017)
3. Cummings, M.L.: Of shadows and white scarves. C4ISR J. **7**, 17–18 (2008)
4. Cummings, M.L., Gao, F., Thornburg, K.M.: Boredom in the workplace: a new look at an old problem. Hum. Fact. **58**(2), 279–300 (2016)
5. Cummings, M.L., Mastracchio, C., Thornburg, K.M., Mkrtchyan, A.: Boredom and distraction in multiple unmanned vehicle supervisory control. Interact. Comput. **25**(1), 34–47 (2013)
6. Dalton, B.H., Behm, D.G.: Effects of noise and music on human and task performance: a systematic review. Occup. Ergon. **7**(3), 143–152 (2007)
7. Daly, T., et al.: Moving vigilance out of the laboratory: dynamic scenarios for UAS operator vigilance training. In: Schmorrow, D.D., Fidopiastis, C.M. (eds.) AC 2017. LNCS (LNAI), vol. 10285, pp. 20–35. Springer, Cham (2017). https://doi.org/10.1007/978-3-319-58625-0_2
8. Davies, D.R., Parasuraman, R.: The Psychology of Vigilance. Academic Press (1982)
9. Dye, M.W., Green, C.S., Bavelier, D.: The development of attention skills in action video game players. Neuropsychologia **47**(8–9), 1780–1789 (2009)
10. Eastwood, J.D., Frischen, A., Fenske, M.J., Smilek, D.: The unengaged mind: defining boredom in terms of attention. Perspect. Psychol. Sci. **7**(5), 482–495 (2012)
11. Fitzgerald, C.: Perception of, subjective feeling and physiological responses to emotion in music according to affective state (2013)
12. Fontaine, C.W., Schwalm, N.D.: Effects of familiarity of music on vigilant performance. Percept. Mot. Skills **49**(1), 71–74 (1979)
13. Gravetter, F.J., Wallnau, L.B.: Statistics for the Behavioral Sciences, 7th edn. Thomson Wadsworth (2007)
14. Helton, W.S., Russell, P.N.: Rest is best: the role of rest and task interruptions on vigilance. Cognition **134**, 165–173 (2015)
15. Mackworth, J.F.: Performance decrement in vigilance, threshold, and high-speed perceptual motor tasks. Can. J. Exp. Psychol. **18**, 209 (1964)
16. Malhotra, D.: The desire to win: the effects of competitive arousal on motivation and behavior. Organ. Behav. Hum. Decis. Process. **111**(2), 139–146 (2010)
17. Manly, T., Robertson, I.H., Galloway, M., Hawkins, K.: The absent mind: further investigations of sustained attention to response. Neuropsychologia **37**(6), 661–670 (1999)
18. McKinley, R.A., McIntire, L.K., Funke, M.A.: Operator selection for unmanned aerial systems: comparing video game players and pilots. Aviat. Space Environ. Med. **82**(6), 635–642 (2011)
19. Mori, F., Naghsh, F.A., Tezuka, T.: The effect of music on the level of mental concentration and its temporal change. In: CSEDU (1), pp. 34–42 (2014)

20. Navarro, J., Osiurak, F., Reynaud, E.: Does the tempo of music impact human behavior behind the wheel? Hum. Factors **60**(4), 556–574 (2018)
21. North, A.C., Hargreaves, D.J.: Music and driving game performance. Scand. J. Psychol. **40**(4), 285–292 (1999)
22. Oken, B.S., Salinsky, M.C., Elsas, S.M.: Vigilance, alertness, or sustained attention: physiological basis and measurement. Clin. Neurophysiol. **117**(9), 1885–1901 (2006)
23. Parasuraman, R.: Memory load and event rate control sensitivity decrements in sustained attention. Science **205**(4409), 924–927 (1979)
24. Per, X., Zhang, J.: The influence of strategy video game and its background music on cognitive control. EC Psychol. Psychiatry **2**(1), 15–25 (2017)
25. Robertson, I.H., Manly, T., Andrade, J., Baddeley, B.T., Yiend, J.: Oops!': performance correlates of everyday attentional failures in traumatic brain injured and normal subjects. Neuropsychologia **35**(6), 747–758 (1997)
26. Schmidt, T.N., Teo, G.W., Hancock, G.M., Amicarelle, Z., Szalma, J.L., Hancock, P.A.: Action video game players and vigilance performance. In: Proceedings of the Human Factors and Ergonomics Society Annual Meeting, vol. 57, no. 1, pp. 1450–1454. SAGE Publications, Sage, September 2013
27. Thomson, D.R., Smilek, D., Besner, D.: Reducing the vigilance decrement: The effects of perceptual variability. Conscious. Cogn. **33**, 386–397 (2015)
28. Tiwari, T., Singh, A.L., Singh, I.L.: Task demand and workload: effects on vigilance performance and stress. J. Indian Acad. Appl. Psychol. **35**(2) (2009)
29. Trisolini, D.C., Petilli, M.A., Daini, R.: Is action video gaming related to sustained attention of adolescents? Q. J. Exp. Psychol. **71**(5), 1033–1039 (2018)
30. Watson, D., Clark, L.A., Tellegen, A.: Development and validation of brief measures of positive and negative affect: the PANAS scales. J. Pers. Soc. Psychol. **54**(6), 1063 (1988)
31. Weinger, M.B.: Vigilance, boredom, and sleepiness. J. Clin. Monit. Comput. **15**(7–8), 549–552 (1999)
32. Wokoun, W.: Vigilance with background music. Human Engineering Lab Aberdeen Proving Ground Md (1963)
33. Zeytinoglu, I.U., Denton, M.: Satisfied workers, retained workers: effects of work and work environment on homecare workers' job satisfaction, stress, physical health, and retention. Research Institute for Quantitative Studies in Economics and Population, McMaster University, Ottawa (2006)

Quantifying Survivability via Measurement of Bodily Exposure During Simulated Combat Engagements

Stephanie A. T. Brown(✉), Clifford L. Hancock, and K. Blake Mitchell

Combat Capabilities Development Command Soldier Center, Natick, MA 01760, USA
{stephanie.a.brown88.civ,clifford.l.hancock4.civ,
katherine.b.mitchell.civ}@mail.mil

Abstract. Protective equipment is designed to help shield against enemy fire, however, such cumbersome gear increases user-burden resulting in performance degradations during combat operations. This paper outlines the development and application of an automated approach to quantify bodily exposure to enemy threat in order to assess the impact of human factors interventions (i.e., protective equipment, tactical training, etc.) on survivability during simulated combat engagements. An image processing algorithm application is presented here, utilizing military personnel ($N = 20$) who ran an integrated combat obstacle course in three equipment configurations with varying encumbrance levels. Compared with baseline, encumbered equipment configurations led to increased exposure duration and cumulative area exposed across time, resulting in an effective implementation of the methodology for quantifying exposure. Quantification of approach behavior and protective cover during combat will result in better understanding of physical tradeoffs of protective equipment.

Keywords: Exposure · Personal protective equipment · Clothing and individual equipment · Test and evaluation methodology · Training · Marksmanship · Military · Survivability · Lethality · Human factors · Human systems integration

1 Introduction

Military Warfighters are at risk of serious and incapacitating injuries due to the nature of combat operations in hostile environments. Typically, injury prevention discipline starts with understanding, predicting and preventing the occurrence. This can be done through interventions that either persuade risk behavior change (e.g., what we do, how we move, etc.), regulate behavior change through law, or provide automatic protection through product and environmental design (e.g., what we wear, where we work, etc.) [1]. In the case of the military, the job requirements include a significant level of inherent risk, so the primary means of injury reduction is through mandated personal protective equipment (PPE) and training. However, it is important for all stakeholders to understand how the protective clothing and individual equipment (CIE) affects a Warfighter's tactical movements both physically and behaviorally. The physical vulnerability tradeoff

J. L. Wright et al. (Eds.): AHFE 2021, LNNS 264, pp. 19–27, 2021.
https://doi.org/10.1007/978-3-030-79763-8_3

between mobility, lethality and survivability has been long debated during the development of military CIE, such as body armor. A recent review of the impact of load carriage on mobility concluded that as load increases, mobility decreases resulting in reduced survivability due to prolonged exposure to enemy fire [2]. Additional bulk and weight may provide improved physical protection against lethal rounds, but reduces the ability for a Warfighter to move quickly and tactically.

Warfighters are trained to move in specific strategic manners, making maximum use of available cover when acquiring hasty firing positions during enemy attack, to be protected and to return fire [3, 4]. Typically, during both product evaluation and training, the effects of CIE's physical encumbrance on general mobility and lethality is assessed objectively via combat obstacle course completion times [5–7], road marches [8], and marksmanship courses of fire [9, 10]. However, assessment of the ability of the Warfighter to achieve tactical positions to promote survivability are primarily evaluated through observation, largely due to lack of a standardized method to capture and quantify such behavioral data. Previous attempts to characterize exposure vulnerability focused solely on exposure duration during combat movement, rather than during enemy engagements [11, 12]. Many other studies have quantified the effects of load carriage on combat mobility [13, 14], but have not related the results to increased risk of injury from exposure to enemy fire. Furthermore, no research has been conducted to quantify cumulative bodily exposure while under cover during combat engagements.

This current research aims to develop an automated method to quantify the effects of equipment on the ability to achieve protective cover during enemy fire. In an effort to advance human factors assessments, this method will provide additional metrics that highlight equipment effects on tactical movements. It is hypothesized that this method will be able to capture and quantify differences in bodily exposure duration and area during shooting engagements as impacted by equipment configurations. Thereby, the method provides more meaningful output measures to support data-driven decisions.

2 Methods

2.1 Measures and Materials

A GoPro Hero 5 Black (GoPro, San Mateo, CA) was attached to a target facing towards the obstacles, in order to capture the participant's exposed body from the vantage point of the enemy combatant. The camera recorded 4k video (30 fps) of the participants engaging targets while traversing obstacles.

Calculated metrics consisted of duration, area, and integrals (cumulative exposure). Duration metrics included duration between the exposure (appearance of participant in camera's frame) and the initial shot, duration between the exposure and the final shot, and duration between the initial and final shot measured in seconds (s). While exposed body area was calculated for every frame of the exposure in order to calculate cumulative exposure area across those frames, area metrics were specifically calculated at the initial and final shot frames. Area (pixels^2) was calculated by summing the number of pixels containing exposed parts of the body and equipment on each frame. The area was also converted to cm^2 using a known distance within a single video frame. The cumulative exposure area metrics, in both units of pixels^2*s and cm^2*s, were estimated using the trapezoidal rule (approximating the region under the curve with trapezoids) of the area curve over identical time intervals as the duration metrics.

2.2 Participants

Twenty active-duty male Warfighters ranging in age from 18 to 35 years ($M \pm SD =$ 24.9 ± 5.11 yrs) volunteered to participate in the field demonstration. The participants were primarily infantry Soldiers (90%) with associated qualifications and training in dismounted tactics. Their years in Army service ranged from less than a year to thirteen (4.5 ± 3.48 yrs), with five (25%) experienced in a combat deployment.

2.3 Test Configurations

All participants executed the field demonstration test scenario in three CIE encumbrance configurations: baseline, minimal or min-encumbrance, and maximal or max-encumbrance. The baseline (unloaded) included a duty uniform, helmet, and a demilitarized rifle. In addition to the unloaded components, the min-encumbrance configuration included a body armor system, with baseline kit for the rifleman duty position (i.e., magazines, grenades, water, and first aid kit), approximately 22 kg. The max-encumbrance configuration included the components of the min-encumbrance and additional grenadier duty position items (i.e., 40 mm grenades), with a total weight of 26 kg. This configuration had the maximum bulk around the torso of 164 cm waist circumference (76 cm greater than baseline, 16 cm greater than min-encumbrance) [15].

3 Procedures

3.1 Laboratory Validation

After creation of the new processing algorithm technique, a validation test was conducted in a controlled laboratory setting to compare the new technique's output against actual known measurements and the previously used method. This test consisted of a data collection on 26 video segments utilizing an International Practical Shooting Confederation (IPSC) target of known area, varying exposed area systematically. For this test, a standard IPSC target was utilized (45-cm by 75-cm), simulating a human sized stimulus for the algorithm to detect. A window was created (50-cm by 50-cm), with black coloration in the foreground and a dark orange background to simplify the detection of the white target by the algorithm. The target was translated within the horizontal and vertical planes of the window in a randomized pattern, using increments of 2-cms. An unbiased assistant who had no knowledge of the actual measurements was employed to collect data utilizing the old methods in real-time. To measure exposure duration, the assistant used a standard stop watch, manually depressed the button at the perceived exposure initiation and stopped the time at the sound of the shot engagement. To measure exposure area, the assistant took visual observations utilizing two measuring sticks attached to the sides of the window, with lines at known increments of 5-cm, similar to a previously used training method [16]. Since previous methods did not utilize area of exposure as a measure, a simple area calculation was derived from the horizontal and vertical distances observed. Videos were post-processed utilizing the new automated technique. Cumulative exposure across time was not measured in the validation since it could not be achieved manually.

3.2 Field Demonstration

A repeated measures test design was used to evaluate the algorithm's ability to detect exposure differences across varying CIE encumbrance levels, where the participants ran through the scenario one time in each test configuration (three times total, quasi-randomized with a counterbalance). The full scenario included a variety of shooting tasks integrated within a combat obstacle course [5]. The low-window and high wall obstacle engagement points were utilized for this methodology assessment (Fig. 1). A set of threatening targets were placed 5-m beyond the obstacles, and could only be seen while traversing the structures. The participants were instructed to maintain tactical discipline in their movements and accuracy during threat engagement, ensuring that proper protection and cover was taken when detecting and identifying a target that simulated a threat. Following engagement of the threat, participants would continue to the next obstacle in the course. Training sessions occurred one-day prior, providing instruction on tactical movements during approach and shooting.

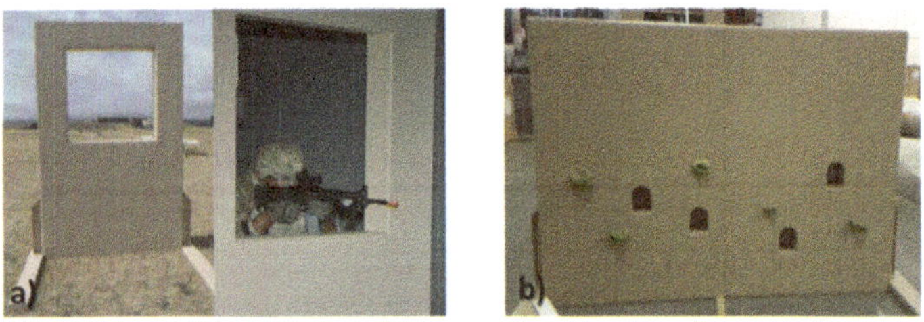

Fig. 1. A window and wall obstacle were used in the field demonstration. The window (a) consisted of a 1.5 m × 3 m × 0.2 m wall with a 0.9 m × 0. 9 m cutout situated 1.2 m from the ground. The wall (b) consisted of a 2.4 m × 1.8 m × 0.45 m obstacle with 9 toe holds to aid in traversal.

4 Data Processing and Analysis

4.1 Image Processing Algorithms

Videos from each trial were analyzed with custom algorithms written in Matlab R2018a (Mathworks, Natick, MA). The goal was to isolate the participant from the foreground and background, and calculate metrics relating to survivability attributes.

Prior to processing the video files, camera-specific parameters were first corrected for lens distortion in order to ensure accurate exposed area estimates. Next, a baseline frame containing the obstacle and empty backdrop was selected. All subsequent frames were compared to the baseline frame, and if differences beyond certain thresholds were detected between the two frames, the regions containing differences were considered exposed body areas. Several major steps were taken in an attempt to isolate the participant within each frame and disregard background noise (Fig. 2):

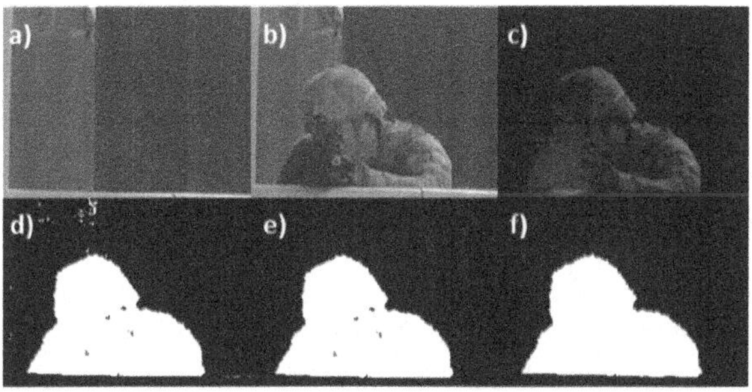

Fig. 2. Exposure area video processing showing: a) baseline frame; b) exposed participant; c) absolute difference between a) and b); d) filtering and converting to grayscale for pixels above a specified luminance threshold; e) removing pixel areas that are less than specified size threshold; and f) final image after filling holes.

1. The baseline frame was identified.
2. For each subsequent frame, the absolute difference between that frame and the baseline frame was calculated.For each subsequent frame, the absolute difference between that frame and the baseline frame was calculated.
3. 2-D median image filtering was applied to reduce noise and preserve edges.
4. The color image was converted into a binary black and white image by making all pixels with luminance greater than a specified value white (value determined trial by trial, from 0.03 to 0.05 in our application) and the remainder black.
5. Connected white pixel areas that were less than a cumulative size of 100-300 pixels were removed to reduce background noise (value determined trial by trial).
6. Holes in the white regions of the binary image were filled.

Shot timings were automatically identified by filtering the audio signal and determining when a specified threshold was met. The audio signal, sampled at 48 kHz, was filtered with a second-order band-pass Butterworth filter with low- and high-pass cut-off frequencies of 13 kHz and 15 kHz, respectively. Subsequently, the signal was rectified and filtered again to create a linear envelope using a second-order low-pass Butterworth filter with a cut-off frequency of 25 Hz. A threshold value of 8E-4, a value 3 standard deviations above the mean of example relatively quiet rectified baseline audio, was selected as the threshold value for shot detection. Shot timing was detected when the audio signal's linear envelope was increasing through this threshold.

4.2 Statistical Analyses

Field demonstration data were analyzed using repeated measures analysis of variance (ANOVA) tests, comparing three equipment configurations (i.e., baseline, min-, max-encumbrance) and two obstacles (i.e., window, wall). Post hoc analysis utilized Tukey tests. Confidence intervals were set at 95% (alpha = .05). Data is presented as $M \pm SD$.

5 Results

5.1 Laboratory Validation

Multivariate correlations were utilized to compare the new automated technique to the old manual technique and actual known measurements. For both exposure duration and area, all combinations were highly correlated, but the automated technique outperformed the manual (*Duration:* auto - known: $r(25) = .9994, p < .0001$; manual – known: $r(25) = .9959, p < .0001$; *Area:* auto - known: $r(25) = .9996, p < .0001$; manual – known: $r(25) = .9308, p < .0001$). Simple linear regression was computed to investigate whether the automated technique predicted the actual measurements. For exposure duration, the new technique (2.24 $\pm$ .43 s) significantly predicted the actual measurements (2.26 $\pm$ .43 s), $F(1, 25) = 20252.91, p < .0001$, adj. $R^2 = 1.00$. Analysis of exposure area showed that the new technique (545.25 $\pm$ .244.62 cm^2) also significantly predicted the actuals (543 $\pm$ 249.45 cm^2), $F(1, 25) = 27246.94$, $p < .0001$, adj. $R^2 = 1.00$.

5.2 Field Demonstration

Exposure Duration. Time from initial exposure to first shot showed a main effect of encumbrance configuration, $F(1, 38.62) = 6.59, p = .003$, and obstacle type, $F(1, 18.89) = 70.64, p < .001$, and an interaction between the two, $F(1, 38.88) = 6.51, p = .004$. Post hoc analysis for the window indicated that baseline (2.9 $\pm$ 1.2 s) and min-encumbrance (3.6 $\pm$ 1.5 s) were statistically different, where min-encumbrance had significantly longer exposure duration times (all other combinations, $p > .05$). For the wall obstacle, baseline (7.6 $\pm$ 3.9 s) and max-encumbrance (11.9 $\pm$ 4.9 s) were statistically different, where max-encumbrance had significantly longer exposure duration times (all other combinations, $p > .05$) (Fig. 3). Similar results were found when analyzing duration of exposure from initial exposure up until the final shot. Exposure from first to last shot showed no main effects of CIE configuration ($p > .05$).

Exposure Area. Analysis of the area of the body exposed to the enemy threat at time of initial shot revealed only a main effect of obstacle type, $F(1,19.46) = 8.57$, $p = .009$, where the wall (1572 $\pm$ 1483 cm^2) had significantly greater exposure area than the window (848 $\pm$ 341 cm^2). Analysis at time of final shot showed similar results.

Cumulative Exposure. Analysis of cumulative bodily exposure area up to initial shot showed a main effect of encumbrance configuration, $F(1, 38.88) = 4.88$, $p = .02$, and obstacle type, $F(1, 19.24) = 18.30$, $p < .001$, and an interaction between encumbrance

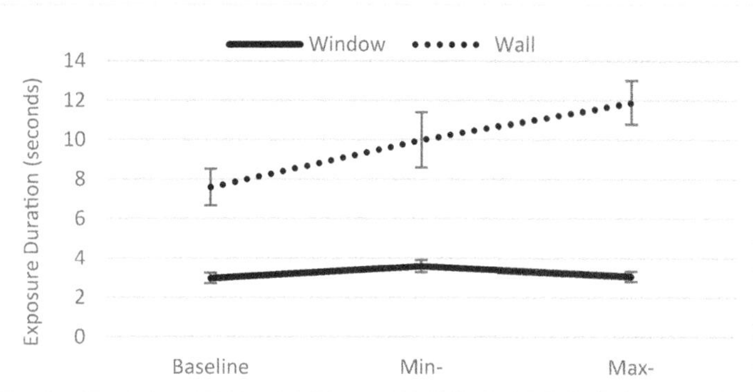

Fig. 3. Exposure duration from initial exposure to first shot across CIE level and obstacle type.

and obstacle which was trending towards significance, $F(1, 39.02) = 2.93$, $p = .06$ (Fig. 4). Post hoc analysis for the window obstacle indicated that baseline (1876 $\pm$ 1004 cm^2*s) was significantly different than min-encumbrance (2245 $\pm$ 989 cm^2*s) and max-encumbrance (2243 $\pm$ 1155 cm^2*s) ($p < .05$ for each). However, min- and max-encumbrance configurations were not significantly different. For the wall obstacle, baseline (4955 $\pm$ 3643 cm^2*s) and max-encumbrance (8687 $\pm$ 8622 cm^2*s) were statistically different, where max-encumbrance had significantly higher cumulative exposure ($p < .01$) (all other combinations, $p > .05$). Cumulative exposure through the final shot of the engagement found only a main effect of obstacle type.

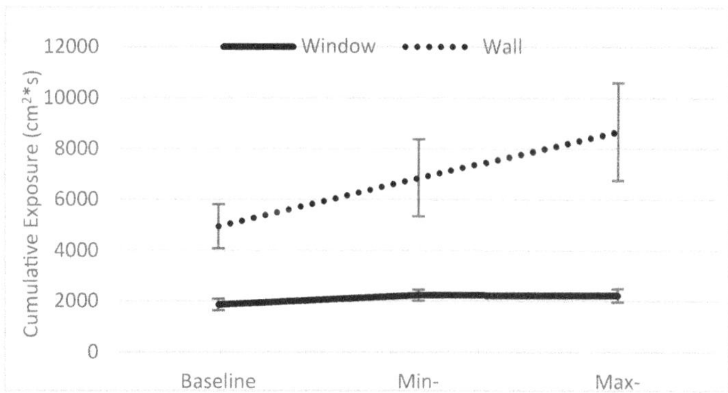

Fig. 4. Cumulative exposure from initial exposure to first shot across CIE level and obstacle type.

6 Discussion

The current methodology was able to detect differences in exposure during enemy engagements across encumbrance levels which would have otherwise been difficult to objectively quantify. The new methodology can capture exposure at a variety of cover locations and approach styles (i.e., side of a wall, through a window, peaking over a place of cover, etc.). This method provides a more comprehensive look at the impacts that CIE or other potential factors (i.e., combat experience, advanced training, etc.) can have on protective behavior and bodily exposure during combat engagements.

Differences in exposure area were not statistically different across configuration. This is not surprising for the window obstacle due to the bulk differences between the kits falling below the window opening, thus not affecting the approach or exposure amount during engagement. At the wall obstacle, a variety of approaches were taken to traverse the obstacle while engaging the target which resulted in a large amount of variability. Differences in exposure duration were found between configurations from initial exposure to first and final shot, but not between the shots. This matches our hypothesis that the increased load and bulk of the CIE would reduce mobility and range of motion, thus translating to increased time to approach, sight, and identify threats, resulting in extended exposure durations during initial stages of the engagement.

Some areas of improvement for future development should be considered. The tarp backdrop behind the window opening was not rigid which allowed for movement, resulting in detected motion by the algorithms. Future studies should employ rigid backdrops with a color that is easily distinguished from the participant, equipment, and weapon, to limit background motion detection and enhance participant detection by the algorithm. Future iterations of the algorithm may attempt to automatically segment the exposed area into categories (e.g., lethal vs. non-lethal regions, weapon vs. arm, gear vs. torso). Additionally, future iterations may incorporate a survivability calculation based on protection level of the equipment worn. Finally, research utilizing this method should consider a feedback system of consequences to force tactical behavior.

References

1. Hedlund, J.: Risky business: safety regulations, risk compensation, and individual behavior. Inj. Prev. **6**(2), 82–89 (2000)
2. Joseph, A., Wiley, A., Orr, R., Schram, B., Dawes, J.: The impact of load carriage on measures of power and agility in tactical occupations: a critical review. Int. J. Environ. Res. Public Health **15**(1), 88 (2018)
3. Headquarters Department of the Army: Combined Arms Operations in Urban Terrain. (Army Tactics, Techniques, and Procedures No. 3–06.11). Washington, DC (2011)
4. Headquarters Department of the Navy: Military Operations on Urbanized Terrain (MOUT). Marine Corps Reference Publication 12–10B.1, Washington, DC (2016)
5. Brown, S.A.T., Mitchell, K.B.: Preliminary development of an integrated mobility, lethality, and survivability soldier performance testing platform. In: Nunes, I.L. (ed.) AHFE 2019. AISC, vol. 959, pp. 153–164. Springer, Cham (2020). https://doi.org/10.1007/978-3-030-20040-4_14
6. Bossi, L.L., Jones, M.L., Kelly, A., Tack, D.W.: A preliminary investigation of the effect of protective clothing weight, bulk and stiffness on combat mobility course performance. Proc. Hum. Factors Ergon. Soc. Annu. Meet. **60**(1), 702–706 (2016)
7. Mitchell, K.B., Batty, J.M., Coyne, M.E., DeSimone, L.L., Bensel, C.K.: Reliability analysis of time to complete the obstacle course portion of the load effects assessment program (LEAP). Technical Report NATICK/TR-17/002, Natick, MA (2016)

8. Carlton, S.D., Orr, R.M.: The impact of occupational load carriage on carrier mobility: a critical review of the literature. Int. J. Occup. Saf. Ergon. **20**(1), 33–41 (2014)
9. Brown, S.A., McNamara, J.A., Mitchell, K.B.: Dynamic marksmanship: a novel methodology to evaluate the effects of clothing and individual equipment on mission performance. Proc. Hum. Factors Ergon. Soc. Annu. Meet. **61**(1), 2020--2024 (2017)
10. Krueger, G.P., Banderet, L.E.: Effects of chemical protective clothing on military performance: a review of the issues. Mil. Psychol. **9**(4), 255–286 (1997)
11. Billing, D.C., Silk, A.J., Tofari, P.J., Hunt, A.P.: Effects of military load carriage on susceptibility to enemy fire during tactical combat movements. J. Strength Cond. Res. **29**, S134–S138 (2015)
12. Hodgdon, J.A., Hesslink, R.L., Hackney, A.C., Vickers, R.R., Hilbert, R.P.: Norwegian military field exercises in the arctic: cognitive and physical performance. Arch. Med. Res. **50**, 132–136 (1991)
13. Dempsey, P.C., Handcock, P.J., Rehrer, N.J.: Impact of police body armour and equipment on mobility. Appl. Ergon. **44**(6), 957–961 (2013)
14. Loverro, K., Brown, T., Coyne, M., Schiffman, J.: Use of body armor protection with fighting load impacts soldier performance and kinematics. Appl. Ergon. **46**, 168–175 (2015)
15. Mitchell, K.B., Choi, H.J., Garlie, T.N.: Anthropometry and range of motion of the encumbered soldier. No. NATICK/TR-17/010, Natick, MA (2017)
16. Taverniers, J., De. Boeck, P.: Force-on-Force handgun practice: An intra-individual exploration of stress effects, biomarker regulation, and behavioral changes. Hum. Factors **56**(2), 403–413 (2014)

A User Interface Prototyping Tool for Exploring Supervisory Control of Automation During Event-Paced Scenarios

Lisa Guo[1](✉), David Kellmeyer[1], and Glenn Osga[2]

[1] Naval Information Warfare Center Pacific, San Diego, CA, USA
lisa.guo@navy.mil

[2] Leidos Inc., San Diego, CA, USA

Abstract. Usability testing is a critical step in the process of agile iterative development. Static and lower-fidelity user-interface mock-ups can suffice during early phases of design or for user-paced environments, but higher-fidelity test beds are required to evaluate event-paced environments especially those exploring the supervisory control of automation. Increased fidelity is desirable as early as possible during the design process to explore the supervisory control design paradigm and generate more numerous and higher-quality user feedback. Complex AI-based systems are more difficult to test during early development phases. Static and low-fidelity prototypes typically cannot adequately represent the complex automation behaviors making usability more difficult to measure. A high fidelity user-interface prototyping approach was employed to simulate the capabilities of pending algorithms that are under development by multiple laboratories for a future intelligent air-defense system. The prototyping tool supported usability data collection by simulating the proposed automation and an initial supervisory control user-interface design within dynamic, group, air defense scenarios. This approach yielded high-quality user feedback early in the system design process even while being subjected to the constraints imposed by remote testing requirements due to COVID-19.

Keywords: User-interface prototyping · Usability testing · Dynamic simulation · Artificial intelligence · Supervisory control

1 Introduction

Usability testing for complex systems with long development cycles presents a challenge to the user-interface (UI) development process of engaging users with a working, dynamic representation of the system interface. Typically, early design evaluations involve the presentation of static images for screen designs, and storyboards can be used to show users the intended information flow and possible features for the end-system [1]. These static representations limit the types of tasks and workflows possible during the user evaluation, and they do not give the user a perception of system behavior, nor do they dynamically present incoming information. This limitation is significant in

J. L. Wright et al. (Eds.): AHFE 2021, LNNS 264, pp. 28–35, 2021.
https://doi.org/10.1007/978-3-030-79763-8_4

dynamic shipboard air defense systems (ADS) where the user-system performance goals relate to ongoing decisions made in response to rapidly changing threats [2] and constrained engagement windows (and opportunities). UI complexity is further increased with the addition of algorithms and automation functions that provide supervisory control options for users under varying degrees of automation assistance. Users must choose how to apply their limited attention resources to observe automation through each step in the air-defense process, and choose when to interdict manually. It is vital the user can quickly understand the anticipated implications of these interdictions. Further UI complexity is added as Group Air Defense is a multiple-user, team decision-making task. Conceptually, a user could direct local automation and responses for their own ship, and also receive direction from higher-level command applied to multiple ships and aircraft. These conditions of varying automation and multi-user command directions can change rapidly within a complex air defense scenario. Also, the pacing of threat actions can present high levels of user workload with an increasing number of threats. The intelligent ADS is designed to present dynamic decision-support information, to enable the protection of friendly forces, using both sensor and weapon resources most efficiently. The system first provides recommended response actions, and automatically processes the responses under user supervision. Users can agree with the recommendations, investigate and validate them, edit or cancel them as desired. A major UI challenge arises as user edits to change or delete system recommendations can invalidate or alter other engagement solutions for own or other ships. Given the complexity of response actions, a prototype that represents these dynamic system qualities is required to support meaningful usability analysis. The prototype included the simulation of a dynamic AI-based ADS UI to support usability analysis, during the system design period when the ADS AI-based algorithms were still being developed. A goal was to not only evaluate man-machine collaboration but also to influence the development of these algorithms.

2 Approach

The Naval Information Warfare Center (NIWC) Pacific, Human-Factors Engineering (HFE) team has been supporting the UI design of an ADS with dynamic automation capabilities to be used in an air-defense team distributed across multiple ships and aircraft. The HFE team used system concepts and algorithm requirements to estimate a variety of threat responses that the algorithms could recommend, and designed a UI to reflect the prospective workflow of the ADS engagement process. A job description for various naval ADS roles was conducted to capture the ADS workflow in the form of Journey Maps, Use Cases and Activity Diagrams [3, 4]. The Journey Maps showed the decision points of primary ADS operators and the state of the system automation (e.g., constrained, not constrained) as each workflow path within the maps were followed. The Use Cases describe major engagement paths such as accepting, rejecting or modifying an AI-proposed engagement. The Activity Diagrams illustrated task activities needed to accomplish the use case. The initial diagrams focused on the current procedures for ADS operations with the integration of AI-based recommendations and automation. Then, low-fidelity UI wireframes were created to illustrate the UI across various user activities. A UI storyboard was created to provide a description of the types of air-defense situation and ADS information that may be needed to inform and guide the users. A "task-centered" UI design approach was used in which the UI directly reflected the ADS workflow with fixed display areas containing lists of ADS engagements during each stage of processing including: Monitoring (tracks awaiting solutions), Recommended (threats with AI-recommended solutions) and Engage (threats in the weapon-control system engagement process). The HFE team reviewed the static images with users and initial prototyping of concepts was done using a Unity [5] game-engine software prototype. Following user review and approval, the concepts were included in the overall User-Interface Toolset as shown in Fig. 1.

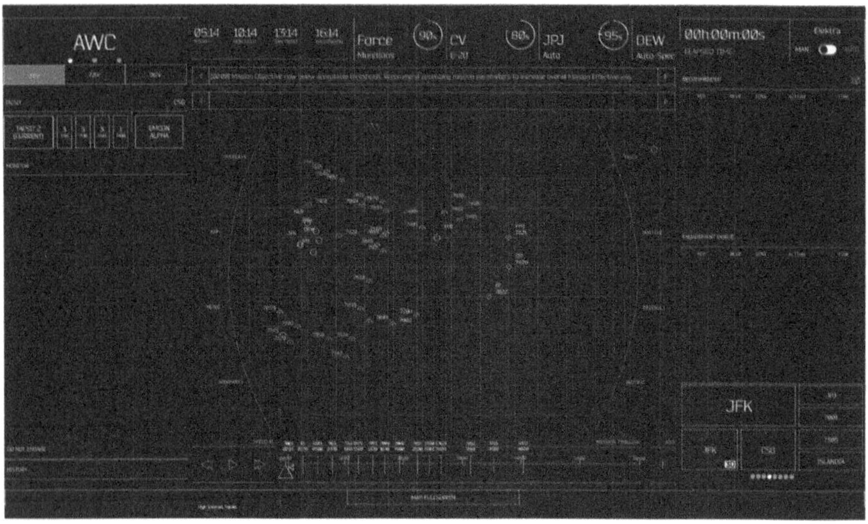

Fig. 1. ADS user-interface display concept design in the UI unity toolset

The static UI toolset became dynamic through the introduction of a scenario script and event driver which directed events to occur following a scenario clock that could be run in real-time or at various speeds to adjust workload via event pacing. The scenario events were defined in a spreadsheet, allowing non-programmers the ability to create, edit, and de-bug scenario events. A track entity (ship, plane, missile, etc.) could be defined giving it an identity, classification, and data description. A track scenario entry time, entry location, direction, and speed defined when and where the track appeared on the scenario map at a designated location and time. The track ID, speed or direction can be easily changed using the spreadsheet. Other rules were incorporated to increase the realism of the simulated algorithm behavior. A simple set of rules simulated the output of the Threat Assessment algorithm score define as Capability, Opportunity and Intent (COI). A COI score of (0) indicated that no factors were threatening while a score of (3) indicated that all three factors were threatening. The UI designers used the track's ID to determine which tracks to display for user attention in the "Monitor Bin" and then used the simulated Threat Assessment score to determine for which tracks a recommended engagement solution would be generated. A user selected automation mode and simulated engagement expiration windows were used to simulate which recommended engagements would be automatically promoted to a weapon control system solution. Small delays were added to simulate real-world processing delay times. The combination of scenario events with simulated automation results allowed exploration of proposed supervisory control rules with Navy air defense operators.

Given a number of friendly forces and potentially hostile threats in the scenario, the engagement pairing data simulating AI-system output described which shooter, sensor and weapon should be used for a particular threat. The actual AI system would parse all the possible pairings and optimize the solution for the entire set of threats, creating the most efficient pairings to minimize risk, optimizing use of less vs. more expensive ordnance, and providing protection for high-value targets. The simulation provided a representation of these pairings, however they were not actually optimized for the dynamic real-world factors. Hence the simulation followed some of the AI system rules but did not actually produce solutions that were based on the algorithms.

3 Event-Paced Scenario Test Plan

A usability test plan was created that included: an introduction briefing, UI familiarization, guided tasks, directed tasks, and a freeplay scenario. Following an introduction describing the purpose and format of the usability test, the administrator demonstrated basic UI functions in the familiarization phase by following the life-cycle of a threat from inception to final engagement. Next, the user was given guided instructions to complete tasks. The administrator would ask the user to perform a task and describe how to perform the task such as "swipe left to put track 7001 into the recommend bin". In the Directed phase, the user was given a task, but was not explicitly guided through the task, with an instruction such as "Engage 7001". In the Freeplay phase, the user was asked to respond to the scenario events without administrator direction using a think-aloud protocol. The users' primary supervisory control tasks included the approval and movement of threats through the engage process by allowing them to proceed in automated mode or by manually moving them from one list to the next. The user could also modify, reject and alter the threat shooter, sensor or weapon. Users were asked to verbally report certain scenario events if tracks met one or more of six conditions. These conditions were described to users as "standing orders" to report the events. These included conditions such as: report any Long-Range Missiles (e.g., high-cost weapon) or if Depth-of-Fire equals 1 (e.g., higher risk shots). The six conditions were based on critical events such as: expensive weapons, urgent shots, high-speed missiles, etc. The user reports indicated if the user was maintaining awareness of important threat and engagement criteria. The administrator would also introduce situation awareness questions such as "what is the COI of Track 7010?" or "why isn't track 7020 being engaged".

4 Test Administration

Two operators who were experienced air-defense instructors participated in the usability event. Users sat at a workstation with a Microsoft Surface Studio 2, 28″ PixelSense™ display and built-in microphone, using a mouse, keyboard and Windows 10 with Unity

Program installed. The prototype used a full-screen configuration on the display as depicted in Fig. 1. Users were permitted to use the mouse to navigate and interact with the scenario. The keyboard was used solely by an on-site administrator in charge of pausing and starting the scenario as needed, or as directed by the administrator. The scenario events were played via a timer run in real-time or accelerated time as required by the events scheduled. The built-in microphone was used to relay voice comments made by users as they interacted with the prototype. MS Teams web-conferencing was used to broadcast the user screen to remotely-located administrators and observers. The recording feature of Teams captured on-screen user actions and audio conversations. The Introduction Phase was 10 min followed by Familiarization and Guided Events for 55 min. The Directed Phase lasted 45 min during which participants alternated interacting and monitoring the other user as directed by the administrator. The monitoring participant was allowed to assist the other user and provide feedback. Each participant performed a 30-min freeplay scenario.

5 Analysis

The video recording captured the on-screen activity and audio commentary of the users and administrators. The video was used to assess user actions taken to complete tasks and measure user response times. The following observations were recorded and counted:

- Bug: An unintended software response as determined by the test administrator or observer.
- Slip: The user appeared to understand how to complete the task but there required multiple attempts to complete (e.g., clicked in the wrong area).
- Standing Order Responses: A correct standing order event verbalization with average and maximum response times as measured from appearance of threat event until user report. Or, a missed or incorrect response to a standing order event.
- Situation Awareness Questions: Correct and incorrect answers to situation awareness questions.
- Threat Processing Actions: A successful threat-card swipe action when achieved on the first attempt and in a timely manner. A Missed Swipe was recorded if the threat-card swipe required multiple attempts or involved a long delay (over 60 s).

6 Results

Results are discussed below for each type of usability observation event. There were 13 software bugs identified. Slips were recorded for various events that involved errant user selections, e.g., errant mouse clicks, noted as UI format issues due to small or unclear clickable areas, or errant movement of UI features contributing to missed selections. Standing orders were correctly done for 80% of the possible reportable events. Users correctly answered 17 questions with 2 incorrect responses. There were 141 correct threat-card swipes and 20 missed-swipe actions. Further details are presented as follows.

7 Standing Orders Reporting

Critical events were reported for all standing orders with an overall average of 80% correct and average response time of 14 s across all events and 10 s average if outliers where ignored. The few "outlier" times were larger than 1 min, in these occasions a threat event may have been overlooked for several minutes. Threat-cards that incorporated color coding to indicate a significant event (e.g., teal and orange color indicating a critical threat) had 87% to 94% accuracy respectively.

8 Administrator Observations

Observations related to the main UI features revealed the following events:

1. Threat list and threat-card manipulation & information (16 items)
2. Map usability (5 items)
3. Decision tools usability and information drill-down (2 items)
4. Symbology and graphics (1 item)

These UI features were noted for each of the major user tasks defined for the ADS. These tasks included:

1. Inspection: user scans, reviews and drills-down to location needed information.
2. Approval/Acknowledgement: user reviews and accepts or denies engagement recommendations.
3. Edit/Correction (of an entity): user alters threat information, such as changing classification, weapon or shooter, or denying an engagement.
4. Request for automation: user nominates a threat for engagement recommendation.

Observations on "inspection" tasks indicated that users were able to view and process the overall ADS workflow, individual threats, and high-priority threats. We observed a high level of user visual scanning during inspection tasks as threat-cards moved from left to right list columns on the display and estimate that visual scanning workload could be reduced by co-locating lists on the display. Inspection was sometimes impeded by discrepant UI results across types of threat-cards which caused user confusion. Other coding issues were identified and UI options to present weapon resource information more efficiently were noted.

The "approve" task observations indicated that users were generally able to approve an engagement, and did so in both manual and automated modes. Thus, they remained actively involved in the approval tasks in concert with the automated processing of threat events.

The "edit" task was conducted successfully, however some usability issues were noted due to a software bug which erroneously re-shuffled the threat-cards in reverse order from the final engage list backwards into the recommend list. Usability issues with editing tasks were observed due to normal movement of threat-cards and edit modes being turned on and off due to errant user selections. Thus, an unplanned secondary task was added to user workload in that they would recall and not re-approve engagements that they had previously approved and which had re-cycled. Users demonstrated an ability to distinguish between threat-cards which they had previously approved and those which were newly added to the list.

Users' interaction with automation were observed as they most often chose to use manual mode, and manual actions to nominate threats to Recommend List occurred frequently even when the system was placed in auto mode. The user would typically switch to auto-mode when the quantity of threats increased.

9 Situation Awareness Questions

The situation awareness and automation awareness (SA) questions posed by the administrator were intended to gauge users' ability to understand a track's status in the automated workflow and to understand automated recommendations via visual or drill-down inspection of the interface, as well as to prioritize their attention. Examples of SA questions involved users identifying threats in the Monitor list that were not being moved along the workflow by the AI, or threats of a different variety such as those only having bearing information and a different threat-level scale. Other SA questions involved users finding "high priority" threats in the Recommended list such as expensive weapons or risky shots, which were denoted with different colors. The majority of user responses were correct for automation and SA questions associated with the Monitor and Recommend Lists indicating that the information and its coding were able to be understood by the users. Incorrect responses were generally related to lack of drill-down information with some types of threats and the subjective decision threshold between engage and do-not-engage for threats with a COI of Level 2. Users demonstrated SA beyond the prescribed questions. In one instance, a user quickly and correctly pointed out when a recommendation did not make tactical sense. In the real system (not simulated) a user decision to process a threat with medium criteria could be the correct path, given that the threat would have quickly crossed a threshold to a higher level. Further analysis with actual algorithms and outcomes involving realistic track dynamics (true speeds and distances x type of air platform) would better determine decision accuracy.

10 Conclusions

Users indicated that the simulated ADS scenario and automation provided an adequate representation of an automated ADS with embedded threat analysis and weapons allocation algorithms. By simulating multiple representative automation use cases we were able to assess and explore how well a user could monitor and respond to anticipated automation even before actual algorithms where available for testing. By coupling this with an event-paced scenario and "standing orders" ancillary probe tasks we were able to obtain an initial interpretation of the user interface's ability to focus attention under different workload conditions. The usability results revealed some UI format and selection usability issues related to UI design. Task time information indicated little variability across types of threat-cards, indicating consistent results across each type of threat. The ability of users to answer questions accurately and respond to standing order reports with a high degree of accuracy indicated that the UI concepts were both understood and usable within the context of the scripted scenario and typical air defense threat processing.

The Unity tool-based simulation of a dynamic scenario and simulated algorithms allowed the team to obtain user feedback and capture video interaction of users with the simulation, resulting in a list of possible design improvements. Had the team relied on lower-fidelity usability tests alone, we would not have been able to simulate time pressure and event pacing. While these usability results do not validate the future algorithms, they indicate that the presentation of results by algorithms are both understood and usable by trained current system ADS operators. It also provides an initial positive indication that these types of automation would be useful and the current CONOPS for its implementation is consistent with user needs and expectations. The UI toolset allows further iterations of design concepts before the algorithms are ready, increasing the likelihood of improved system usability and user acceptance for the system when algorithms are incorporated with the UI.

References

1. Truong, K.N., Hayes, G.R., Abowd, G.D.: Storyboarding: an empirical determination of best practices and effective guidelines. In: Proceedings of the 6th Conference on Designing Interactive Systems, pp. 12–21, June 2006
2. Joint Publication 3-01 Countering Air and Missile Threats, 21 April 2017
3. Kellmeyer, D., Guo, L., Osga, G.: Design Use Cases FIRECAPE. Distribution D US Govt and US Govt Contractors Only, March 2019
4. Kellmeyer, D., Guo, L., Osga, G.: Activity Diagrams v1.0. Distribution D US Govt and US Govt Contractors Only, March 2019
5. Wikipedia: Unity Game Engine (2020). https://en.wikipedia.org/wiki/Unity_(game_engine)

Visualizing Team Trust and Cohesion

Exploiting Interdependence in Autonomous Human-Machine Systems to Avoid Disaggregation and Vulnerability

William Lawless(✉)

Paine College, Augusta, GA 30901, USA
w.lawless@icloud.com

Abstract. We review the interdependence theory of complementarity for autonomous human-machine teams (A-HMTs) operating in open systems. In open systems, interdependence theory is state dependent, reactive to every situation, including uncertainty. Reactiveness to uncertainty explains why the best teams are highly interdependent, outperforming the same members of a team when they are acting independently as disaggregated individuals. A key ingredient we argue is that highly-effective team members are in orthogonal roles, reducing the degrees of freedom and entropy. In contrast, we also briefly review the competing, rational theories of interdependence that promote disaggregation: social interdependence (cooperation theory), games (applied social interdependence theory), innovation (the atomistic, rational "homo economicus" of economics theory), and system dynamics (the interdependence of feedback under certainty); in conjunction, we review the problems with generalizing, applying or scaling rational but closed system theories to the design and operation of A-HMTs and systems facing uncertainty.

Keywords: Human machine teams · Autonomy · Complementarity · Systems

1 Introduction

In this chapter, we review the interdependence theory of complementarity for autonomous human-machine teams (A-HMTs) and systems like organizations operating as, or in, open systems. In open systems, interdependence theory is state dependent, reactive to every situation, especially uncertainty. Reactiveness to uncertainty explains why the best teams are highly interdependent, outperforming the same members of a team when those members are acting separately as disaggregated individuals. A key ingredient we argue is that highly-effective team members are in orthogonal roles, reducing their structural degrees of freedom (e.g., for a 3-person restaurant: cook, waiter and cashier). A key distinction: as the entropy produced by the structure is reduced, the team can direct more entropy produced by the team's function or mission [1].

In contrast, we also briefly review the competing, rational theories of interdependence that promote disaggregation: social interdependence (cooperation) theory, games

J. L. Wright et al. (Eds.): AHFE 2021, LNNS 264, pp. 39–45, 2021.
https://doi.org/10.1007/978-3-030-79763-8_5

(applied interdependence theory), innovation (the atomistic, rational "homo economicus" of economics theory), and system dynamics (the interdependence of feedback under certainty); in conjunction, we briefly review the problems with generalizing, applying or scaling rational but closed system theories to the design and operation of A-HMTs and systems facing uncertainty. Next, we briefly outline our supporting evidence for the interdependence theory of complementarity and its mathematical application to A-HMTs, notably for metrics and tradeoffs (e.g., least entropy production, LEP, in a team or system structure can occur as part of a tradeoff that promotes maximum entropy production, MEP, in a system's performance as seen when engaging in the widest possible search of a solution space for, say, innovation). Derived from interdependence theory, we also briefly review our discovery of a new field of research on structural vulnerability in teams, organizations and systems; how pursuing the natural motivation to reduce structural vulnerability unexpectedly produces organizational resilience to social shocks; or how the seeking of structural vulnerability in opponents motivates decisions in opposing forces that lead to mergers and alliances to create more order across a system in accordance with the second law of thermodynamics. We close with a brief review of future research.

Interdependence is an umbrella term for the phenomena that transmit all social effects in the form of interference (e.g., the synergism that produces emergence; the dysergism that produces divorce, splits and conflict; and the asynergesism that destabilizes opponent defenses; together, these phenomena combine into forces that drive local change, organizational restructuring, or even political and possibly social evolution).

In contrast to our theory of the interdependence of complementarity, social interdependence theory [2] has a long and hopeful history, culminating in homilies of peace and harmony in an attempt to supplant Darwin's [3] "survival of the fittest" with new ageism's "survival of the friendliest" [4]. Not surprisingly, the results of this new age theory cannot be generalized nor scaled. However, the primary weakness of social interdependence theory and other rational theories, developed largely with aggregations that sum the rational choices of individuals (e.g., [5]), is their inability to predict outcomes in natural social settings involving competition or uncertainty, the very situations to be addressed to establish fundamental science and engineering relationships for the design, operational guidance and metrics to prepare for the rapidly approaching age of autonomous human-machine teams and systems that will be used in the real world [6].

The reality of robots today [7] has exceeded expectations. In 2019, there were over 500,000 robots in industry and service, not counting those employed for data collection and analyses, commercial drone and private use. But robots in autonomous human-machine teams are new. Until now, many of the advancements with robots have been made using rational approaches [8]. We have argued, however, that facing uncertainty, the design and operation of teams performing in competitive situations against other teams cannot be solved with rational approaches alone; instead, since interdependence entails state dependent effects, a quantum-likeness approach is recommended [9].

2 Research Findings

In contrast to social interdependence theory [2], by relying on the interdependent effects found in human-team studies (e.g., the best science research teams are highly interdependent; in [10]; see also [11]), by relying on field studies, and by adopting state-dependent effects in theory (quantum-like; see [9]), including Schrödinger's and Lewin's [12] separately derived concept of the "whole being greater than the sum of its parts," which has, as an overarching philosophy, been adopted by modern Systems Engineering [13], our revised theory for the interdependence of complementarity has guided us to make several predictions along with supporting findings in the field. Making this chapter more of a brief inquiry of what has been found and what is planned, we proceed with only our briefest review of both starting next; we provide ample citations for readers who wish to study the topics in more detail.

Redundant team members in teams and organizations impede performance including the responses by teams and organizations to social and resource shocks [9], thereby increasing the likelihood of corruption and accidents [14, 15].

We can mathematically distinguish Shannon's information theory for two factorable situations or two independent agents in factorable social situations represented by Eq. (1):

$$S_{A,B} \geq S_A,\ S_B. \tag{1}$$

Equation (1) can be revised as shown in Eq. (2):

$$S_{A,B} = S_A + S_B. \tag{2}$$

However, when two social agents are interdependent, known as subadditivity, we adopt Von Neumann's [16] equation for non-factorable entities:

$$S_{A,B} < S_A + S_B. \tag{3}$$

Equation (3) helps us to account for Schrödinger's,and Lewin's speculations about the whole being greater than the sum of its parts, and why boundaries around teams and organizations are required to isolate their states of interdependence from outside interference (as an example of the value of geopolitical boundaries today, see, for example, [17]).

The best teams are interdependent [17] and composed of orthogonal relationships [18]. Characterizing an orthogonal relationship, an intelligent organism can act independently as an individual, or as the member of a group, but not both simultaneously, accounting for the failure of complementarity theory (e.g., p. 207, [19]) in social science and in game theory (e.g., [20]), forcing traditional game theory to be restructured by adapting implicit preferences [21], but, consequently, reducing its ability to generalize, especially for applications to autonomous human-machine teams.

Orthogonal relationships in highly interdependent teams of humans (e.g., the 3-person restaurant discussed above; a marriage between two people in different roles; or a business relationship between, say, a boss, a secretary and an employee) operate on the fly in a sea of communication, with each member of the team making necessarily orthogonal contributions to the team's output yet all the while being instantly aware of a problem when one occurs with the performance of others on the team ([22]; see also [9]).

Were it possible to interrupt a member of a team composed of members acting in orthogonal roles, however, to ask for evidence from the perspective of other team members, they cannot provide an objective one since they each report as a disaggregated individual, not as the member of a team; this tangential aspect of reporting from orthogonal relationships is likely the very reason the theory of complementarity has failed in social science. Orthogonal contributors cannot perceive the same reality simultaneously, one reason for cross-training as a means to educate other teammates about all of the different roles in a team [9].

The value of intelligence reflected by a nation's higher education for all of its citizens increases the nation's ability to promote innovation as indicated by the number of productive patents it produces [18, 23].

Most situations considered by researchers are either well-defined, defined beforehand, or clearly defined for competitors, but what steps should be taken when all paths going forward are uncertain? Facing uncertainty, humans weigh the choice of a path going forward by engaging in debates [9], supporting the claims by AI scientists that machines must be able to express their intentions and actions in a causal language that both humans and machines are able to understand from each other and from their leaders (viz., using artificial intelligence, or AI, for machines; see [24, 25]).

How should a team determine context? Suppose a team's mission has become blocked by unforeseen circumstances; how should the team proceed (see [6])? We have argued that context cannot be determined by a single individual, a single team, or a single organization; that to determine context, a state of competition must exist between at least two social units and judged by an audience of observers ([14]; also, see [15] and [9]). We have further argued that debate and other forms of competition are needed to determine the best path moving forward when facing uncertainty ([22], see also [9]).

With our theory of the interdependence of complementarity, we have discovered an entirely new field of research [9], that a sense of vulnerability motivates teams and organizations to pursue avoidance behaviors (e.g., mergers or spin-offs), to engage in exploitative behavior (e.g., direct attacks among competitors), or to create vulnerability in opponents (e.g., with the use of deception). This discovery of the sense of vulnerability in the "self," or promoted in the "other," appears to be key to the survival of individuals, teams and systems in nature by promoting resilience, leanness, and, unexpectedly, the ability to adapt (e.g., from the *Wall Street Journal*, the leanest hospitals were able to adapt to covid-19 compared to rural hospitals; in [26]).

For future research, we propose to advance the theory of interdependence by providing a mathematical model of vulnerability in a team or system, how it is identified or created, how it is exploited, and how to avoid the vulnerability arising with a false sense of security derived from relying on convergence processes alone, both computationally and socially ([18, 22]; see also [9]).

By devaluing an opponent's reasoning, the over-reliance on convergence processes for autonomous systems, especially for computational systems (e.g., from the feedback in complex system dynamics; in [27]), can be counter-productive, leading to poorer decisions, misleading conclusions, or producing more accidents [22].

3 Future Research Activities

We plan to continue to develop and test metrics for the performance of teams and autonomous human-machine teams (e.g., [14, 15]). For example, if the best teams are fully interdependent, Eq. (3) governs; if team interdependence is reduced or eviscerated, as happens with the presence of redundant (ineffective, unneeded) personnel, or with social suppression, Eqs. (1) and (2) govern.

We have separated mathematically the structure of a team from its performance [1]. This separation allows us to follow the effectiveness of a team, organization or system. For this, we plan to model tradeoffs in teams between structure and performance [22]; see also [9]).

We plan to continue to explore models of scaling between teams, organizations, systems and systems of systems (e.g., especially with redundancy; in [14, 15]), and extend models to systems facing unexpected shocks (e.g., [23]; see also [9]).

We plan to continue to explore models of non-factorability to incorporate randomness in models of tradeoffs as teams face uncertainty [22].

We plan to explore convergence processes, both computationally and socially, which assume that social situations are factorable. In contrast, and mindful of the nonfactorability arising from interdependence, we plan to explore the inability of rational models to predict when facing uncertainty (e.g., the limits of the rational model in social and uncertain situations; in Mann, 2018; and the failure of the superforecasters championed by Tetlock and Gardiner; in [28]).

We plan to provide a mathematical model of vulnerability in teams [9]. We plan to test vulnerability mathematically characterized by a reduction in a team's operational or functional entropy production and an increase in its structural entropy production.

We plan to provide a mathematical model of debate for systems facing uncertainty to explore tradeoffs. We plan to explore decision quality, decision power, and decision speed. We plan to explore a model of trust in human-machine systems [29].

Finally, in the review of quantum biology by McFadden and Al-Khalili [30], unexpectedly, recent work has demonstrated in biological systems that the retention of quantum effects (e.g., tunneling; entanglement) is intricately connected with the environmental fluctuations regularly taking place at biologically relevant length and time scales. These results are unexpected because it was thought that in living systems, decoherence would preclude quantum effects. By extrapolating to social systems, this finding suggests that the quantum-like effects of interdependence may be kept coherent from the noise and tumult afforded by the competition that normally occurs in real life.

References

1. Bisbey, T.M., Reyes, D.L., Traylor, A.M., Salas, E.: Teams of psychologists helping teams: the evolution of the science of team training. Am. Psych. **74**(3), 278–289 (2019)
2. Johnson, D.W., Johnson, R.T.: Cooperation and competition. In: International Encyclopedia of the Social & Behavioral Sciences, 2nd edn, pp. 856–861 (2015). https://doi.org/10.1016/B978-0-08-097086-8.24051-8
3. Darwin, C.: The Descent of Man. Appleton, New York (1973)
4. Hare, B., Woods, V.: Survival of the Friendliest. Penguin Random House, New York (2020)
5. Mann, R.P.: Collective decision making by rational individuals. PNAS **115**(44), E10387–E10396 (2018)
6. Lawless, W.F., Mittu, R., Sofge, D.A., Hiatt, L.: Editorial, AI, autonomy and human-machine teams: interdependence, context and explainable AI. AI Mag. **40**(3), 5–13 (2019)
7. Mims, C.: On the 100th Anniversary of 'Robot,' They're Finally Taking Over. Wall Street Journal (2021). https://www.wsj.com/articles. Accessed 23 Jan 2020
8. Rees, M.: Dreams of a rational world. A long-term look at the future of humanity. IAI News (2021). https://iai.tv/articles/dreams-of-a-rational-world-auid-1732?
9. Lawless, W.F.: Quantum-like interdependence theory advances autonomous human–machine teams A-HMTs. Entropy **22**(11), 1227 (2020). https://doi.org/10.3390/e22111227
10. Cummings, J.: Team Science Successes and Challenges. NSF Sponsored Workshop Fundamentals of Team Science & Science of Team Science, Bethesda, MD (2015)
11. Cooke, N.: Effective human-artificial intelligence teaming. In: AAAI-2020 Spring Symposium, Stanford, CA (2020). https://aaai.org/Symposia/Spring/sss20symposia.php#ss03
12. Lewin, K.: Field theory of social science. D. Cartwright (ed.). Harper & Brothers, New York (1951)
13. Walden, D.D., Roedler, G.J., Forsberg, K.J., Hamelin, R.D., Shortell, T.M. (eds.) Systems Engineering Handbook, 4th edn. INCOSE-TP-2003-002-04. Wiley, Hoboken (2015)
14. Lawless, W.F.: The entangled nature of interdependence. Bistability, irreproducibility and uncertainty. J. Math. Psychol. **78**, 51–64 (2017)
15. Lawless, W.F.: The physics of teams: interdependence, measurable entropy and computational emotion. Front. Phys. **5**, 30 (2017). https://doi.org/10.3389/fphy.2017.00030
16. von Neumann, J., Wheeler, N.A. (ed.): Mathematical Foundations of Quantum Mechanics. New Edition. R.T. Beyer (Tr.). Princeton University Press (2018)
17. Friedman, G.: U.S.A. Russia's Search for Strategic Depth, Geopolitical Futures (2020). https://geopoliticalfutures.com/russias-search-for-strategic-depth/
18. Lawless, W.F.: Towards an epistemology of interdependence among the orthogonal roles in human–machine teams. Found. Sci. **26**(1), 129–142 (2019). https://doi.org/10.1007/s10699-019-09632-5
19. Berscheid, E., Reis, H.T.: Attraction and close relationships. In: Gilbert, D.T., Lindzey, G., Fiske, S.T. (eds.) Handbook of Social Psychology 2, 4th edn. Erlbaum, London (1998)
20. Jones, E.E.: Major developments in five decades of soc psych. In: Gilbert, D.T., Fiske, S.T., Lindzey, G. (eds.) The Handbook of Social Psychology, vol. 1, pp. 3–57. McGraw-Hill, Boston (1998)
21. Amadae, S.M.: Rational choice theory. Encyclopaedia Britannica (2017). https://www.britannica.com/topic/rational-choice-theory
22. Lawless, W.F.: The interdependence of autonomous human-machine teams: the entropy of teams, but not individuals, advances science. Entropy **21**(12), 1195 (2019)
23. Lundvall, B.A.: The Learning Economy and the Economics of Hope. Anthem Press (2016)
24. Pearl, J.: Reasoning with Cause and Effect. AI Mag. **23**(1), 95–111 (2002)

25. Pearl, J., Mackenzie, D.: AI Can't Reason Why. Wall Street J. (2018). https://www.wsj.com/articles/ai-cant-reason-why-1526657442
26. Gold, R., Evans, M.: Why Did Covid Overwhelm Hospitals? A Years long Drive for Efficiency. Wall Street J. (2020). https://www.wsj.com/articles/hospitals
27. Forrester, J.W.: The Beginning of System Dynamics. MIT, Cambridge (1989)
28. Tetlock, P.E., Gardner, D.: Superforecasting: The Art & Sci of Prediction, Crown (2015)
29. Lawless, W.F., Sofge, D.A.: The intersection of robust intelligence & trust. In: Lawless, W.F., Mittu, R., Sofge, D., Russell, S. (eds.) Autonomy & AI: A Threat or Savior? Springer, New York (2017)
30. McFadden, J., Al-Khalili, J.: The origins of quantum biology. In: Proceedings Royal Society A, vol. 474, p. 20180674 (2018). rspa.royalsocietypublishing.org

Visualizing Human-Autonomy Team Dynamics Through the Development of a Global After-Action Review Technology

Michael Taberski[1(✉)], Kristi Davis[2], Kristin E. Schaefer[2], and Ralph Brewer[2]

[1] US Army JSOC, Ft Bragg, NC, USA
michael.j.taberski.mil@mail.mil
[2] DEVCOM Army Research Laboratory, Aberdeen Proving Ground, MD, USA
{kristi.l.davis26.civ,kristin.e.schaefer-lay.civ,
ralph.w.brewer.civ}@mail.mil

Abstract. This paper describes the importance of visualization to support after-action review technologies to develop appropriate human-autonomy team dynamics. Key to developing accurate visualizations in this context requires the capability to quantify critical team dynamics by parsing large, diverse sensor data in offline, low-latency environments. Here, we describe the use of open-source technology and scalable web-based applications for the quick ingestion of these massive datasets and timely visualizations for the AAR process for the development of the Global After-Action Review Technology. This technology explores human-autonomy team interaction problems by identifying key features needed for computer vision, including image classification, object recognition, scene understanding, gait recognition, pose estimation, tracking, and behavior recognition.

Keywords: Visualization · Human-autonomy teaming · After-action review · User interface · Training

1 Introduction

After-action reviews (AARs) have a long-standing history with the US Army as a method to improve training and team effectiveness. New technologies are needed to allow both humans and machines to receive and provide feedback. Visualization is a major component to providing a deep understanding of "what happened" during an exercise, where assets were on the battlefield, and the information conveyed during the mission. Additionally, visualizations provide us with a better mechanism to communicate what decisions the autonomous systems are making and why. This, combined with the knowledge and data associated with what the human team members are doing provides insight into how improvements can be made in teaming.

J. L. Wright et al. (Eds.): AHFE 2021, LNNS 264, pp. 46–53, 2021.
https://doi.org/10.1007/978-3-030-79763-8_6

1.1 After-Action Reviews

Anytime there is a mission or exercise performed by units in the Army, leadership conducts an AAR. This review gives the unit insight into what occurred, including where things ran smoothly, but more importantly where things went awry in order to discuss and alter future outcomes. The key principles in an AAR include:

1. Allow team members to engage in debriefs in order to integrate lessons learned from past performance [1, 2];
2. Provide opportunities to practice resolving conflicts in a productive manner [3, 4] in order to improve coordination during future performance;
3. Allow team members to develop accurate mental models of problems, tasks, spaces, and shared knowledge [5, 6]; and
4. Provide a constructive process of reviewing performance to improve a team's ability to monitor their performance, recognize errors, and self-correct [7, 8].

Historically these AARs include only human voices, but current technologies on the battlefield require robotic assets be afforded a voice. These AAR technologies provide a method for evaluating mission-specific, environmental, and social context. This is critical because understanding these contexts can improve team comprehension and understanding, thus improving performance and collaborative decision-making [9, 10].

1.2 AARs for Manned-Unmanned Teaming

Manned-unmanned teaming (MUM-T) is the synchronization of Soldiers, manned and unmanned vehicles, and sensors for increased situational understanding, greater lethality, and improved survivability during military operations [11]. The Wingman Joint Capabilities Technology Demonstration focused on one manned command-and-control (C2) vehicle coupled with an unmanned robotic vehicle outfitted with a remote weapon station mounting a 7.62 mm machine gun [12]. This team conducted live fire gunnery exercises to test the crew's ability to engage targets using the system's autonomy. The Warfighter Machine Interface (WMI) provided interactive customizable displays for the crew to share situation awareness (SA), data, map, sensor, and alerts. The Global AAR Technology (GAART) was developed to support this scenario [13].

More recently, this MUM-T was expanded to integrate an array of manned and unmanned systems which provide real-time SA to the commander in order to generate options across all domains and enable maneuver forces to achieve military objectives. This team consists of one manned autonomous C2 vehicle outfitted with two teams of lethality and mobility operators, one set for each robotic combat vehicle (RCV). Future phases will bring together larger groups, up to a company size. Therefore, the GAART must be able to handle multiple sets of vehicles for platoon and company size exercises.

1.3 GAART

GAART is a way to look at the mission from a global top-down view. This technology allows Soldiers to view exactly what happened in their mission during a replay, accompanied by statistics to see what changes could be made and what actions to replicate when they are faced with the same scenario in the future. This technology adds visualization to support advanced understanding of a prior mission. Now the robotic team member can be considered when showing their field of view and whether their autonomy was being utilized to the maximum benefit of the mission. The initial GAART design used Google Earth with animations using keyhole markup language to depict the environment, vehicles, targets, and firing data (Fig. 1; green block). This was complimented by visualizing performance metrics (blue block); mobility and weapon autonomy use (yellow block); and event log (red block).

Fig. 1. Version 1 of the GAART showing the dynamic map (green), performance metrics (blue), autonomy use (yellow), event log (red), events (orange) (Color figure online)

2 Background: Appropriate Visualization Techniques

Data visualization, or the techniques used to communicate data and information through the use of a visual object, is an incredibly powerful way to quickly and efficiently obtain pertinent information from large sets of data. This is especially relevant in today's world of Big Data comprised of sensors, satellites, and networks all delivering millions of pieces of information every minute. No human is capable of deriving relevant facts from that amount of data but the use of data visualizations can help humans parse and quickly see important facts from the massive amount of information at our fingertips. Visualization tools have pushed data analysis from combing through archaic tables to interacting with visual charts, graphs, and maps in real-time as the data is being collected. Learning and thinking through visualizations makes complex information accessible and cognitively tractable which allow Soldiers to perceive and draw relations among items which would be difficult to comprehend without [14].

Particularly challenging in the realm of data analytics is the comprehension of temporal and spatial data. However, these two components make up the vast majority of information leveraged for analysis during an AAR. "Where were you?" and "When were you there?" are often key pieces of information that have direct impacts on the success of the mission. As for the spatial component, maps and graphs are examples of external representations for learning by organizing that are easily interpretable [15]. Therefore, the GAART tool has been constructed with a map-centric approach to ensure the focal point of the application communicates important data to users in a way they understand. As for the temporal component of GAART, the goal was to develop a feature that allowed users to analyze their data over time in an effort to compare discrete points in time to others over the course of the mission. However, aggregating a large amount of data with varying time-based fields into a comprehensive data visualization is challenging as the time intervals, measures, and the value of those intervals can change over different variables [16]. To facilitate the visualization of time, especially how it relates to space, the GAART leverages an interactive brushing tool that allows users to interact with and control where in time they are observing the data. Additionally, all time intervals are maintained in seconds and minutes to ensure continuity across variables. This interactive component is key to the analysis of complex interactions within an exercise over time and space.

To support appropriate recall, GAART leverages research on visual memory, visual imagery, and knowledge representations [15] to help soldiers understand each team member's role, responsibility, and intent. The dashboard joins information from many different sensors and data to visualize both the manned vehicle as well as the unmanned robotic vehicle. To support the Soldier's ability to see the "whole picture" of the battlefield or training environment, the development of context in relation to machine, time, and space is important. Therefore, the fidelity of our visualizations supports various states of both human and machine over time. Although often overlooked, words that support context by addressing outliers, explaining equations, or defining the data facilitate the analysis of the data by the user [17].

In an environment where one teammate (the machine) is unable to verbally communicate in real-time, the visual aspect of communication is even more important and conversely, miscommunication is much more likely. Tufte [17] notes that the distortion of data as it relates to visualizations can severely impact an individual's ability to ascertain information from the visualization. This distortion not only impacts the analysis of the data but also an individual's trust in the data and the tool being utilized, especially in a situation where the tool is dependent on artificially intelligent systems. Soldiers are not going to inherently trust machine learning (ML) algorithms. Instead, ML is coupled with human guidance displayed visually for Soldiers to understand both sides of the MUM-T equation.

3 GAART Visualization

Open-source technology is being leveraged throughout the GAART to provide a scalable, flexible, and customizable environment to produce meaningful visualizations. Open-source technology refers to a technology where its source code is freely accessibly to copy, redistribute, or modify based on the needs of the user [18]. This technology was constructed utilizing the Python programming language to support data processing and data visualization for detailed, flexible, scalable, and interactive charts, graphs, and maps that convey key metrics about the event.

3.1 Mobility and Maneuver Visualization

The overall goal of the GAART is to allow users to visually inspect what happened during an exercise in support of an AAR. Understanding the mobility and maneuver component of the exercise and communicating that visually to users is the largest feature in the technology both in terms of real estate and importance to users. Everything else is constructed in support of what happens on the map and provides context for what is occurring at all points in time. For a minimally viable product that provides users with enough functionality to support the AAR process while not overdeveloping the technology, the following features were focused on:

1. Vehicle Maneuver: The vehicle icons depict heading and latitude/longitude coordinates moving with the manned and unmanned vehicles drawing a path behind them.
2. Visualizing Paths: The robotic vehicles show projected paths from the path planner. Each vehicle shows a field of view for the weapon and camera systems.
3. World Feature Visualizations: Visualization of the targets are as human or vehicle icons according to the type of target exposed during the exercise.
4. Visualizing Planning: Phase lines, target reference points, battle positions, and any additional planning visualizations can be added to the map through the use of geospatial data sets and toggled for increased fidelity.

3.2 Key Metrics Visualization

For gunnery operations, the goal for the crew is to detect, identify, decide, engage, and assess potential enemy targets in a field setting [13]. Performance is rated on 10 engagements in both offensive and defensive postures against single or multiple stationary and moving targets during both daylight and night operations. Scoring is calculated based on timing, accuracy, and effective team communication. Each engagement is worth a maximum of 100 points with the standard of qualification at a minimum of 70 points for all targets. The crew must qualify at least 7 of the 10 engagements with a minimum score of 700 points total and qualify at least one night engagement [19]. Additional visualizations need to be included for the unmanned vehicles in the AAR.

Visualizing the performance of individual crews is important for the AAR but it is helpful to also showcase how all the crews did overall. Here leadership can see trends in specific engagements that may need additional training for the crews. This performance data is analyzed with the Soldier's technology use to determine how the scoring relates to autonomy. The event log's global view shows all of the engagements while the local view drills down into each engagement.

Screen capture video of each Soldier's user interface is synchronized to the global visualization of the run to look into crew interactions with the unmanned system. Crew interactions with the screen and the controllers are how the manned operators work with the robotic vehicle's mobility and lethality. Malfunctions or anomalies help understand why and how it affected the crew performance. Future work will include looking at physiological and behavioral indicators as well as communication dynamics [20].

3.3 Autonomy Performance Visualization

Visualizing the use of autonomy during the mission is crucial to gaining a perspective of the vehicle and weapons systems. As with any new technology Soldiers are apprehensive at first in using it and do not want to give up control. Combining the autonomy visualization with other event alerts that impact a mission such as malfunctions and anomalous behaviors can highlight issues that otherwise might not be found.

A map overlay provides a medium for outlining the paths showing when and where the robotic vehicle was in autonomous or teleoperation mode. Future efforts look to synchronize the sensor views of the RCV as well as planning algorithms into the map view. Soldiers will have an opportunity to provide input to discuss team trust, justification of the decisions made by the RCV and continued learning opportunities for human robot interaction.

Assigning a value to the quality of performance of the team can show improvement and give encouragement when changes are made that raise metric scores. This tool can reinforce trust building between the Soldier and robot. During a replay, autonomy visualization will outline when the systems are engaged in their respective modes. This will enable the crew to understand their autonomy use over time as well as links to system malfunctions and anomalous behaviors to illustrate potential areas that can cause performance issues. Being able to show causation over the use or lack of use of these systems will allow team trust to build. By having a numeric value associated with this metric allows the Soldiers to see exactly how their interactions with the robot are affecting the overall performance of the crew.

3.4 Interactive Capabilities of the Technology

After-action visualizations should be interactive and allow the user to dig deeper into the underlying data if desired. Typically, when users want to analyze temporal data, they look to compare data at various points in time with second-order goals that those observations will lead to insights [18]. GAART leverages several interactive components so users can inspect and analyze explicit data points to facilitate an after-action review. Multiple user-centric features were added to GAART to focus the user and allow them to interact with what is relevant to them. Initially, the focus was on small interactive capabilities that provide the most benefit, while outlining several interactive capabilities for future consideration.

Manipulating the time component through use of a brushing toolbar is often the most characteristic way to provide users the ability to interact with their time-series data [18]. The major feature is the user's ability to inspect and analyze data at different time points. The toolbar controls the playback of the AAR as well as updates the corresponding charts, tables, and supporting data based on the user-defined point in time.

The corresponding graphs and charts that showcase autonomy metrics, team performance metrics, and other supporting data are updated automatically and are interactive. Users can hover over data points depicted in the charts and graphs to inspect the data with more fidelity. This feature allows users to better understand anomalous data points or areas for sustainment and improvement in relation to the overall AAR.

4 Conclusion

Although the goal of this technology was to provide users with a comprehensive visual dashboard showcasing highly dimensional data from many sensors, there was also a need to ensure that the technology is scalable as needs arise over time. Creation of a framework of customizable visualizations that are adjustable, modifiable, or alterable to fit any environment or leadership requirement was made possible through leveraging highly adaptable, open-source technology. The current technology provides the machine with a voice and allows users to understand what is happening during different stages of the training exercise but it does not allow users to interact with the machine or provide it feedback.

Future work will include the ability to capture and store data across multiple teams, training exercises, and timeframes to provide a robust baseline for team success metrics as well as provide data for further analysis in predicting team cohesion and human-autonomy usage during training. This coupled with users inputting self-reported data at time points during the AAR, help the machine learn creating a feedback loop. The robotic asset should learn on hundreds of training exercises to provide a visual outline of how to be most successful during training. During the mission the robot should call out anomalous data points and compare across a database full of similar exercises. Scoring of team performance should be done directly in the dashboard decreasing data aggregation time. Lastly, provide users a voice to make comments, rate the usefulness and reliability of the autonomous system, and allow users to provide their improve and sustain comments for the autonomous system through built-in, survey-like features.

Acknowledgments. The views and conclusions contained in this document are those of the authors and should not be interpreted as representing the official policies, either expressed or implied, of the Army Research Laboratory or the U.S. Government. The U.S. Government is authorized to reproduce and distribute reprints for Government purposes notwithstanding any copyright notation herein.

References

1. Smith-Jentsch, K.A., Johnston, J.H., Payne, S.C.: Measuring team-related expertise in complex environments (1998)

2. McNeese, N.J., Demir, M., Cooke, N.J., Myers, C.: Teaming with a synthetic teammate: insights into human-autonomy teaming. Hum. Factors **60**, 262–273 (2018)
3. De Dreu, C.K., Weingart, L.R.: Task versus relationship conflict, team performance, and team member satisfaction: a meta-analysis. J. Appl. Psychol. **88**, 741 (2003)
4. Marks, M.A., Mathieu, J.E., Zaccaro, S.J.: A temporally based framework and taxonomy of team processes. Acad. Manag. Rev. **26**, 356–376 (2001)
5. Blickensderfer, E., Cannon-Bowers, J.A., Salas, E.: Theoretical bases for team self-correction: fostering shared mental models. Adv. Interdisc. Stud. Work Teams **4**, 249–279 (1997)
6. Endsley, M.R.: Toward a theory of situation awareness in dynamic systems. Hum. Factors **37**, 32–64 (1995)
7. Salas, E., Sims, D.E., Burke, C.S.: Is there a "big five" in teamwork? Small Group Res. **36**, 555–599 (2005)
8. Marks, M.A., Panzer, F.J.: The influence of team monitoring on team processes and performance. Hum. Perform. **17**, 25–41 (2004)
9. Schaefer, K.E., Aksaray, D., Wright, J., Roy, N.: Challenges with addressing the issue of context within AI and human-robot teaming. In: Computational Context, pp. 80–96. CRC Press (2018)
10. Schaefer, K.E., Oh, J., Aksaray, D., Barber, D.: Integrating context into artificial intelligence: research from the robotics collaborative technology alliance. AI Mag. **40**, 28–40 (2019)
11. Brewer, R.W., Cerame, E., Pursel, E.R., Zimmermann, A., Schaefer, K.E.: Manned-unmanned teaming: US army robotic wingman vehicles. In: Cassenti, D.N. (ed.) AHFE 2018. AISC, vol. 780, pp. 89–100. Springer, Cham (2019). https://doi.org/10.1007/978-3-319-94223-0_9
12. Schaefer, K.E., Baker, A.L., Brewer, R.W., Patton, D., Canady, J., Metcalfe, J.S.: Assessing multi-agent human-autonomy teams: US Army Robotic Wingman gunnery operations. In: Micro-and Nanotechnology Sensors, Systems, and Applications XI, p. 109822B. International Society for Optics and Photonics (2019)
13. Brewer, R.W., Walker, A.J., Pursel, E.R., Cerame, E.J., Baker, A.L., Schaefer, K.E.: Assessment of manned-unmanned team performance: comprehensive after-action review technology development. In: Chen, J. (ed.) AHFE 2019. AISC, vol. 962, pp. 119–130. Springer, Cham (2020). https://doi.org/10.1007/978-3-030-20467-9_11
14. Uttal, D.H., O'Doherty, K.: Comprehending and learning from 'visualizations': a developmental perspective. In: Gilbert, J.K., Reiner, M., Nakhleh, M. (eds.) Visualization: Theory and Practice in Science Education, pp. 53–72. Springer, Dordrecht (2008). https://doi.org/10.1007/978-1-4020-5267-5_3
15. Rapp, D.N., Kurby, C.A.: The 'ins' and 'outs' of learning: Internal representations and external visualizations. In: Gilbert, J.K., Reiner, M., Nakhleh, M. (eds.) Visualization: Theory and Practice in Science Education, pp. 29–52. Springer, Dordrecht (2008). https://doi.org/10.1007/978-1-4020-5267-5_2
16. Aigner, W., Miksch, S., Müller, W., Schumann, H., Tominski, C.: Visualizing time-oriented data—a systematic view. Comput. Graph. **31**, 401–409 (2007)
17. Tufte, E.R.: The Visual Display of Quantitative Information. Graphics Press, Cheshire (2001)
18. Von Krogh, G., Von Hippel, E.: The promise of research on open source software. Manag. Sci. **52**, 975–983 (2006)
19. Command, U.: Training and qualification, crew. US), DotA (ed.) Training Circular No.: TC 3-20.31 (2015)
20. Baker, A.L., Schaefer, K.E., Hill, S.G.: Teamwork and Communication Methods and Metrics for Human-Autonomy Teaming. CCDC Army Research Laboratory Aberdeen Proving Ground United States (2019)

Development of a Neural Network Algorithm to Detect Soldier Load from Environmental Speech

Angelique Scharine(✉)

U.S. Army Research Laboratory, Human Research and Engineering Directorate, 7101 Mulberry Point Rd, Aberdeen Proving Ground, MD 21005-5425, USA
angelique.a.scharine.civ@mail.mil

Abstract. The objective was to develop a model based on speech input that can identify when team members need adaptive autonomous assistance. Human teams often adjust their behavior to work cohesively and effectively as a team. Similarly, it is beneficial for autonomous agents to be able to adaptively adjust to team needs. We constructed a convolutional recurrent neural network model based on those developed for the recognition of emotion from speech. Audio recordings from a recent field exercise were used to train and validate the model. These data were labeled according to whether the speech occurred during an engagement (engaged, neutral, or no-speech). The model classified more than 99% of the training, validation, and test sets correctly. This information will allow us to design systems in which autonomous agents can prioritize, assist with, and take autonomous control of tasks.

Keywords: Human-robot teaming · Neural-network modeling · Adaptive autonomous assistance

1 Introduction

As part of the US Army modernization priorities, the Army proposes to integrate autonomous systems, with the warfighter to augment and enhance Soldier and team performance in increasingly complex environments [1]. Ideally, autonomous systems will provide several advantages, such as reducing threats to Soldiers, providing additional distributed sensor capabilities to increase situation awareness, and enabling faster, more-informed decision making. While this will increase the lethality of such teams, it will also increase the responsibilities of its crewmembers. As such, a goal of the Combat Capabilities Development Command Army Research Laboratory's Human–Autonomy Teaming (HAT) Essential Research Program is to address the complexities of teaming humans and autonomous systems within an already complex tactical environment in order to create synergistic teams that function effectively and adapt to the dynamic nature of combat.

J. L. Wright et al. (Eds.): AHFE 2021, LNNS 264, pp. 54–62, 2021.
https://doi.org/10.1007/978-3-030-79763-8_7

2 Background

The emotional intelligence of a team predicts the efficacy of its performance [2]. Specifically, team members with high levels of emotional intelligence are able to modulate their behaviors by perceiving and adapting to the emotional state of other team members [3]. We learn to implicitly perceive the emotions of others by reading subtle cues, including facial expression and changes to the voice [4]. This allows us to adapt our own communications and actions. If a team member is overloaded, another team member can assist with tasks, thus dividing the workload more evenly. If a team member is angry, the other team member may act in ways to appease the other; this ability to adapt to others is a useful human skill. Good teamwork often consists of knowing when to pass along information and knowing when this information will not be helpful due to the constraints of the recipient's state [5].

Operational platforms are being designed today that can provide adaptive autonomous assistance (AAA), taking over elements of tasks to reduce the operator's operational load. Although autonomous agents may be reliable in the tasks they are programmed to do, they are not currently programmed to adapt their actions based on those of other team members. Therefore, to facilitate cohesive human–robot teamwork, it is necessary to identify when assistance through automation and cognitive streamlining would be beneficial and when it would decrease performance.

2.1 Models of Speech Emotion Recognition (SER)

Machine learning (ML), deep neural network (DNN), and artificial intelligence techniques have been used successfully to identify the emotional state of human talkers using the acoustical features of their voices [6–8]. These methods, which have proved useful in numerous ways, including automatic speech recognition and image analysis, extract the mathematical probability of a category or event from relationships in a given dataset [9]. In the simplest version, a classifier is trained on a database of labeled speech items. Features are extracted by dividing the speech waveform into overlapping segments (windows) and extracting statistical information about the signal [10]. These features usually include information about the spectral content, the log-energy content, the pitch period, and statistical information about their averages, minima, and maxima. Derivatives of these features provide information about feature changes over time [11, 12]. Thus, the task of the algorithm is to estimate the model parameters from the data, determining the weight of each feature with respect to the probability of an outcome, and by optimizing these weights, speech can be classified into output categories.

Recurrent neural networks (RNNs) models incorporate connections to the layer's previous state, thus allowing the model parameters to incorporate successive inputs, making them ideal for speech [13–15]. The use of RNN helps the model to account for individual differences. One form of RNN is that of long short-term memory (LSTM), which incorporates a forgetting parameter, sometimes called momentum, which determines the weight of the previous input on the current parameter. In this way, the parameters are more dependent on recent speech since the influence of earlier speech is reduced over time. Taking these elements a step further, convolutional neural networks (CNNs) transform data using a convolutional kernel followed by a pooling layer. The convolutional kernel acts as a filter, extracting different levels of features. The pooling layer reduces the number of similar or correlated features by subsampling.

3 A 3-D CRNN (LSTM) Model

The model structure is a 3D convolutional recurrent neural network (CRNN) model [15, 16]. The feature input consists of three vectors: the static log-Mel filterbank energies and their derivatives, delta (magnitude of change), and delta-delta (rate of change). This input provides both spectral and temporal information about the speech signal, effectively reducing the effect of emotion-irrelevant factors, such as the style of speaking, speech contents, and the environment. This part of the model uses code developed by James Lyons that computes the filterbank energies from the signal [17].

These log-Mel filterbank energies contain information about the power spectrum of speech. The signal is passed through a pre-emphasis filter, gets sliced into (overlapping) frames, and is filtered with a windowing function. Each frame is Fourier transformed, and the power spectrum is computed. From these, the filterbanks are computed. Essentially, the log-Mel filterbank process is designed to capture elements of the speech signal while accounting for human perception. The derivatives capture the change in this content, giving relative information about the changes in speech over time.

This gives a 3-D feature representation, $\in R^{t \times f \times c}$, as input to the CNN model, where t denotes the frame length, f denotes the Mel-filter bank number, and c denotes the number of feature channels.

$$m_i = \log(p_i) \tag{1}$$

$$m_i^d = \frac{\sum_{n=1}^{N} n(m_{i+n} - m_{i-n})}{2\sum_{n=1}^{N} n^2} \tag{2}$$

$$m_i^{dd} = \frac{\sum_{n=1}^{N} n\left(m_{i+n}^d - m_{i-n}^d\right)}{2\sum_{n+1}^{N} n^2} \tag{3}$$

The CRNN includes four CNN layers, one 3-D max-pooling layer, one linear layer, and one LSTM layer. Each direction contained 128 cells, resulting in a 256-dimensional representation of high-level features. The first convolutional layer had 128 feature maps, and each remaining convolutional layer had 256 feature maps. The filter size of each convolutional layer was 5 × 3, where 5 corresponded to the time axis, and 3 to the frequency axis. The pooling size for the max-pooling after the first convolutional layer was 2 × 2. To reduce the number of model parameters, the CNN output was flattened into a linear layer with 1768 outputs before passing it to the LSTM layer. Finally, the output from the LSTM layer was fed into a bidirectional RNN with long short-term memory cells (BLSTMs) for temporal summarization. These utterance-level representations were then passed to a fully connected layer with 64 output units to obtain higher-level representations that help the softmax classifier map the utterances into one of the three different emotion classes (nsp, neu, eng). Batch normalization was applied to each layer accelerated training and improved the generalization performance.

Table 1 gives a summary of the model layers. The CRNN architecture was implemented with the TensorFlow 2.0 toolkit [18] and Python 3.7 [19, 20]. The parameter weights of the model were optimized by minimizing the sparse categorical cross-entropy loss function, using the Adam optimizer [21]. The initial learning rate was set to 10^{-4} and the momentum was set to 0.9. A dropout factor of 0.2 was used to prevent the model from overtraining.

Table 1. Summary of compiled model

Block	Layer	Filter	Size	Params	Output	Activation
1	Conv2D	5 × 3	128	5888	(:, 296, 38, 128)	re_lu
	BatchNorm			512	(:, 296, 38, 128)	
	Leaky_re_lu					(Leak = 0.1)
	MaxPool2D	2 × 2			(:, 148, 19, 128)	
	Dropout					(p = .02)
2	Conv2D	5 × 3	256	491,776	(:, 144, 17, 256)	re_lu
	BatchNorm			1024	(:, 144, 17, 256)	
	Leaky_re_lu					(Leak = 0.1)
	Dropout					(p = 0.2)
3	Conv2D	5 × 3	256	983,296	(:, 140, 15, 256)	re_lu
	BatchNorm			1024	(:, 140, 15, 256)	
	Leaky_re_lu					(Leak = 0.1)
	Dropout					(p = 0.2)
4	Conv2D	5 × 3	256	983,296	(:, 136, 13, 256)	re_lu
	BatchNorm			1024	(:, 136, 13, 256)	
	Leaky_re_lu					(Leak = 0.1)
	Dropout					(p = 0.2)
5	Reshape				(:, 256, 1768)	
	Bidir LSTM	...	128	1,942,528	(:, 256, 256)	
	Flatten				(:, 65,536)	
	Dense			4,194,368	(:, 64)	re_lu
Class	Dense	...	...	195	(:, 3)	Softmax

4 Method

4.1 Field Exercise

Speech recordings were made during the October 2018 Wingman JCTD event held at Fort Benning, Georgia. The Wingman program developed prototype manned and unmanned vehicle platforms (high-mobility multipurpose wheeled vehicles [HMMWVs]) for use in joint manned–unmanned team gunnery tasks. The command and control (C2) vehicle carried the crew, while the robotic combat vehicle (RCV) was unmanned and weaponized, controlled by two crew members in the C2 vehicle. One crewmember was responsible for RCV mobility, and the other crewmember operated the weapon. During the JCTD event, two five-man crews (one Marine crew and one Army Soldier crew) conducted the live-fire gunnery evaluation exercises [22, 23].

The field exercises followed the standard engagement process, known as DIDEA (detect, identify, decide, engage, and assess). This required the teams to acquire targets, identify them as potential threats, decide to engage, engage the target(s), and then assess the effects of each firing action. The crews were trained on the command terminology. The training process included a dry run (no ammunition) and multiple blank-fire runs (blanks). After training, the Soldiers and the Marines (all male) took part in multiple stationary live-fire engagement runs. These runs were like the training runs but with two differences: live ammunition was used and, to comply with safety requirements, the robotic combat vehicle was chocked to prevent vehicle movement while the weapon was loaded.

4.2 Speech Input

Audio recordings were made during all runs; however, due to technical issues, some of these recordings were not clear enough to use. We used the recordings from one live-fire run for both the Soldiers and Marine [24].

The original recordings were made in stereo at a 44.1 kHz sampling rate with 16-bit quantization. Each track was down sampled to 16 kHz and split into two monaural tracks. Segments from both tracks were used for training and testing. Segments were formed that contained a single speech utterance; the shortest segments were 300 ms. Each recorded segment was manually labeled as nonspeech, neutral, or engaged. Nonspeech segments were segments with no speech for longer than 2 s. An engaged segment meant that the team was actively engaged in one of the target runs. Neutral segments contained casual conversation and sometimes discussion of issues experienced during the test. Descriptive statistics for the wave-files are shown in Table 2.

The wav-files for each field exercise were divided into segments, each labeled with a single state category. Text files contained the wav-file segment names and their labels. This larger dataset was divided into three subsets, 70% of which were used for training, 15% for cross validation, and 15% for testing. The items within these sets were randomly chosen from the overall set to contain equivalent proportions of nonspeech (nsp), neutral (neu), and engaged (eng) items (Table 2).

Table 2. Wingman speech data

	Test M	Test A	Total
Type	Live run	Live run	...
Participants	Marines	Army	...
Duration (mm:ss.msec)	31:12.000	29:29.000	1:00:41.000
Count nonspeech segments	280	220	500
Count neutral segments	148	146	294
Count engaged segments	176	92	268
Count all segments	604	458	1062
Mean duration nonspeech	00:07.617	00:08.766	00:08.123
Mean duration neutral	00:04.192	00:04.931	00:04.559
Mean duration engaged	00:03.646	00:05.079	00:04.138

5 Results

The reasoning behind the use of validation sets is to avoid overfitting the data to the training set. By maximizing both the fit to the training set and to the validation set, it increases the probability of generalizability to an independent test set. Here it was possible to fit parameters to the training set, getting accuracy scores of nearly 100%. Although much of the training occurred in the first 20 epochs, small incremental changes were observed throughout and additional training led to greater accuracy for the test set. The summary of the first 10 and last 5 epochs are shown in Table 3. The trained model classified the test set with more than 99% accuracy.

6 Discussion

Despite the high performance of the model in detecting these classes, the generalizability of this model is limited. Probably the most significant weakness for this model presently is that the size of the dataset is so much smaller than the number of parameters used to fit to it. Another potential weakness is that the training, validation and test sets might not have been completely independent. Because each track of the stereo input was used separately, there is a chance that items in the validation and test sets matched items in the training set. Hopefully, overtraining was avoided by using a drop-out factor.

It will be important to the development of this project to obtain more data and test the model's performance, controlling for factors such as potential overtraining. Plans include testing the model's performance on other speech datasets. A similar set of data was collected for the Wingman project in 2019. As with the 2018 study, it incorporated a team of one manned and one unmanned vehicle. By evaluating the accuracy of the model predictions with new data, we can measure the extent to which the model generalizes. Additional plans include testing the model's efficacy for larger teams using data from teams consisting of two manned and four unmanned vehicles.

Table 3. Model history for the first 10 and last 5 epochs

Epoch	Training loss	Training accuracy	Validation loss	Validation accuracy
1	0.7268	0.7142	0.6358	0.7848
2	0.3707	0.8447	2.9755	0.1646
3	0.2518	0.9163	1.6121	0.2447
4	0.1714	0.9385	0.7501	0.7637
5	0.0720	0.9778	0.8656	0.7806
6	0.0603	0.9861	0.0578	0.9831
7	0.0332	0.9924	0.1073	0.9620
8	0.0147	0.9981	0.0338	0.9873
9	0.0050	0.9975	0.0070	0.9958
10	0.0027	0.9987	0.0035	0.9958
996	1.15E-06	1.0000	5.50E-09	1.0000
997	2.50E-07	1.0000	5.00E-09	1.0000
998	7.70E-06	1.0000	4.00E-09	1.0000
999	3.30E-06	1.0000	3.00E-09	1.0000
1000	1.50E-04	1.0000	1.50E-09	1.0000

Note: Performance of the trained model on the test set was also 1.0000

7 Conclusions

Using speech to predict Soldier load to trigger AAA is a novel application of NN modeling. As such, it represents a significant contribution to the Wingman, Next-Generation-Combat-Vehicle (NGCV) and Mission Enabling Technologies Demonstrator (MET-D) programs because it is not reliant on additional body-worn sensors. As such, it is less vulnerable to contact and transmission issues. It can work with anyone in the operational environment without outfitting them with specialized sensors. NN-derived models have several potential applications to the combat vehicle environment, from use in speech recognition to target detection to navigation. Thus, it is a tool we may see often in future operations.

The model easily distinguished the three categories of speech (nonspeech, neutral, and engaged), making it a promising option for use as an algorithm to detect when Soldiers are engaged or are cognitively loaded. Although some work needs to be done before it can be implemented in a test vehicle, these results are encouraging.

References

1. South, T.: From unmanned teaming to new capabilities, the Army's Next Generation Combat Vehicle will be a leap forward - Reader Mode. Army Times (2019)

2. Elfenbein, H.A.: Team emotional intelligence: what it can mean and how it can affect performance. In: Linking Emotional Intelligence and Performance at Work: Current Research Evidence with Individuals and Groups, pp. 165–184. Taylor and Francis (2013). https://doi.org/10.4324/9780203763896
3. Waldinger, R.J., Hauser, S.T., Schulz, M.S., Allen, J.P., Crowell, J.A.: Reading Others' Emotions: The Role of Intuitive Judgments in Predicting Marital Satisfaction, Quality, and Stability (2004). https://doi.org/10.1037/0893-3200.18.1.58
4. Neubauer, C., Chollet, M., Mozgai, S., Dennison, M., Khooshabeh, P., Scherer, S.: The relationship between task-induced stress, vocal changes, and physiological state during a dyadic team task. In: ICMI 2017 - Proceedings of the 19th ACM International Conference on Multimodal Interaction (2017). https://doi.org/10.1145/3136755.3136804
5. Lingard, L., et al.: Communication failures in the operating room: an observational classification of recurrent types and effects. Qual. Saf. Heal. Care. **13**, 330–334 (2004). https://doi.org/10.1136/qshc.2003.008425
6. Casale, S., Russo, A., Scebba, G., Serrano, S.: Speech emotion classification using machine learning algorithms. In: Proceedings - IEEE International Conference on Semantic Computing 2008, ICSC 2008 (2008). https://doi.org/10.1109/ICSC.2008.43
7. Stuhlsatz, A., Meyer, C., Eyben, F., Zielke, T., Meier, G., Schuller, B.: Deep neural networks for acoustic emotion recognition: raising the benchmarks. In: ICASSP, IEEE International Conference on Acoustics, Speech and Signal Processing - Proceedings (2011). https://doi.org/10.1109/ICASSP.2011.5947651
8. Koolagudi, S.G., Rao, K.S.: Emotion recognition from speech: a review (2012). https://doi.org/10.1007/s10772-011-9125-1
9. Bishop, C.M.: Pattern Recognition and Machine Learning. Springer, New York (2006)
10. Ververidis, D., Kotropoulos, C.: Emotional speech recognition: resources, features, and methods. Speech Commun. (2006). https://doi.org/10.1016/j.specom.2006.04.003
11. Schuller, B., Rigoll, G., Lang, M.: Hidden Markov model-based speech emotion recognition. In: Proceedings - IEEE International Conference on Multimedia and Expo (2003). https://doi.org/10.1109/ICME.2003.1220939
12. El. Ayadi, M., Kamel, M.S., Karray, F.: Survey on speech emotion recognition: Features, classification schemes, and databases. Pattern Recognit. (2011). https://doi.org/10.1016/j.patcog.2010.09.020
13. Badshah, A.M., Ahmad, J., Rahim, N., Baik, S.W.: Speech emotion recognition from spectrograms with deep convolutional neural network. In: 2017 International Conference on Platform Technology and Service, PlatCon 2017 - Proceedings (2017). https://doi.org/10.1109/PlatCon.2017.7883728
14. Lim, W., Jang, D., Lee, T.: Speech emotion recognition using convolutional and recurrent neural networks. In: 2016 Asia-Pacific Signal and Information Processing Association Annual Summit and Conference, APSIPA 2016 (2017). https://doi.org/10.1109/APSIPA.2016.7820699
15. Chen, M., He, X., Yang, J., Zhang, H.: 3-D convolutional recurrent neural networks with attention model for speech emotion recognition. IEEE Signal Process. Lett. **25**, 1440–1444 (2018). https://doi.org/10.1109/LSP.2018.2860246
16. Huang, C.W., Narayanan, S.S.: Attention assisted discovery of sub-utterance structure in speech emotion recognition. In: Proceedings of the Annual Conference of the International Speech Communication Association, INTERSPEECH, pp. 1387–1391. International Speech and Communication Association (2016). https://doi.org/10.21437/Interspeech.2016-448
17. Lyons, J.: python_speech_features Documentation Release 0.1.0 (2017)
18. Abadi, M., et al.: TensorFlow: a system for large-scale machine learning (2016)
19. van Rossum, G., Drake, F.L.: Python 3 Reference Manual (2009)

20. Scharine, A.A., Schaefer, K.E.: Adapting a Model of Emotional State Recognition to Detect Stress in a High-Noise Environment (ARL-TR-9137), Aberdeen Proving Ground, MD (2021)
21. Kingma, D.P., Ba, J.: Adam: A Method for Stochastic Optimization (2014)
22. Schaefer, K.E., et al.: US Army Wingman Joint Capability Technology Demonstration (JCTD): Initial Soldier and Marine Feedback on Manned-Unmanned Gunnery Operations (ARL-TR-8663). Aberdeen Proving Ground, MD (2018)
23. Brewer, R.W., Cerame, E., Pursel, E.R., Zimmermann, A., Schaefer, K.E.: Manned-unmanned teaming: US army robotic wingman vehicles. In: Advances in Intelligent Systems and Computing (2019). https://doi.org/10.1007/978-3-319-94223-0_9
24. Baker, A.L., et al.: Team Trust in Human-Autonomy Teams: Analysis of Crew Communication during Manned-Unmanned Gunnery Operations (ARL-TR-8969). Aberdeen Proving Ground, MD (2018)

Incorporating Key Leader Engagement into Tactical Simulations

Jacob Ziadeh(✉), Jordan Blackman, Morgan Chewning-Kulick, Mazie Jones, Anna Loar, James Ness, Arno Harholt, and Jacob Whisenhunt

Department of Behavioral Sciences and Leadership, United States Military Academy, 601 Cullum Rd., West Point, NY 10996, USA
{Jacob.Ziadeh,Jordan.Blackman,Morgan.Chewning-Kulick, Mazie.Jones,Anna.Loar,James.Ness,Jacob.Whisenhunt}@westpoint.edu

Abstract. Following the recent changes in conventional warfare throughout the Middle East and the importance of "winning hearts and minds," the military has placed emphasis on establishing bonds with the local populace and leadership through Key Leader Engagements (KLEs) [1]. Virtual reality (VR) technologies such as the Virtual Human Toolkit [2], allow for simulations of KLEs to be contained and disseminated with ease. However, coding the natural language processor for meaningful immersive learning requires the language processor to engage the user in a story. Capturing the experiences of those with direct KLE experience and personifying an Intelligent Agent with the essentials of that experience are key to an engaging story. The process begins with front-end analysis (FEA) [3]. This paper describes that process pursuant to a simulation designed to train awareness of critical features of the KLE.

Keywords: Virtual reality · Military · Simulation development · Afghanistan · Thurstone scaling · Front-end analysis · Key leader engagement

1 Introduction

"To its credit, the institutional Army learned these hard lessons, from successes and failures, during and after each of the cases examined in this study. However, they were often forgotten as the Army prepared for the next major conventional conflict.

These lessons are still important and relevant today." [4].

Lessons are often lost in large organizations because they are rarely applied within all levels of the organization. The lessons are documented at the higher echelons of the Army and archived. However, as archives, the lessons learned are typically not effectively pushed down to the operator in the form of relevant stories and practice. Doing so would allow the Army to not only fully embrace each lesson, but also establish a tradition of diffusing those lessons across echelons and cohorts.

J. L. Wright et al. (Eds.): AHFE 2021, LNNS 264, pp. 63–70, 2021.
https://doi.org/10.1007/978-3-030-79763-8_8

The Army does have an existing means to pass on stories: mentorship. Mentorship is fundamental to accelerating life experience through the wisdom of those who come before us or as Sir Isaac Newton remarked, "if I have seen further it is by standing on the shoulders of Giants [5]." Mentors are the "Giants" who add perspective to issues confronting a leader, put feedback in context, guide resolution, and mitigate repeating past errors. However, access to mentors is limited by the ratio of those seeking mentorship to those who have relevant experiences. It is also limited by the loss of contact with mentors as soldiers disperse during cyclic rotations to other assignments.

The question is then how to leverage principles of mentorship to accelerate learning and mitigate forgetting, as an institution, the lessons gained from those with real-world experience. Knowledge gained from real-world experience can take years to form, and cohorts within an organization may not encounter the situations in training or combat that would allow them to reach a given lesson independently. That experience is difficult, if not impossible to convey from archives beyond recounting steps of a process and post hoc assessments of the outcomes. In the archive, the logic of the event is conveyed, often through a secondary source, and thus the learner does not directly experience the situation as conveyed through a mentor. As an immersive story, simulations are means to mitigate the loss of real-world experience.

Simulations are proven in fields such as aviation and medicine to provide users with relevant experiences to gain, redintegrate, and improve skills. Within a simulation, one can gain insight into the strengths and weaknesses of a particular course of action and explore the breadth of possible decisions the simulation has to offer to become fully aware of the nature and effects of such decisions [6]. In this paper we present a method to build a training simulation to familiarize the learner with Key Leader Engagements (KLE) through capturing the experiences of those with direct experience, personifying an Intelligent Agent with the essentials of that experience.

The quintessential human means for diffusing lessons is through passing on stories [7]. Representations of experiences, as in cave paintings and storytelling, are the oldest traditions of recounting events, imparting lessons, and projecting affect. These formats structure information in part-whole relations that supply the listener or reader with schematic frameworks to interpret past, present, or future analogous events. The diffusion of stories using technology-mediated means, in this case a virtual reality environment, imparts the lessons within these stories in a rapid and salient manner that is resistant to the fleeting nature of humanity's collective memory.

Since all permutations of experiences in the deployed environment cannot be known in advance, the prescribed intentional memorization of facts or scripted sequences are likely to be of limited value. Thus, a narrative approach, as found in the traditions of storytelling, is taken to orient the observer to recurrent themes [8]. Examples of themes are the "clash of cultures," "struggle with moral dilemmas," or "tempering of frustration" - the latter being the tempering of frustration of individual achievement motivation in deference to slower, systematic change of a nation developing to provide security to its citizens. To capture relevant themes of the KLE, we began with a Front-End Analysis which helped determine the essential elements to constructing an immersive experience. From this we were able to select mentors that would inform the story. A storyboard technique was then used to codify the story pursuant to building the simulation.

The simulation blended two software systems. Virtual Battlespace [9] was used to build the tactical context and Virtual Human Toolkit [2] was used to build the Intelligent Agent for the KLE. The remainder of this paper provides an overview of the process of building cultural training, how to incorporate an Intelligent Agent, and lessons learned.

2 Cultural Training Building

> "Afghan culture is founded on personal relationships. Earning the trust of the people is a large part of our mission. Build relationships with tribal, community, and religious leaders. Success requires communication, collaboration, and cooperation."
>
> General Stanley McChrystal [10].

With the mentorship of the 10th Mountain Division in the Webisode Narrative Themes, we are provided with insight into KLEs at the tactical level [11]. The 10th Mountain Division embraces several themes regarding tensions between locals, showing the amount of cultural difference, moral confrontation, and aggravation.

Without delving into documentation that is only for official use, we can best understand the current status of KLE training through the Commander's Handbook (2010). The handbook mentions the process of teaching commanders through the use of KLE Cells, KLE Plans, KLE Assignments, and later, KLE Assessments. KLE Cells are groups of trained individuals (Civil Affairs and Public Affairs) who help to modify face-to-face engagements and messages so that they reach the local audience in the best manner. That said, though these Cells often contain good information, they are often poorly manned, hard to reach, and rarely available. KLE Plans and Assignments are implemented engagement plans that assign local leadership to units so that issues of the same nature are not discussed or resolved by different units with different aims. Through this, we understand the level of thought implemented into military KLEs, but we notice how they may fail to achieve the end goal of building meaningful engagements with local leadership.

Although the KLE Plans and Assignments help to improve synergy, the cultural experts in KLE Cells are few and incredibly hard to reach, which provides motivation for the development of a virtual system. A virtual system could easily and comprehensively offer and disseminate all that KLE trainings have to best equip commanders with the tools for managing a cultural exchange, misunderstandings, or whatever else may come about from exchanges with KLE Cells.

2.1 The Systematic Process

To correctly build a virtual simulation with a culturally cognizant Intelligent Agent, a systematic process is required, one that helps track goals, assess completion and quality, and provide a full analysis of what is expected. A Gantt chart was composed to guide the process and keep the team engaged. This chart helps to guide the entirety of the developmental process by creating a detailed roadmap, discussing the priorities of work and their associated timelines. By using this chart, it enables the team to readily track completion and analyze the needs of the project.

2.2 Front-End Analysis

Collecting the entirety of an important face-to-face interaction can be difficult and carries a variety of different variables. For instance, being able to properly represent the cultural differences, social issues, and political tensions within the environment poses a breadth of complexities. By using the Front-End Analysis (FEA) method, the developmental process is streamlined. The analysis itself is typically used to condense training down for businesses, resulting in fewer costs and less time wasted [12]. In the case of this research, the FEA provides a framework with which we may understand and capture the complexity of the persona that the simulation is centered around. In doing so, the research team can minimize the costs of developing and providing a quality persona. This FEA was composed of several levels, focusing on demographics, culture (religion, food, customs, etc.), environmental implications, regional history, and interest analysis. Through this and an analysis of the functions and tasks that the training is meant to purvey, the FEA model allows for the creation of a holistic persona for military personnel to interact with and learn from on demand. Moreover, the framework could be applicable to building other personas within other regions and situations. That said, the FEA in this case helps to realize what exactly is important to run the simulations and understand exactly what servicemen may need to understand when venturing abroad and having day-to-day interactions with locals.

2.3 Use of Thurstone Scaling

No matter how accurate of a representation the Intelligent Agent can create, if the trainee is not able to properly identify the character, then the training is a failure. There are specific audiovisual cues which the trainee relies on, which can be discovered with Thurstone scaling [13]. A Thurstone scale begins with a focus group representative of the population of interest. The group is asked to generate statements that are most like, least like, and neutral on an issue in question. A separate group is asked to rank each statement along an 11-point scale from least like to most like in representing the issue. A cumulative frequency of responses to each statement is generated to determine the scale score and agreement among the group. The technique can be used to personify an Intelligent Agent.

For example, Fig. 1 shows the results of the Thurstone Scaling technique to capture the US Military Academy Corps of Cadets' impressions of an Arms Room officer. The Corps of Cadets is composed of four distinct classes, each of differing class rank and status. The ranks from least status to highest are Plebes or Freshman, Yuks or sophomores, Cows or juniors, and Firsties or seniors. As Fig. 1 indicates, there are striking similarities and differences of opinions concerning the Arms Room officer. For example, Q2-14 shows a difference in the Plebes versus the other classes in comparative judgment of harshness on those who do not follow rules. The context is that the Arms Room has a sign, posted in a non-obvious place, that states that the forms will be filled out in pen. The Arms Room has a table with forms and pens and pencils laying on the table. If you use a pencil, the form is shredded, the individual is berated and sent to the end of the line. Plebes do not see this as harsh as opposed to the other class ranks. The discrepancy points to a personification to represent in the Intelligent Agent to help the Plebes understand the difference between enforcing standards and hazing.

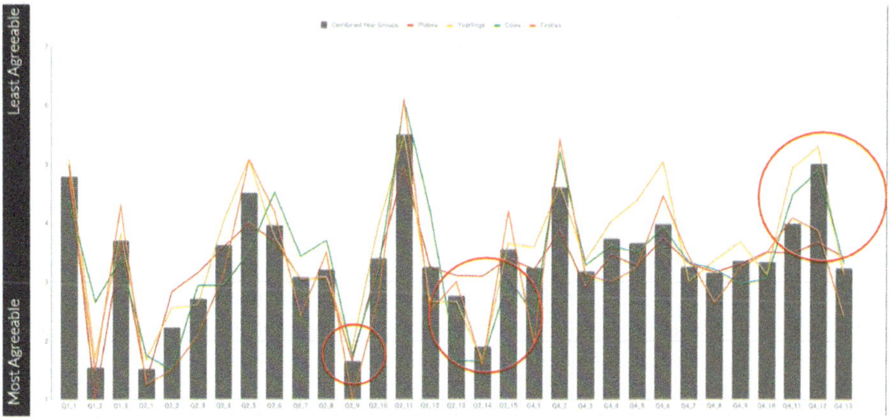

Fig. 1. Results of a thurstone scaling

2.4 Building Intelligent Agents

Simulating KLEs requires a substantial amount of data, driven by insight from experienced military leaders. However, data and algorithms alone will not suffice for the development of an Intelligent Agent [14]. For this reason, it is important to have military leaders experienced in hostile negotiation environments as a part of the data collection process. The experience of a military leader is an element that cannot simply be quantified, which is why it is so unique. The quality of KLE training depends on the scenario. The strength of the scenario ultimately drives the data and creates an efficient training environment.

Tolman's three dynamisms: regression, fixation, and displacement of aggression onto outgroups all possess relevance to the development of KLE training [15]. Regression describes the event in which an individual resorts to impetuous behavior in the face of an overwhelming situation [16]. This is important to consider as the development of the Intelligent Agent occurs, as the intent of the training is to put participants in uncomfortable positions to evaluate their behavior and responses. Fixation describes what is commonly known as expert bias, an instance in which an individual that possess a significant amount of knowledge on a given subject refuses to consider other scientific reasoning [14]. Human behavior is subject to bias when a certain amount of knowledge is acquired, which can be detrimental in a KLE because modern-day leaders operate in an ambiguous geo-political climate. Displacement of aggression onto outgroups describes the process of an individual directing frustration to others when they feel threatened, which often occurs in hostile environments [14]. This is relevant as leaders who encounter aggression must learn to counter it in a tactful manner to successfully de-escalate any hostile situation they encounter. Significantly, incorporating elements of Tolman's dynamisms into the development of this Intelligent Agent provide an exceptional KLE training experience.

3 Building the Interface

3.1 The Virtual Human Toolkit

The Virtual Human Toolkit (VHToolkit) allows for a realistic and immersive environment in which operators can naturally interact with the system and receive instant feedback. Virtual reality training simulations have become organic to models where iterative real-life experiences are not possible or potentially harmful. Current sentiment undervalues the level of nuance which is possible while simulating human interactions with the VHToolkit. The version used in these training engagements accurately simulates non-verbal communication such as posture, facial expression, and natural movement [17]. Earlier imprecise immersion systems left significant doubt in the veracity of using virtual reality in training, so long as realistic conditions are set there is no reason in which the VHToolkit cannot be used. Proper preparation and understanding of the target outcome allow for the Toolkit to augment human to human interaction in a training environment.

The depth of the simulated engagement is based upon the predicted interactions and templated responses created by the Toolkit operator. Using a method of storyboarding ideal interactions, the creator then deviates from the ideal and approaches the simulation from the perspective of those undergoing training. The storyboard or a series of graphics which represent the dialogue, environment and nonverbal communication which is presented to the trainer. This level of control allows for specific triggers or stimuli to promote the trainee to the ideal interaction or highlight real-life pitfalls which have been identified. The storyboard allows the VHToolkit to utilize previous scenarios and allow for the trainee to make mistakes without real-life consequences. The dialogue follows the natural progression initiated by the trainee. The Toolkit accounts for slang, inconsistent speech and continues to provide feedback. In order to prevent "gaming" of the system, levels of randomness and a constantly evolving dialogue tree prevents the trainee from trying every option. As the dialogue tree progresses, the interaction moves along with the storyboard and every response made by the trainee is weighed.

3.2 Developing the Intelligent Agent

The main focus when developing an Intelligent Agent should be the main characteristics and fine details of the key leaders (KLs) so that once the construction of the Intelligent Agent in the VHToolkit begins there is an accurate representation of the leader to go from. Without having experienced a KLE, we interviewed various professionals who led KLEs in Afghanistan, were Afghan Interpreters and were Afghan KLs assigned to the Afghan Consulate in D.C. Employing the Thurston scaling technique to the interview data, we generated an Intelligent Agent that reflects past common experiences. The interviews and cultural research allowed us to create multiple different scenarios to generate an Intelligent Agent to train those unfamiliar with KLEs on salient features of the exchanges.

One of the main areas of focus was ensuring that these Intelligent Agents were realistic and provided adequate training to soldiers before deploying. The VHToolkit provides the ability to use a natural language processor and gestures. It is crucial to use both tools to ensure that the Intelligent Agent does not come off as robotic and is as humanly accurate a depiction of those who are KLs in Afghanistan. Users can utilize the Intelligent Agent to talk to or type out questions. From there the Intelligent Agent reacts and the situation plays out, however there is not just one way for the Intelligent Agent to react, therefore making it a training environment. Each user can run through the simulation and see what methods or interactions work the best when it comes to these interactions.

4 Limitations

Afghanistan is an ethnically diverse tribal country. Afghan key leaders represent the breadth of the ethnic diversity of which we only have experiences of those from the Kabul area. Thus, the simulation is only valid for encounters in Kabul and the immediate surrounding area.

In building the scenario, the team used DoD owned software to maximize cost/benefit. The tactical scenarios were developed in Virtual Battle Space (VBS). The KLE was developed in Virtual Human Toolkit. The systems are not integrated and so a "fade to black" transition was adopted to switch between the VHT governed KLE and the VBS governed tactical scenario. The transition did disrupt the immersion. Improved transitions or better integration of the software is something the team is working on.

Scoring the KLE to render a tactical consequence was not automated. This required an additional operator to manage the simulation. Workloads for operating the simulation need to be assessed and distributed such that the simulation is easily run with no more than 2 operator/trainers managing the training simulation.

5 Discussion

Following a review of literature and research regarding program design, product development, and procedural analysis in development, the research team was able to field a worthwhile product on a strict time schedule and budget. By utilizing several tenants and tools of product design, namely Front-End Analysis, the Thurstone Scale, an interviewing process, and frameworks for storyboarding and scripting, the development of the KLE situation was streamlined. Being able to identify the process of developing and reviewing such a complex product up front and devising a strict schedule for the research team to adhere to was critical in coming up with a worthwhile product in the short time span of half a year.

From the processes, the team was able to retrieve a breadth of cultural intricacies and implications. The interviewing process gave the team important data to consider in building the situation. The availability of experienced combat veterans was weighed on heavily to conduct an analysis of cultural behaviors from the perspectives of servicemen and women.

Building the scenarios and integrating VHToolkit and Virtual Battlespace required a basic level of understanding of the programs. The programs are relatively easy to script but do require some time to learn each system. The scripts and accompanying user guides, once built, can be saved and distributed to others for use in their unit's training. The ease of dissemination and its fostering of communication across training elements mitigates the loss of lessons learned and facilitates across the DoD a culture of lessons learning.

References

1. Lapping, B.: End of Empire. St. Martin's Press (1985)
2. Hartholt, A., et al.: All together now. In: Aylett, Ruth, Krenn, Brigitte, Pelachaud, Catherine, Shimodaira, Hiroshi (eds.) IVA 2013. LNCS (LNAI), vol. 8108, pp. 368–381. Springer, Heidelberg (2013). https://doi.org/10.1007/978-3-642-40415-3_33
3. Harless, J.H.: An ounce of analysis (is worth a pound of objectives). Harless Performance Guild, Newnan (1970)
4. Ramsey, R.D.: Advising indigenous forces: American Advisors in Korea, Vietnam, and El Salvador; Global War on Terrorism Occasional Paper 18. Combat Studies Institute Press, Fort Leavenworth, Kansas (2006)
5. Newton, I., Hooke, R.: Isaac Newton letter to Robert Hooke (1675)
6. Ness, J., Kolditz, T. Lewis, P., Lam, D.: Development and fielding of the US army leader self-development portfolio. In: Bartone, P.T., Pastel, R.H., Vaitkus, M. (eds.) The 71F Advantage: Applying Army Research Psychology for Health and Performance Gains. Center for Technology and National Security Policy, National Defense University, Washington, D.C. (2010)
7. Lord, A.B.: The Singer of Tales. Atheneum, New York (1971)
8. Campbell, J.: Myths to Live By. Bantam Books, New York (1972)
9. Virtual Battlespace 3. Bohemia Interactive Simulations. Orlando, FL
10. Commander's Handbook for Strategic Communication and Communication Strategy Version 3.0, US Joint Forces Command Joint Warfighting Center
11. Self-directed Immersive Readiness Tool (ASIRT): Developed for TRADOC, Combined Arms Center-Training, Ft Leavenworth, KA. https://asirt.ict.usc.edu/
12. Schlegel, M.J.: A Handbook of Instructional and Training Program Design (1995)
13. Thurstone, L.L.: Attitudes can be measured. Am. J. Sociol. **33**, 529–554 (1928)
14. Wasdyke, C., Ness, J.: Lexicon of mood: ethical implications for IA design. DoD Human Factors Technical Advisory Group Meeting 73, Aberdeen, MD, 15–19 April 2019
15. Tolman, E.C.: Cognitive maps in rats and men. Psychol. Rev. **55**(4), 189 (1948)
16. Belani, H., Vukovic, M., Car, Ž.: Requirements engineering challenges in building AI-based complex systems. In: 2019 IEEE 27th International Requirements Engineering Conference Workshops (REW), pp. 252–255. IEEE, September 2019
17. Hartholt, A., Fast, E., Reilly, A., Whitcup, W., Liewer, M., Mozgai, S.: Multi-platform expansion of the virtual human toolkit: ubiquitous conversational agents. Int. J. Seman. Comput. **14**(03), 315–332 (2020)

Computational Modeling Approaches in Politics and Economics

Assess Electric Vehicle Adoption Through an Agent Based Approach

Yi Ling Chang(✉), Yuan-Yuan Lee, and Zining Yang

School of Social Science, Policy and Evaluation,
Claremont Graduate University, Claremont, CA, USA
{yi-ling.chang,yuan-yuan.lee,zining.yang}@cgu.edu

Abstract. With technological innovations and support from government agencies, the auto industry for electric vehicle (EVs) has now gained a renowned popularity among consumers who are in search for emission-free substitutes to replace diesel and petrol-fueled vehicles. This paper employs an Agent-Based Modeling (ABM) technique, providing us the opportunity to capture interactions among agents (consumers), each with different socio-economic levels and unique utility function. The simulated result allows us to analyze micro-level EV adoption and macro-level impact from changing environment. The aim of our research is to investigate interconnected factors, especially on decision making process, behind the evolving preference towards EVs. Under each tick (tick = day), agents are designated to move from home, to work and have the opportunity to other locations of their choice. Hence, each agent has distinct daily travel distance to mimic reality. In addition, we input data collect from official government agencies including average gas price, EVs rebate program offering and vehicle range per charge to empirically analyze the comparison between electric and gas vehicles. Furthermore, we leverage cost benefit analysis using willingness to pay (WTP) and affordability as key thresholds in the decision-making process; only when both affordability and willingness thresholds are met, agents are then able to consider adopting the optimal vehicle type to maximize their utilities. In three scenario analysis, external macro-level shocks that can impact consumer's affordability or/and willingness: number of available public charging stations, price of EVs, and EV rebate programs are considered. Simulated result presents the aggregated emergent behaviors and illustrates a calibrated multi-agent framework that enables us to address issues related to purchase incentives and consumer price sensitivity in decision-making process.

Keywords: Agent-based modeling · Cost benefit analysis · Electric vehicle adoption

1 Introduction

Electric vehicles (EVs) have been identified as a potential mitigation method to reduce energy consumption and carbon dioxide (CO2) emissions in the transport sector. The positive long-term implication has generated a widespread interest among researchers to

J. L. Wright et al. (Eds.): AHFE 2021, LNNS 264, pp. 73–83, 2021.
https://doi.org/10.1007/978-3-030-79763-8_9

identify possible drivers and barriers against consumer adoption of EVs. Consequently, to encourage the adoption of EVs, various entities including both federal and state government, electric utility agencies across the United States have offered incentives as an effort to positively influence consumer preference and adoption behaviors. Yet, adoption of EVs at the local level remain a challenging topic for social science research. Given the complexity of human interaction in socio-economic systems, assessments made through conventional approach poses shortcomings. Under the conventional approach, which analyze individual behavior from an aggregate level and assumes unified adaptive level, can only capture part of complex relationship between human decision making and societies.

Literature on adoption of EVs have been studied and experimented in various aspects including econometrics and psychology. However, extensive research typically has taken a myopic viewpoint of the issue, such as the consumer's perspective [1, 2] and policy effectiveness [3, 4]. In this paper, we offer an alternative framework using an agent-based modeling (ABM) technique to assess interdependencies between adaptive human interactions and adoption of EVs. In particular, we explore the impact of attitudes on preferences using rational choice theory as foundation and investigate the conflicting forces of maximizing utility while minimizing cost that comes play during individual decision making through cost benefit analysis. This study adopts ABM, which encompass the strength to investigate strategies stimulating behavioral changes (e.g. decision to switch from gas vehicles to EVs), allowing systematic exploration of environmental changes over time and providing insights to calibrate effective incentive policies. To increase model validity, we parameterize empirical data for average gas price, price of EVs, vehicle range per charge, and rebate offered.

This paper is organized as the following: Sect. 2 we summarize related work on consumer behavior change to EV adoption, from theoretical conventional approach to dynamic interactive agents. Our model design framework and calibrated parameters are described in detail in Sect. 3. Under Sect. 4, we present simulation experiment results under various scenarios. Validation process illustrated through sensitivity analysis follows in Sect. 5. Finally, policy implication and discussion for future extension is concluded in Sect. 6.

2 Theoretical Background

Researchers from multiple disciplines (most prominently in social science, economics, psychology and neuroscience) have been utilizing converging theories and methods to study the role of decision rules and emergent social norms with the goal to better understand human decision-making process [5]. They found that distinct differences between the behavior of the system and the behaviors of agents within that system. The underlying dependencies component leading agents have the ability to continue to revise their rules for interactions (changing behaviors); whereas independent agents do not [6, 7]. One of the key insights observed from the lens of complex dynamic systems is that local interacting parts, since governed by its own rules, can collectively impact the overall outcome and/or influence actions of others. A familiar example is self-organization, a spontaneous emergence of new behavior pattern that has arisen from a combined activity of multiple dynamically linked components following simple rules [8].

Influence of Social Norms in Decision Making: Empirical findings from experiments designed and analyzed through conventional approach, which commonly assumes individuals as independent acting agents, reveals its limitation with inconsistent results in specifying relevance factors related to local interactions that are driving changes in agents' behavior patterns [9]. Computational focused researchers have explored micro-foundations of global patterns from a bottom-up perspective, leaning on ABM technique to incorporate the interdependency among agents [10]. More precisely, the interdependency can generate behaviors such as imitation and sanctioning, which are actions agents may potentially illustrate in response to the influence they receive from other individuals. On the other hand, there can also be indirect interdependency; when individual behavior changes elements of the environment, which in turn impact the behavior of other agents, here each agent's decision is then dependent in part of the choices of others [11].

In this paper, individual agents are the unit of analysis, we model their interactions and observe the emergent properties of change in preference between gas and electric vehicles over time. More specifically, agents follow a set of interaction rules with each other and with the environment, calculating affordability and willingness to purchase, to optimize their own utilities through the use of cost benefit. Hence, the complexity within the model stems from agent's individual utility maximization can have a direct impact to other agents' decision making and consequently have an indirect influence to the overall environment. Furthermore, previous literature and case studies which have identified issue of a "stated preference-action" gap of EV adoption from inconclusive results related to effectiveness of incentives [12]. In general, EV adoption and change in preference involve socio-economical difference between individuals, information spread through communication, and knowledge update. These processes, though not viable to specify through the conventional equilibrium methods, are suitable for agent-based modeling techniques. We aim to address the stated challenges through our ABM, providing valuable inferences on the adoption of EVs.

3 ABM Development

The goal of our ABM is to investigate the range of emergent behaviors on EV adoption, the designed micro-level framework (Fig. 1) is focused on agent interactions; each agent has its own unique complex cognition process of evaluating and choosing between the types of vehicle to optimize their utilities. The experiment includes agents with two types of vehicle preferences, i.e. gas and electric vehicle, and each individual is randomly assigned a "home" (not shared with other agents) and "work" location (can be shared with others). Each tick, agents will than move among different locations: home, work and potentially other locations. The inclusion of additional locations is one of our efforts in building a realistic model, while it also allows us to test public charging station availability and vehicle range may have on agents' preferences.

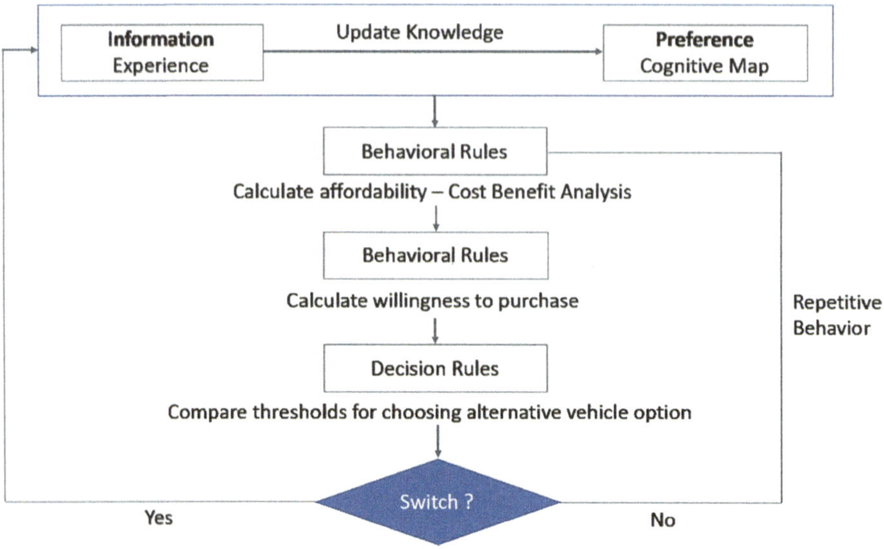

Fig. 1. Flowchart diagram of our designed agent behavior path.

3.1 Descriptive Information

To accurately capture the changing preferences between gas and electric vehicles, we incorporate real data for average cost for gas and electricity per mile [13], average car price for both types of vehicles, average vehicle charge range by types of charger (e.g. home, work, and public), as well as the maximum number of hours can be spent to charge by location. We also include rebate incentives offered by clean vehicle assistance program [14], only agents who satisfy the income qualifications may be considered. The probability of approval for the rebate offer is set to be normally distributed. On a macro-level, we use data from United States Environmental Protection Agency by vehicle type [15] to illustrate the overall greenhouse gas emission trend with the change in EV adoption over the observed time frame.

3.2 Behavior and Decision Rules

As shown in Fig. 1, agents' movement are designed to follow a set of behavior rules to maximize individual utilities. We use empirical data as input for agents to compute their affordability using benefit-cost ratio (BCR), which depends on income, and external differences including expected savings of fuel spend by vehicle types, potential cost incurred from switching (sell their current vehicle) and expected benefit from rebate program:

$$BCR = \sum[PV(Benefits)] \Big/ \sum[PV(costs)] \tag{1}$$

For willingness to purchase, we adopt a multinomial logit approach to estimate purchase probability for agent i choosing vehicle $j \in (G, EV)$ and vice versa:

$$P_{ij} = exp(G_{ij}) / \sum (exp(\alpha_{ij} * G_{jv})) + (exp(\beta_{ij} * EV_{jv})) \tag{2}$$

Then, agent *i* are designed to follow intuitive mechanism to maximize utilities *uij* for vehicle *j*, which depends on the characteristics *v* including distance, fuel type, miles per gallon with its respective pricing, and vehicle range per charge.

The applied algorithm behind the simulated process modeling agents' interactions, we then are able to observe agents' emergent change in preference that can have an effect on their decision making.

```
To setup
clear-all
set total-CO2 initial emission amount
setup-locations
setup-drivers
create-drivers-network
ask drivers to create-links-with other drivers in-radius closeness with
[source? = 0]
reset-ticks
End

To go
If day in 365 tick => stop the model
ask links  to set thickness 0
ask drivers  link to work places from home locations
ask drivers link other locations to home locations
if drivers do not have sufficient chargers will receive penalty
ask drivers spread-information
ask drivers calculate-willingness
ask drivers calculate-cost-benefit
ask drivers calculate-rebate
  if income < 51040 ] ;; clean vehicle assistance program  offers $5000
  income qualification for single person less than $51,04
  [ if random-float 1.0 > rebate-approval
  [ set rebate rebate + 5000 ]
ask drivers adoption
    let diff ([ closeness ] of myself - [ closeness ] of one-of link-
   neighbors )
   ifelse diff > 0
   [set adopted? True
    set adopted? false]
ask drivers calculate-CO2
tick
```

4 Scenario Analysis

In this section, we present scenario analysis exploring adoption of EV with changing one parameter each time to observe change in the micro-level under different scenarios:

4.1 Baseline

Table 1 includes initial input parameters for calibrating the model, setting population number N = 50 with 20% of population already driving electrical vehicles. Agents are designed with heterogeneous socio-economic attributes including income, age, living expenses and costs of their current car (see Table 1 for model parameter). Figure 2–3 summarizes how the simulated decision-making process change over time. At each tick, agents follow the described set of rules then compare thresholds between affordability and willingness to decide action: to adopt electric vehicle or not. The thickness of the lines in the interface represent frequency of drivers passing through the route (e.g. the more frequent drivers commute on the path, the route then becomes thicker).

At t = 1, majority of the agents have higher affordability and willingness for gas vehicle supporting their final choice. An interesting trend surfaced when t = 10, that even though agents' willingness to purchase still exemplifies gas vehicle as a superior choice, their affordability has changed from gas to electric vehicle. When t = 30, we notice agents' willingness for gas vehicle starts decreasing with a converging trend line, and the gap in affordability by vehicle type continue to enlarge leading to more drivers have switched to adopt EVs. Finally, at t = 60, there is 41 eV drivers with only 9 drivers who remain using gas vehicle; the monitor record shows that all of the agents in the experiment have at one point changed their preference. The simulated non-linear result echoes challenge previous scholars may face in predicting or estimating adoption of EVs since micro-level interactions and preferences exemplifies dependencies in decision making process.

Table 1. Table parameters - baseline

Parameters	Description	Base value
Population	Number of population in this model	50
Initial percentage of using EV	Percentage of population already driving electrical vehicles	20%
Price of EV	Price of electrical vehicles	$31,000
Home charge	Percentage of home which using electrical vehicles charge	30%
Shared work places	Percentage of work places in this model	20%
Workplace charge	Percentage of work place which using electrical vehicles charge	40%
Public charging stations	Setting initial public charging station for electrical vehicles	4
Rebate approval rate	Percentage of individual approved for clean vehicle assistance program	50%
Percentage of connections	Percentage of individual to connect to each other within the network	50%
Percentage of friend informed	Percentage of individual could receive electrical vehicles information from friend	50%

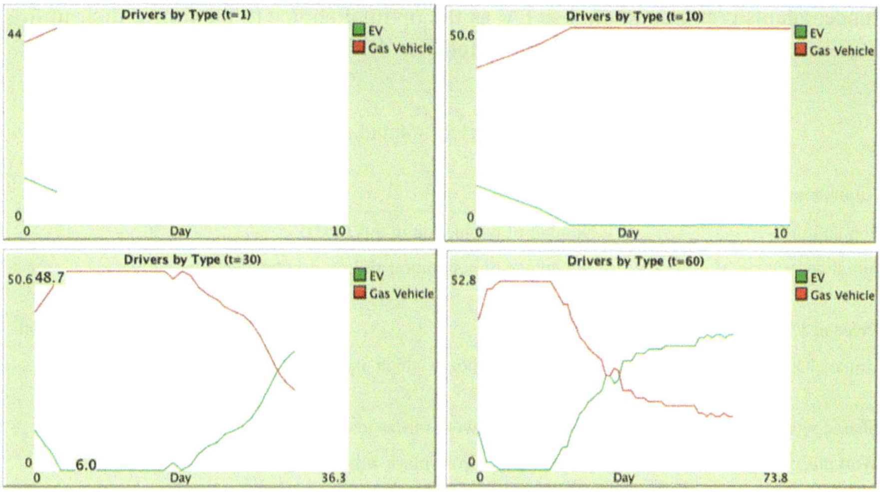

Fig. 2. Drivers by type at different time frame under baseline.

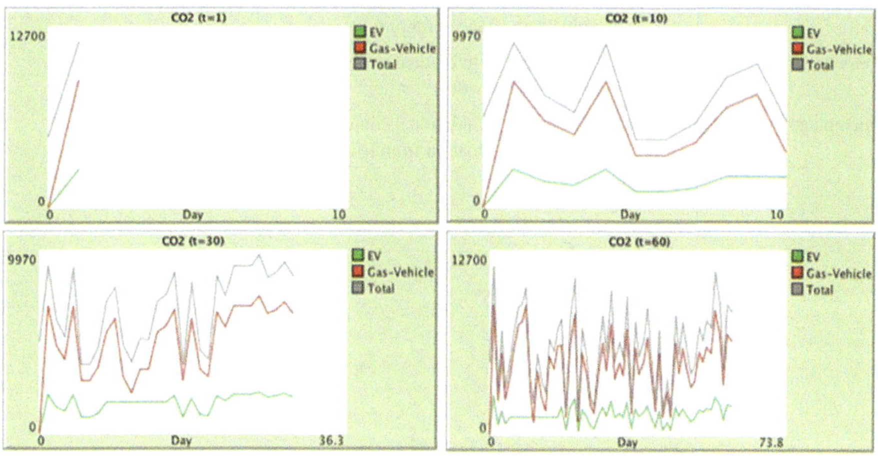

Fig. 3. Simulated CO2 emission at different time frame under baseline.

4.2 Reduction in Price of EV

In the second scenario, we want to explore the effect of reduced price for electric vehicle has on adoption. By recalibrate our model to reduce EV price to $21,000 and leaving the remaining parameters unchanged, we observe significant changes in adoption of EV through agents' interactions. With the reduced EV price, agents' willingness and affordability for both types of vehicles at $t = 1$ displays close to identical trend lines and the total number of gas vehicle has already started to decrease. Shortly after $t = 1$, the number of drivers for electric vehicle exceeds the number of drivers for gas vehicles.

Hence, agents continue to choose EV as the optimal choice to maximize their utilities as shown in t = 10 and t = 30 (Table 2 and Fig. 4).

Table 2. Table parameters – simulating price of EV

Parameters	Description	Base value
Population	Number of population in this model	50
Initial percentage of using EV	Percentage of population already driving electrical vehicles	20%
Price of EV	**Price of EV**	**\$31,000–\$50,000**
Home charge	Percentage of home which using electrical vehicles charge	30%
Shared work places	Percentage of work places in this model	20%
Workplace charge	Percentage of work place which using electrical vehicles charge	40%
Public charging stations	Setting initial public charging station for electrical vehicles	4
Rebate approval rate	Percentage of individual been approval from clean vehicle assistance program	50%
Percentage of connections	Percentage of individual to connect to each other within the network	50%
Percentage of friend informed	Percentage of individual could receive electrical vehicles information from friend	50%

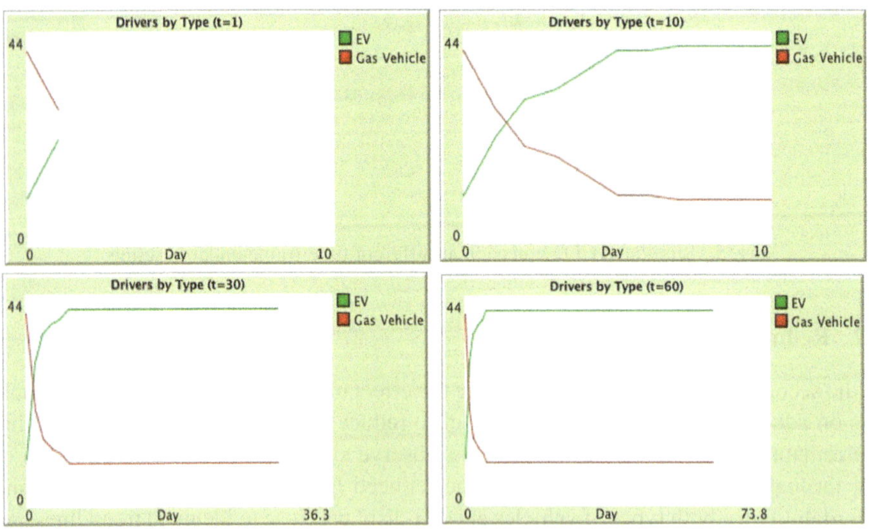

Fig. 4. Drivers by type at different simulated time periods when EV price is reduced to $21,000.

5 Sensitivity Analysis

Table 3 shows results from our sensitivity analysis assessing the effects of independent variables including price of EV, pubic charging stations, and rebate approval rate have on various dependent variables. We opt for pooled OLS regression for the following models:

Table 3. Sensitivity output with pooled OLS

VARIABLES	(1) EV	(2) EV	(3) GAS	(4) GAS	(5) CO2	(6) CO2
EV-Price	-0.699***	-0.699***	0.706***	0.706***	-0.132***	- 0.138***
	(0.0126)	(0.0126)	(0.0125)	(0.0125)	(0.0246)	(0.0248)
Public-charging-stations	0.00243	0.00243	-0.00883	-0.00883	-0.00167	- 0.000684
	(0.0126)	(0.0126)	(0.0125)	(0.0125)	(0.0176)	(0.0176)
Rebate-approval		0.0124		-0.00742		
		(0.0126)		(0.0125)		
Num-EV					-0.143***	
					(0.0246)	
Num-GAS						0.151***
						(0.0248)
Constant	5.32e-09	5.32e-09	-2.92e-09	-2.92e-09	1.26e-09	9.44e-10
	(0.0126)	(0.0126)	(0.0125)	(0.0125)	(0.0176)	(0.0176)
Observations	3,200	3,200	3,200	3,200	3,200	3,200
R-squared	0.489	0.489	0.498	0.498	0.011	0.013

All variables are standardized
Standard errors in parentheses
*** p<0.01, ** p<0.05, * p<0.1

In model 1 and 2, we could see electric vehicle price is statistically significant at 95% confidence interval. The variable has a negative impact on total number of EV drivers (DV). Hence, when there is an increase in electric vehicle price, it would lead a 0.6 unit decrease in EV drivers. Although, public charging stations and rebate approval rates are not statistically significant, these two variables are identified to have a positive influence to increase the adoption of EV.

Next, we use number of gas drivers as DV for Model 3 and 4. Price of EV remains statistically significant at 95% confidence interval, but it now has a positive effect on the number of people who prefer gas vehicle. The output results indicate that with every dollar increase in electric vehicle price increase, there is a 0.706 increase in number of drivers who decides to use gas vehicle. This result supports our findings from scenario analysis that price of electric vehicle is one of the key factors in adoption of EV. Since the change of EV price can have a direct impact to agents' affordability and willingness, therefore, agents will change their strategies in order to optimize their utilities.

Lastly, in model 5 and 6, we are interested in quantifying the variability among the selected IVs have on CO2(DV). Price of EV shows statistically significance at 95% confidence interval with a negative impact on CO2. More specifically, the total number of electric vehicle drivers has a negative effect on CO2 which per electric vehicle would diminish 0.143 CO2 grams/per mile (model 5). On the other hand, we identify the positive relationship between total number of gas vehicle and the amount of CO2 emission.

Taken together, the preliminary statistical provides evidence that our designed experiment as a valid approach to study the adoption of EV. In our ABM, price of electric vehicle has a significant influence on agents' behavior and decision, eventually lead to adoption of EV. Furthermore, in our designed experiment, we observe that with an increase in electric vehicle user there is a reduction in CO2 emission which supports the common belief as one of the major benefits of switching from gas to electric vehicles.

6 Conclusion

In this study, we develop an ABM that combines empirical data collected from government agencies as input to a simulated experiment. The intent of this model is to provide practical insights to adoption of EVs incorporating underlying connectivity among individuals, which is an advantage of an agent-based approach, allowing us to examine agents' interactions and observe emergent properties over the observed time. In our experiment, we discover the most effective change of parameter is the price of electric vehicle that can lead to agents' non-linear change in preferences while comparing thresholds of willingness and affordability.

Limitation and Future Direction for Extension: in this model, we did not consider the degrees agents can be connected, i.e. random v.s preferential network, which can potentially impact the level of information diffusion. In addition, we only focused on micro-level interactions with three scenarios. There are many different types of government policies that can be simulated with an extended multi-level integrated ABM that we are interested in exploring. Despite the mentioned limitations, we hold firm belief that the results obtained from our experiment provides valuable insights towards understanding challenges related to EV adoption. Furthermore, future refinement of the model will only enhance clarity to the complex human decision-making process.

References

1. Beggs, S., Cardell, S., Hausman, J.: Assessing the potential demand for electric cars. J. Econom. **17**(1), 1–19 (1981)
2. Byrne, M.R., Polonsky, M.J.: Impediments to consumer adoption of sustainable transportation. Int. J. Oper. Prod. Manag. **21**, 1521–1538 (2001)
3. Agrawal, A.W., Dill, J., Nixon, H.: Green transportation taxes and fees: a survey of public preferences in California. Transp. Res. Part D: Transp. Environ. **15**(4), 189–196 (2010)
4. Nixon, H., Agrawal, A.W.: Would Americans pay more in taxes for better transportation? Answers from seven years of national survey data. Transportation **46**(3), 819–840 (2018)
5. Sanfey, A.G., Stallen, M., Chang, L.J.: Norms and expectations in social decision-making. Trends Cogn. Sci. **18**(4), 172–174 (2014)
6. Waldrop, M.M.: Complexity: The Emerging Science at the Edge of Order and Chaos. Simon and Schuster, New York (1993)
7. Kenrick, D.T., Li, N.P., Butner, J.: Dynamical evolutionary psychology: individual decision rules and emergent social norms. Psychol. Rev. **110**(1), 3 (2003)

8. Nowak, A., Vallacher, R.R.: Dynamical Social Psychology, vol. 647. Guilford Press, New York (1998)
9. Bettenhausen, K., Murnighan, J.K.: The emergence of norms in competitive decision-making groups. Adm. Sci. Q. **30**, 350–372 (1985)
10. Epstein, J.M.: Agent-based computational models and generative social science. Complexity **4**(5), 41–60 (1999)
11. Macy, M.W., Willer, R.: From factors to actors: computational sociology and agent-based modeling. Ann. Rev. Sociol. **28**(1), 143–166 (2002)
12. Coffman, M., Bernstein, P., Wee, S.: Electric vehicles revisited: a review of factors that affect adoption. Transp. Rev. **37**(1), 79–93 (2017)
13. Gasoline and Diesel Fuel Update - U.S. Energy Information Administration (EIA). https://www.eia.gov/petroleum/gasdiesel/. Accessed 07 Dec 2020
14. Clean Vehicle Assistance Program. https://cleanvehiclegrants.org/. Accessed 07 Dec 2020
15. Greenhouse Gas Emissions from a Typical Passenger Vehicle (10 May 2018). https://www.epa.gov/greenvehicles/greenhouse-gas-emissions-typical-passenger-vehicle. Accessed 07 Dec 2020

The Impact of Housing Programs on Unsheltered Homeless Population: An Agent-Based Approach

Dandan Kowarsch(✉) and Zining Yang

Claremont Graduate University, 150 E 10th Street, Claremont, CA 91711, USA
{dandan.kowarsch,zining.yang}@cgu.com

Abstract. This article presents an Agent-Based Modeling approach to simulate the cascading effect that community-based health care services and two temporary housing programs would have on the unsheltered homeless population in Los Angeles. Our model overcomes the challenge that traditional research using aggregated data isn't sufficient to provide satisfying explanations. We incorporate micro, meso, and macro level components, to uncover the step-by-step changes introduced by integrated program that synchronizes health services and housing programs, including the feedback loop between homeless population and the socio-economic environment they live in. Using key policy levers, we explore the impact of various policies on the unsheltered homeless population. Particularly, we assess the Rapid Re-Housing program, which is one important component of the Housing-First policy, as well as the transitional housing program, which especially aims to support chronically unsheltered homeless people through an aggregate program that conflates the temporary housing and supportive services. To achieve our objective, we use sensitivity analysis to estimate the impact of the number of social workers, effectiveness of policy level, and community-based special care services for mentally ill homeless on macroscopic phenomena. We also conduct two scenario analyses to evaluate two major temporary housing programs on unsheltered homeless. The regression result based on simulation data suggests that the Rapid Re-Housing program is neither effective nor efficient in reducing unsheltered homeless. However, the result illuminates that social workers play vital roles in building relationship with unsheltered homeless people and facilitating chronical unsheltered homeless to receive needed treatment and to be stabilized by housing programs.

Keywords: Agent-based modeling · Unsheltered homeless · Housing-first policy · Rapid-rehousing program · Transitional housing program · Community special care · Mentally Ill homeless · Social workers

1 Introduction

The objective of this article is to evaluate the feasibility and the effectiveness of current housing programs and potential solutions, including integrated housing programs

J. L. Wright et al. (Eds.): AHFE 2021, LNNS 264, pp. 84–92, 2021.
https://doi.org/10.1007/978-3-030-79763-8_10

with health services as well as community-based special care for mentally ill homeless. Homelessness is an inevitable consequence of socio-economic issues. In recent years, the unsheltered homeless populations in Los Angeles, San Francisco, and Portland have increased significantly. In 2019, Los Angeles County's total homeless population was approximately 56,257 people, of which 42,471 were unsheltered.[1] As of 2020, homeless population in LA County was 66,436, approximately 12.7% increase.[2] The unsheltered homeless has become the most challenge to LA County as there is no accessible effective policy that can be implemented to mitigate the homeless crisis.

It becomes urgent and critical for the County to come up with an integrated policy that takes effect. It is also fundamental to understand the difference between chronically unsheltered homeless and non-chronically unsheltered homeless, and to apply two different strategies based upon the characteristics of the unsheltered homeless. The reason for doing so is that the former needs abundant supportive health services to maintain normal conditions before being assigned into any housing programs but the latter only needs temporary housing.

Solving homelessness can be costly and time-consuming. Los Angeles County has increased local sale taxes just for the homeless. The state of California and HUD also allocate funds for the homeless. However, the payoffs for every dollar the government spends on the homeless seem ineffective because the trends in unsheltered homeless population and the spending coincide.

In this article, we focus on the investigation of the viability for a potential approach that LA County proposed in 2019. Particularly, we examine following aspects: On micro level, the impact of social workers on unsheltered homeless; On meso level, the impact of community-based mental cares and housing programs on unsheltered homeless; On meso-micro/meso-macro levels, the impact of administrative or policy efficiency on social worker's capability and unsheltered homeless, respectively.

2 Literature Review

Economic theory predicts that people will have difficulty paying rent and, in some cases, end up homeless. A large body of academic literature confirm that higher home prices are indeed associated with higher rates of homelessness [1–3]. Clifasefi et al. (2016) described Housing-First policy as an approach to end homelessness by providing immediate, permanent, low-barrier, non-abstinence-based supportive housing for individuals with the lived experience of homelessness [4]. According to Clifasefi et al., Housing-First attempts to improve many domains for the homeless in two primary ways: affordable individual housing units and assertive community treatment and other supportive services.

However, despite the number of chronically homeless people living on the nation's streets and shelters dropping to as low as 30% between 2005 and 2007, the Housing-First approach was faced with many critiques [5]. Fowler et al. (2018) argue that an emphasis

[1] Data is extracted from LAHSA 2019 Homeless Count Data by Supervisorial District. https://www.lahsa.org/data?id=41-2019-homeless-count-by-supervisorial-district.

[2] Data is extracted from LAHSA 2020 homeless point-in time count result. https://www.lahsa.org/documents?id=4558-2020-greater-los-angeles-homeless-count-presentation.

on housing services could negatively impact communities' efforts to end homelessness and that communities should focus instead on homelessness prevention and moving homeless people into permanent housing [6].

Undoubtedly, unsheltered chronical homeless are most likely to suffer mental illness. However, Basaglia's Trieste Model of mental health system suggests that people with even the most severe mental illness could live a "normal" life accommodating their condition in the "community" [7–10].

3 Research Design

3.1 Model Entities and Variables

We use ABM implemented in NetLogo[3] to assess the feasibility of the implementation of the Trieste model. The model allows us to capture the adaptive and emerged behaviors through the consistent interactions among individuals. There are two types of agents, unsheltered homeless people, and outreach homeless services social workers. We use attributes in the modeling space to capture special care centers. The function of community special cares is to cultivate an ideal environment which facilitates the mentally ill homeless to have "normal" lives. Since the County generates a 3-year proposal, we set each time step to present a day, and the model stops running at time step at 1095 or if there is not any unsheltered homeless exited.

The attributes for unsheltered homeless are as follows: Mental health condition, which is drawn from a uniform distribution, is an important indicator for survival; Unsheltered homeless' willingness of being housed is a binary variable; The qualification status for any available housing programs is a dummy variable; Whether the client is successfully connected with a social worker or not is a binary variable. In addition, the social worker owns work capability, which is determined by the unsheltered homeless population. The global attributes consist of policy-efficiency level, housing sufficiency, the average lengths of staying in RRH and TH programs, and the total amount of available services across 1095 iterations. Figure 1 below portrays the model's high-level process and multi-scale architecture.

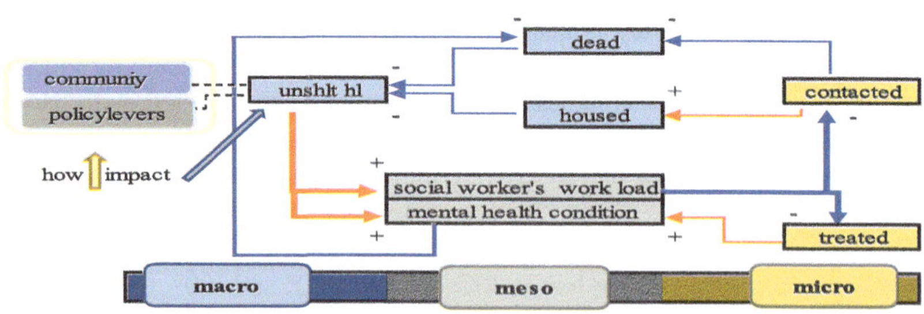

Fig. 1. Simplified flowchart of multi-scale agent-based model

[3] NetLogo, Wilensky, U. (1999).

The flowchart consists of micro, meso, and macro levels. Unsheltered homeless and social workers are two components on micro level. The number of social workers is an important policy lever on this level because if there isn't sufficient number of social workers, the social workers' workloads will be large, and their work capacities will be lower than normal condition. Social worker's work capacity determines the successful contact ratio. Alternatively, the successful contact ratio is an essential cause that affects the meso level behaviors.

On meso level, the policy efficiency level and the number of community-based health centers together determine how many unsheltered homeless can actually get health treatment and housed. For instant, after reaching out the homeless, social workers evaluate the unsheltered homeless' mental condition. If the homeless is selected as a mentally ill client, he or she will get mental health treatment at community-based health care. The length of staying in the health care center relies on the person's mental health condition. In contrast, if the person does not need a mental treatment and is willing to be housed, then the person will be randomly assigned into one housing program. Even though, sometimes, the unsheltered homeless is contacted, there is still a chance which the person would rather sleep in public than be housed. Moreover, the community also affects social workers' capacity and housing sufficiency. For example, if community is supportive, the residents are willing to communicate with the homeless, provide water and food, and contact social workers and report unsheltered homeless location as well as there will be sufficient affordable housing for the homeless. In other words, those indirect supports from the community not only increase social worker's work capacity, but also facilitate policy implementation.

On macro level, the dynamically accumulative effect from lower level arises and the macroscopic phenomena emerge. We specifically assess causal relationships between the various sets of inputs and their corresponding changes in the unsheltered homeless population, the stabilized homeless population, and deaths among the unsheltered homeless, accordingly.

3.2 Parameterization Verification and Simulation Experiment

As we set up the baseline, we calibrate initial values including average length in RRH program[4], the percentage of mentally illness in unsheltered homeless,[5] the level of funding for health services[6] and the average homeless growth rate[7] based on official reports and the relevant publications. We then initialize the community to be not supportive, housing inventory to be insufficient, the probability of being assigned into RRH to be

[4] HUD User Data: https://www.huduser.gov/portal/pdrdatas_landing.html.

[5] Health Conditions Among Unsheltered Adults in the U.S. https://www.capolicylab.org/wp-content/uploads/2019/10/Health-Conditions-Among-Unsheltered-Adults-in-the-U.S.pdf.

[6] Los Angeles County Homeless Initiative Draft Detailed Measure H Funding Recommendations FYs 19-20, https://homeless.lacounty.gov/wp-content/uploads/2019/03/FY19-20/Measure-H-Individual-Funding-Requests.v2_3.21.19.pdf, pp15. We use the findings for RRH as the initial value for special care center.

[7] The arithmetic mean is calculated based on the reported unsheltered homeless population. Data source: PIT Count and HIC Data and Reports. https://www.hudexchange.info/programs/hdx/pit-hic/datareports/.

50%, and the policy efficiency to be neutral. The initial value for unsheltered homeless population is 100, while the number of social workers is 5. Social work capability is 20 homeless per social worker, and the number of special care centers is 35. The initial mental level of each unsheltered homeless is drawn from a uniform distribution. The lower value indicates higher degree of mental illness. If the unsheltered homeless' mental level is below 0, we assume that this person is too weak to stay alive.

The initial value for the percentage of unsheltered homeless who have the desire to be housed is 50% and the treatment length is set to 1095 days, but the actual time for a given individual depends on the individual's own mental health level. The length of being in a temporary housing program is drawn from a normal distribution.

We adjust the number of social workers from initial value 5 to 1 and observe the simulated results. As what we expect, when there is only one social worker who is dealing 100 unsheltered homeless, on average, only 42% unsheltered homeless are willing to be sheltered and nobody can receive supportive treatment and be settled into housing programs. Additionally, the longer unsheltered homeless wanders on the street, the more likely this person is going to die eventually. Contrarily, we increase the number of social workers to 10, meaning that a social worker serves 10 unsheltered homeless. The results indicate that on average, 80% unsheltered homeless are willing to be sheltered. Among them, 55% unsheltered homeless can receive treatment and 77% homeless are housed through a housing program. A snapshot for a sample run is shown in Fig. 2.

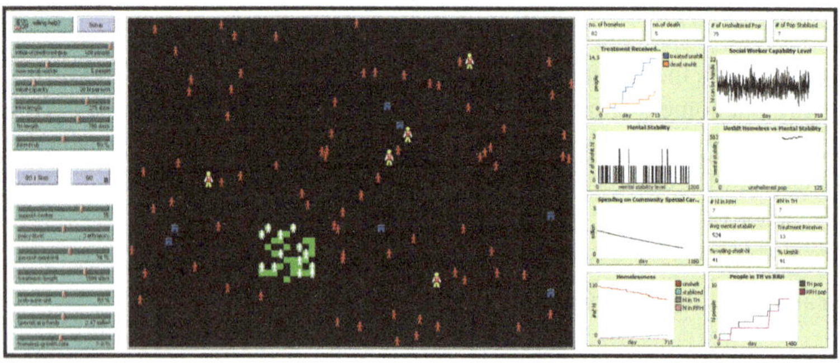

Fig. 2. NetLogo interface with a sample run (t = 713) under initial setup.

4 Sensitively Analysis

4.1 Quasi-Global Sensitivity Analysis

Sensitivity analysis helps identify the parameters that have significant impact on model's output. We set up the simulations as follows: the number of social workers to be 3, 5, and 10 accordingly; policy efficiency levels to be 1, 2, and 3; the number of community special care centers to be 0, 20, 40, and 60. In addition, to distinguish the impact of community's supportive level on dependent variables, the sensitivity analysis is executed under two different environments: the community is supportive, and the community is not supportive, respectively.

We perform pooled OLS for statistical analysis based on simulation data from sensitivity analysis. Table 1 depicts the result in detail.

Table 1. Regression result. P-value = 0: "***"; <0.001: "**"; <0.01: "*"; <0.05: "."; Significance level at 5%. All estimates are standardized, and all standard errors are robust errors.

Pooled OLS	Model 1: lnUnshlt		Model 2: lnTreated		Model 3: lnDead	
(Intercept)	0.000***	0.016	0.000***	0.038	0.000***	0.013
willing.helpTRUE	−0.197***	0.004	0.172***	0.008	−0.115***	0.006
#.SocialWorker	−0.343***	0.0009	0.457***	0.001	−0.006**	0.001
Policy.Level	−0.135***	0.002	0.312***	0.005	−0.064***	0.003
#.SupportCenter	−0.082***	0.0001	0.010**	0.0002	−0.352***	0.0002
Mental level	0.682***	0.00003	−0.100***	0.00006	−0.635***	0.00003
DF	70666		52318		48655	
P (F-statistic)	0		0		0	
RMSE	0.514		0.9		0.62	
R.Adj	0.569		0.351		0.804	

The aim of Model 1 is to examine the correlation between the theoretical variables and unsheltered homeless population. The regression result of model 1 suggests that the community's contribution, total number of social workers, and policy efficiency level are negatively and significantly associated with unsheltered homelessness. We find that the mental health level of unsheltered homeless is positively associated with unsheltered homeless. It is because the unsheltered homeless population is composed of mentally ill and non-mentally ill homeless. The latter have better mental conditions and if the non-mentally ill homeless are settled in a housing program, only those who are mentally disabled are left on the streets.

Model 2 seeks to assess the correlation between the theoretical variables and number of treated homeless. The result of model 2 indicates that all theoretical variables are statistically significant associated with treated homeless. Particularly, we find that the number of social workers has the most significant impact on remaining the necessary mental health level among unsheltered homeless.

Model 3 aims to help understand the relationship between theoretical variables and the number of deaths. The result of model 3 projects the most significant factor for homeless death is associated with their mental condition (std. est. = −0.635), and mental health services (std. est. = −0.352).

Readily apparent, the results from three models illuminate that all variables are highly significant. Indeed, the results indicate that the possible solutions are not only feasible, but also effective for overall unsheltered homelessness mitigation.

5 Scenario Analysis on Two Housing Programs

To evaluate two housing programs, the RRH and TH programs, we perform scenario analysis to investigate the impact of the housing programs on unsheltered homeless. The first scenario aims to examine the impact of RRH program, so we initialize the model with all unsheltered homeless assigned into the RRH program. The second sicario is to test the impact of TH program, so we assign all unsheltered homeless to to the transitional housing program. Holding all other variables to their initial values, we run 100 simulations for each scenario and conduct the descriptive analysis based on outcomes at the final steps.

The simulation result suggests that without the contribution from community, both the RRH program and TH housing strategy produce equivalent impact on unsheltered homeless: The mean of unsheltered homeless for only RRH available is 54.18; While the mean of unsheltered homeless for only TH available is 53.37. However, the TH program surpasses the RRH program on homeless stabilization. It implies the homeless who are assigned in the TH program are more likely to achieve stabilization through the permanent supportive housing program. The detailed descriptive result and histogram plots of the frequencies in unsheltered and treated homeless are shown in Table 2 and Fig. 3.

Table 2. Descriptive Table. The 100 iterations on the last step (step = 1095). The data projects the overall situation by the end of three years.

Scenario I: only RRH available							
Step = 1095	N	Mean	SD	Median	Min	Max	Skew
tot_unshlt_pop	100	54.18	4.73	54	42	64	−0.23
total_stablized	100	9.98	11	4	0	41	0.83
Scenario II: only TH available							
Step = 1095	N	Mean	SD	Median	Min	Max	Skew
tot_unshlt_pop	100	53.37	7.01	54	35	67	−0.33
total_stablized	100	32.02	8.19	32	16	58	0.43

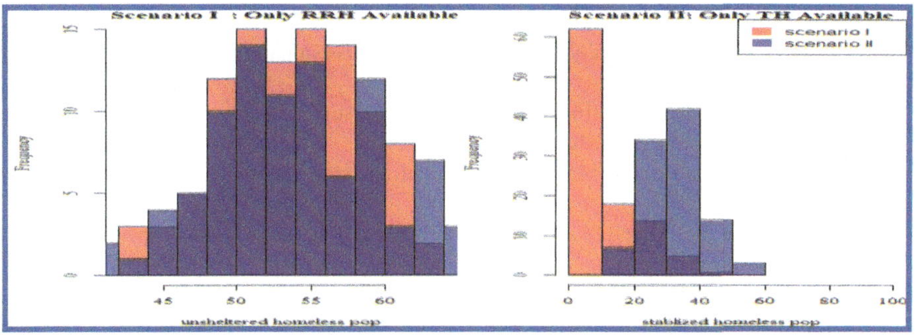

Fig. 3. The histogram plot of unsheltered homeless population (*Left*) and the histogram plot of treated homeless population (*Right*).

6 Conclusion

ABM is a promising approach to understanding the complex dynamics, enable to simulate multi-level nonlinear behavior, and generate abundant data for appropriate statistical analysis. It is particularly useful in testing policies, as real world experiment is almost impossible. The simulation results suggest that social workers have the most impact on decreasing unsheltered homeless population as well as on increasing the number of the homeless in receiving mental treatment. Apparently, social workers are the only people whom the vulnerable homeless can rely on. The work capacity of the social worker determines the velocity of mitigation process of unsheltered homelessness. The community-based special cares play a crucial role in improving the mental condition of unsheltered homeless so that more homeless can successfully be stabilized through housing programs. Last but not the least, based on the result of scenario analysis, comparing to the transitional housing program, the Rapid Re-Housing program does not have a superior impact on reducing unsheltered homeless population.

References

1. The Council of Economic Advisers September 2019: The State of Homelessness in America. https://www.whitehouse.gov/wp-content/uploads/2019/09/The-State-of-Homelessness-in-America.pdf
2. Corinth, K.: The impact of permanent supportive housing on homeless populations. J. Hous. Econ. **35**, 69–84 (2017)
3. Quigley, J.M., Raphael, S.: Regulation and the high cost of housing in California. Am. Econ. Rev. **95**(2), 23–28 (2005)
4. Clifasefi, S.L., Collins, S.E., Torres, N.I., Grazioli, V.S., Mackelprang, J.L.: Housing first, but what comes second? A qualitative study of resident, staff and management perspectives on single-site housing first program enhancement. J. Community Psychol. **44**(7), 845–855 (2016)
5. Stanhope, V., Dunn, K.: The curious case of housing first: the limits of evidence based policy. Int. J. Law Psychiatry **34**(4), 275–282 (2011)

6. Fowler, P., Wright, K., Marcal, K., Ballard, E., Hovmand, P.: Capability traps impeding homeless services: a community-based system dynamics evaluation. J. Soc. Serv. Res. **45**(3), 348–359 (2018)
7. Basaglia, F., Basaglia, O.: Crimini di Pace: Ricerche sugli Intellettuali e sui Tecnici come Addetti all'Oppressione. G. Einaudi, Torino (1975)
8. World Health Organization: Mental health in europe, stop exclusion, dare to care. Geneva (2001)
9. Scheper-Hughes, N., Lovell, A.M.: Breaking the circuit of social control: lessons in public psychiatry from Italy and Franco Basaglia. Soc. Sci. Med. **23**(2), 159–178 (1986)
10. Portacolone, E., Segal, S.P., Mezzina, R., Scheper-Hughes, N., Okin, R.L.: A tale of two cities: the exploration of the Trieste public psychiatry model in San Francisco. Cult. Med. Psychiatry **39**(4), 680–697 (2015). https://doi.org/10.1007/s11013-015-9458-3

Opening Borders. Peru's Expected Utility on Venezuelan Immigration: A Social Network Analysis

Jorge Guerra(✉)

Claremont Graduate University, 150 E 10th St, Claremont, CA 91711, USA
jorge.guerra@cgu.edu

Abstract. An exodus, a consequence of political, economic, and social dissatisfaction, has driven more than 4.6 million of Venezuelans to leave their country since 2016. The spillover effect of this humanitarian crisis has impacted neighboring countries in South America which now face unprecedented migratory influx. Perú's robust economic growth of the last decade has drawn Venezuelans over the border in search of better living standards. At the same time, Peru's recent political scandals with corruption, impeachments, and graft probes have created economic turmoil. Venezuelan immigration has exacerbated this instability, and Peruvians, law makers, media, and international stakeholders have not yet found agreement in how to respond. The application of Social Network Analysis, Spatial Bargaining, and Game Theory reveals the Expected Utility of Peru's political agenda concerning migration. As anticipated, the power of influential states and the potential economic benefit that a developing country such as Peru could gain from migration will influence future political decision making on this issue even if this response refuses to support popular demand.

Keywords: Immigration · Expected utility · Game theory · Spatial bargaining

1 Introduction

Since 2017, nearly 870,000 Venezuelans have emigrated to Peru. The impact of this migration has built up a subtle resentment in the country in recent years. The perception of this complex social issue has generated a division between public opinion and political actors. Peruvian voters call for new laws and policies to block the pathways to legal immigration; these demands for reform are in opposition to policies adopted by the Global Compact for Migration[1]. Although the USA and the EU's financial support is fundamental for the development of the country, a majority of public officials still call for anti-immigration reforms. The fact that on average 57% of working age Venezuelans

[1] Global Compact for Migration. "Refugees and Migrants". https://refugeesmigrants.un.org/migration-compact (2017).

J. L. Wright et al. (Eds.): AHFE 2021, LNNS 264, pp. 93–102, 2021.
https://doi.org/10.1007/978-3-030-79763-8_11

in Peru have a higher education, and of these, half have a university degree[2] has not yet created pro-immigration policies. Facing this socio economic and political challenge, this research explores the economic implications of Peru's political decision-making on Venezuelan's exodus using three formal models. First, the theoretical perspective of Social Network Analysis analyzes relational ties and flows between nodes based on the degree of centrality. Second, the Spatial Bargaining approach aids in finding a legislative equilibrium in the Peruvian Congress by identifying coalitions among stakeholders, especially salient given a dissatisfaction with the status quo. Third, the application of game theory in this case demonstrates the influence of developed countries on developing countries based on calculated actions to obtain the highest payoff.

My findings indicate that the Expected Utility of Peru for offering shelter, harbor, and refugee is determined by financial gains. It also reveals the assumption of most political parties and voters that the immigration of Venezuelan citizens hinders Peru's current economic growth and causes delays and negative effects on public safety, employment, and opportunities for Peruvians. This belief is dominant and represents a major point of contention between Peruvian stakeholders and international stakeholders who are in favor of policy that offers humanitarian assistance to Venezuelans while demanding continued growth of Peru's economy.

2 Literature Review

The Venezuelan migration to Peru has been growing at a fast speed since 2016 reaching, a peak in 2018[3], intersecting with political upheaval in Peru. By 2020, two upper-middle-income economies[4], Colombia and Peru, have hosted 1.8 million and 830,000 people respectively[5]. The economy in Peru seems to be asphyxiated by the instability of the government. It also has been hampered by massive migration. Research in regulatory policy suggests that Anti-immigration initiatives [9], proposed at the Legislative Chamber of Peru, depicts a process presented by Olson [5] in which legislators differ in their policy preferences depending on their constituency's interest, their exposure to different interest's group's concerns, and their own personal ideas.

The flows of migrants are described by Cioffo-Revilla [3] as a social network with directional social relations. Christopolous [2] suggests that Social Network Analysis (SNA) can be employed to: 1) examine the interactions between nodes (Peru, Venezuela); 2) measure the resource flows between nodes, and 3) measure the information flows between nodes. Forecasting future policy decisions is a difficult task because it implies estimated probabilities of events to occur. The Expected Utility model, explained by Bueno de Mesquita [1], has proven to be a flexible and reliable forecasting tool. This analysis identifies the groups, issues, and other variables required by the model (policy preferences held by each group on each issue; the relative power controlled by each group; and the salience, or level of importance, each group attributes to each issue).

[2] The World Bank. "Venezuelan Migration: The 4,500-kilometer Gap Between Desperation and Opportunity" https://www.worldbank.org/en/news/feature/2019/11/26/migracion-venezolana-4500-kilometros-entre-el-abandono-y-la-oportunidad.

[3] Venezuelan Migration in Peru: https://publications.iadb.org.

[4] New country classification by income level https://blogs.worldbank.org.

[5] Venezuelan immigration: crime and fake perceptions: https://www.migrationpolicy.org.

3 Methodology

This research combines Social Network Analysis, Spatial Bargaining, and Expected Utility Theory. The Venezuelan immigration is briefly analyzed in terms of social networks and policy decision process from the Peruvian standpoint. Three considerations are described prior to implementing the models. First, the structure of the unicameral National Parliament and their methods to issuing laws and legislative resolutions. Second, identifying stakeholders, international organizations, and influential states which affect policy outcomes. Third, the identification of relational ties (linkages) between nodes, which are channels for transfer or flow of resources [10]. This permits an examination of countries as our central unit of analysis, observing the centrality and interactions of Venezuela with other countries.

3.1 Unicameral National Parliament

Peru Legislative Chamber is represented by a unicameral National Parliament composed of 130 legislators serving five-year terms. In 2020, members of the Peruvian Congress presented a proposal to block Venezuelan citizens from coming to the country [9]. Figure 1 shows 86–44 votes against immigration. Figure 2 displays 74–56 preference on their attempts to close the border blocking Venezuelans from coming to Peru.

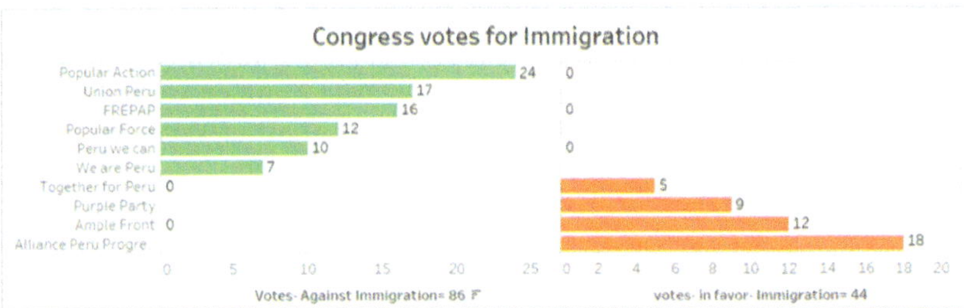

Fig. 1. .

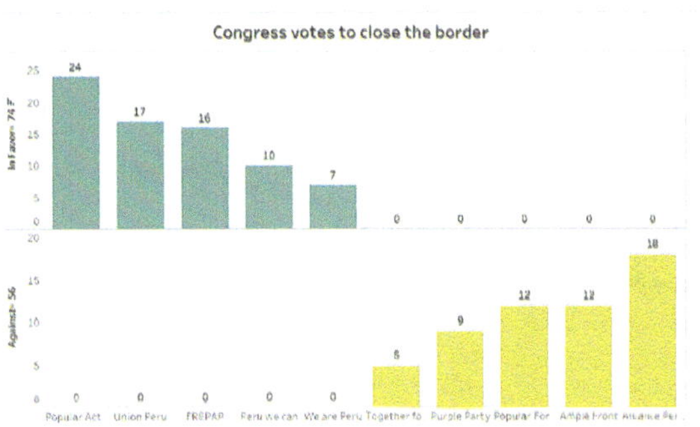

Fig. 2. .

3.2 Stakeholders

The European Union has disbursed €70 million to promote the socio-economic integration of Venezuelan migrants and refugees and supporting their host communities. This support will be distributed between Colombia, Peru, and Ecuador, countries that host most refugees and migrants. In addition, The United States provided more than $75 million in Foreign Assistance funds to Peru in Fiscal Year 2019 to catalyze private sector investment and development. Figure 3 shows the financial assistance received between 2018–2020.

Stakeholders	Assistance	Amount	Year
U.S Embassy	COVID-19 Pandemic	$ 25,800,000	2020
International Narcotics and Law Enforcement	Law Enforcement Security Development relationship	$ 75,000,000	2019
US. Agency for International Development	Cocaine reduction Public integrity-anti-corruption Sustainable environment	$ 49,000,000	2018
European Union	Venezuelan refugees and migrants	€70 million	2019

Fig. 3. .

3.3 Level of Analysis and Directed Network

South America is the major network for Venezuelan citizens, with migration increasing since 2017. Colombia and Peru combined host more than 2 million Venezuelan citizens [1]. Figure 4 demonstrates that Peru is the second major host of Venezuelan refugees. The out-degree value (Fig. 5) is one of the highest in the region.

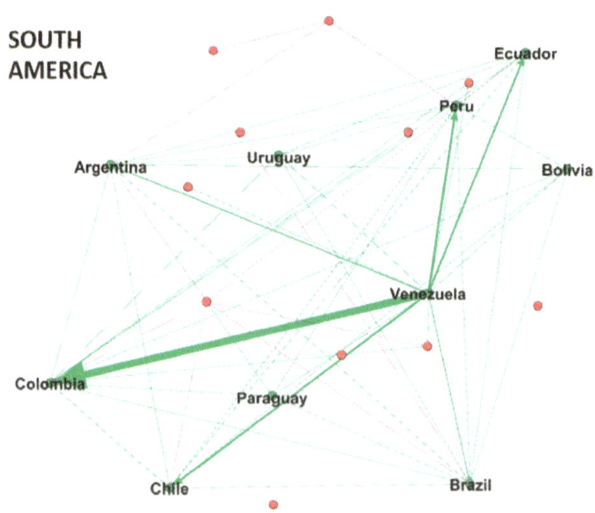

Fig. 4. .

Id	Label	Interval	node.type	In-Degree	Out-Degree	Degree
Colombia	Colombia		South America	1	9	10
Venezuela	Venezuela		South America	1	9	10
Brazil	Brazil		South America	9	6	15
Argentina	Argentina		South America	8	5	13
Ecuador	Ecuador		South America	6	5	11
Peru	Peru		South America	9	4	13
Chile	Chile		South America	6	4	10
Bolivia	Bolivia		South America	2	4	6
Paraguay	Paraguay		South America	4	3	7
Uruguay	Uruguay		South America	5	2	7

Fig. 5. .

4 Model Implementation

4.1 Spatial Bargaining

The Spatial Bargaining model estimates the formation of parliamentary coalitions and predicts their position on two issues: 1) closing the borders to impede a massive migration of Venezuelan citizens; 2) to enact laws in favor or against immigration. Figure 6 shows the list of stakeholders involved. One hundred and thirty (130) lawmakers distributed among 10 political parties. Figure 7 depicts the preference of stakeholders in the two-dimensional issue space using a scale of values from 0–100 being 100 the preferred outcome.

Stakeholders

Name	Category
Willax TV	TV Channel
Pan-American TV	TV Channel
Latin Frequency TV	TV Channel
Channel N	TV Channel
America TV	TV Channel
RPP radio	Radio
We are Peru	Political Party
Union Peru	Political Party
Together for Peru	Political Party
Purple Party	Political Party
Popular Force	Political Party
Popular Action	Political Party
Peru we can	Political Party
FREPAP	Political Party
Ample Front	Political Party
Alliance Peru Progre..	Political Party
The Republic newsp..	Newspaper
Express newspaper	Newspaper
El Comercio newspa..	Newspaper
UN Refugees	International Org
Venezuela	Country
USA	Country
EU	Country

Fig. 6. .

Spatial Bargaining values		
Name	X value- Immigration	Y value- Close the border
The Republic new..	80.0	30.0
Latin Frequency TV	80.0	30.0
FREPAP	10.0	100.0
Channel N	80.0	30.0
Alliance Peru Pro..	80.0	30.0
Union Peru	5.0	100.0
Popular Action	5.0	100.0
Willax TV	80.0	20.0
We are Peru	50.0	50.0
USA	100.0	0.0
UN Refugees	100.0	0.0
RPP radio	80.0	20.0
Purple Party	80.0	20.0
Peru we can	20.0	80.0
Pan-American TV	80.0	20.0
Express newspap..	80.0	20.0
EU	100.0	0.0
El Comercio news..	80.0	20.0
Ample Front	80.0	20.0
America TV	80.0	20.0
Together for Peru	70.0	20.0
Venezuela	30.0	50.0
Popular Force	10.0	20.0

Fig. 7. .

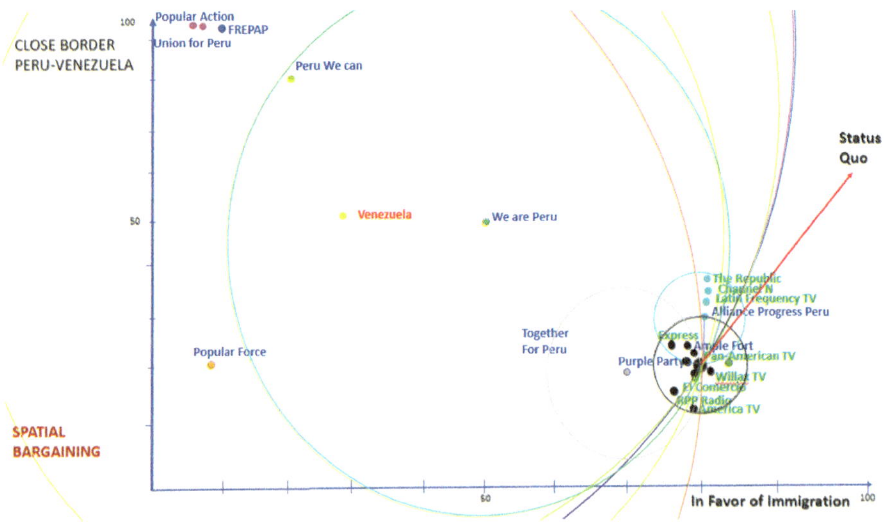

Fig. 8. .

The multipartyism causes legislative gridlocks often resolved by trade-offs among their legislative decisions. Figure 8 depicts Media groups and a few political groups (Alliance, Ample Fort, and Purple Party) seem to have a chance to convince two political groups ("Together for Peru" and "We are Peru") to join their proposals. The formation of a coalition could defeat the status quo which represents support in the interests of Venezuelan citizens. Similarly, Fig. 9 illustrates two powerful stakeholders: 1) the Peruvians and 2) a cluster of stakeholders with high interest in migratory reform. The ideal points of The United States and European Union favor the Venezuelan cause in Peru. Their influence is based on a long-term financial support directed to humanitarian assistance and development programs.

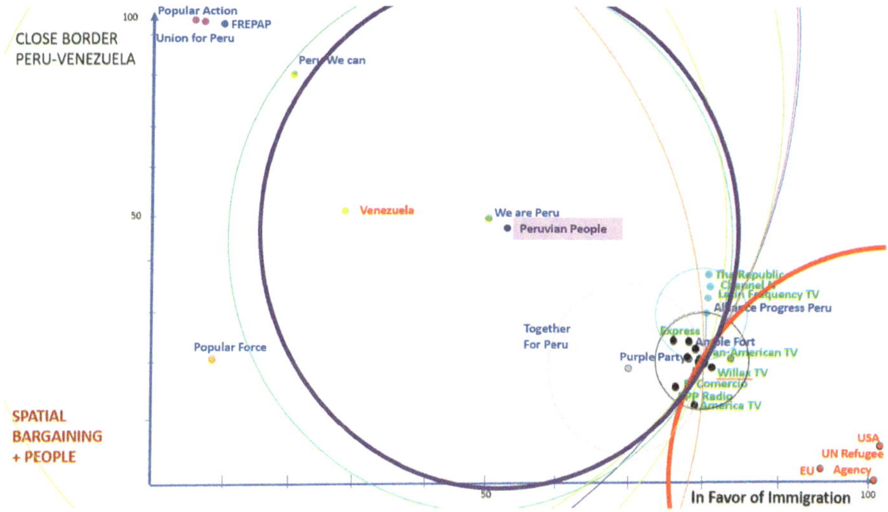

Fig. 9. .

4.2 Expected Utility Theory

Although most political groups in the Peruvian Parliament disfavor immigration, lawmakers will have to contemplate the influence of the United States and EU in their final decision. The financial loss and the risk of jeopardizing future support (and diplomatic relations) will exacerbate the humanitarian and economic crisis. Under the premises of the Expected Utility theory, the model evaluates each actor's willingness to take on risk and estimates what kind of conflictual relationship is likely to emerge between them [1]. Figure 10 displays a game theory using trees with two major actors (players): The United States and Peru with assigned symbolic payoffs from one (1) to six (6), being one (6) the better or highest payoff for each country. They present their optimal strategies. Assuming the players are rational calculators, Fig. 11 throws a payoff of (4, 6) using rollback (backward induction).

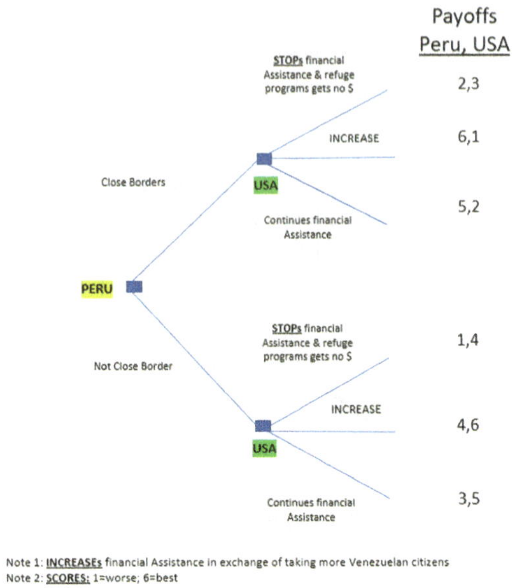

Fig. 10. .

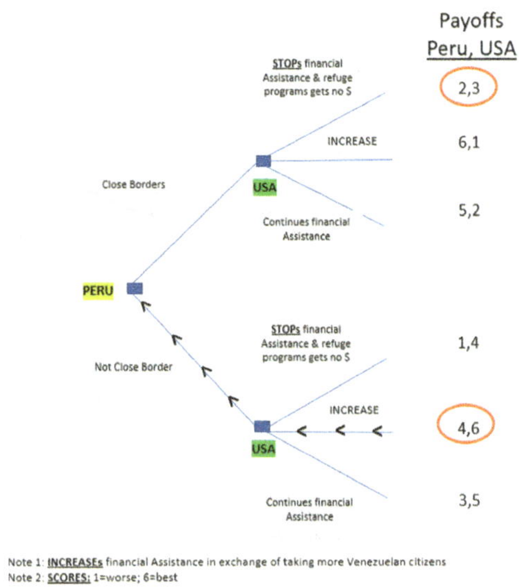

Fig. 11. .

While developed states have the capacity to deter and negotiate at macro scales, developing countries must evaluate their decisions based on economics and financial gains. The optimal strategy for Peru in the long run is not closing the border. It is estimated that the country would benefit with $111 M as the EU model predicts. Probabilities for each action was estimated. Figure 12 displays the values and calculations.

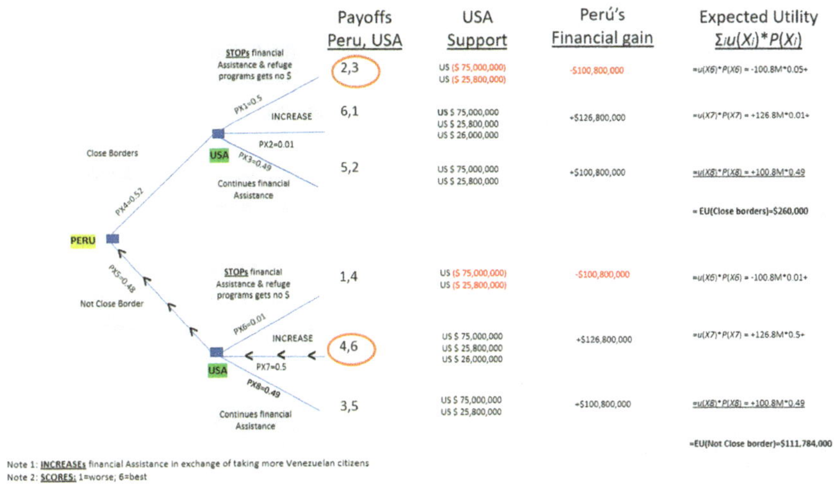

Fig. 12. .

5 Policy Implications and Conclusion

Computational Social Science under the form of information processing can predict policy outcomes and decisions that the Peruvian government should take to maximize its utility. Understanding the complexity of the political system in Peru, the agenda setting, and the actors involved in the decision making, it helps to generate proposals for better governance and more progressive and inclusive policies. The application of advanced formal models in a real scenario shed a light to question multipartyism as an effective system for policy implementation. Venezuela's return to political, social, and economic stability would benefit from Diplomatic actions and multilateral efforts. The migration of Venezuelan's citizens is an opportunity for growth and progress. Priority actions in Perú should be directed toward naturalization, health coverage, investments, and equal opportunities. Transitioning from humanitarian assistance toward development will require cooperation among states that sees immigration as a mechanism to generate prosperity for all.

References

1. Bueno de Mesquita, B.: The contributions of expected utility theory to the study of conflict. In: Manus, M. (ed.) Handbook of War Studies, pp. 143–169. University of Michigan Press, Ann Arbor (1993)
2. Christopoulos, D.C.: Network Analysis in Social & Political Science (2004)
3. Cioffi-Revilla, C.: Introduction to Computational Social Science: Principles and Applications. Springer Science & Business Media, Berlin (2013)
4. Dixit, A., Skeath, S., Reiley, D.H.: Games of strategy (2004)
5. Olson, M.K.: Political influence and regulatory policy: the 1984 drug legislation. Econ. Inq. **32**(3), 363–382 (1994)
6. Tsebelis, G.: Decision making in political systems: veto players in presidentialism, parliamentarism, multicamerism and multipartyism. Br. J. Polit. Sci. **25**, 289–325 (1995)

7. Impact of hosting forced countries in poor countries. https://www.annualreviews.org
8. Venezuelan Migration in Peru: Short-term Adjustments in the Labor Market. https://publications.iadb.org
9. Migratory agenda in the Peruvian Congress. https://idehpucp.pucp.edu.pe
10. Essential Issues of the Peruvian Legal Systems. https://www.nyulawglobal.org
11. Congress composition. https://peru.as.com/peru

Optimal Strategy in International Relations

Dandan Kowarsch(✉)

Claremont Graduate University, 150 E 10th Street, Claremont, CA 91711, USA
dandan.kowarsch@cgu.edu

Abstract. The aim of this article is to explore the impact of power distribution and international trade on world stability by an agent-based simulation modeling approach. The superiority of using ABM is not only to discover non-linear patterns dynamically, but also to capture uncertainties in order to prevent imperceptible risks, such as global wars. There is a large body of studies attempting to address the cause of wars. Theories in international relations portray the disparate mechanisms of war from various perspectives. In this article, however, I primarily consider two power-based theories, balance of power and power transition theory, and the interconnections among power, trade, and war. Proponents of both sides assert seemingly controversial arguments on the cause of wars. According to balance of power theory, the polarity distribution in power enhances peaceful dyadic relationship between two powerful countries because the superpower states will prefer the status quo and cooperation with their allies. In contrast, the supporters of power transition theory claim that parity and the challenger's dissatisfaction with the status quo are essential conditions for war. The simulated result, however, indicates that the two theories are comparable as power parity is positively associated with trade. Alternatively, trade is positively associated with interdependence, and interdependence increases both the costs of war and opportunities of waging war. This endogeneity, integrated with the third dimension, time, results in powerful states most likely and ironically avoiding wars with other powerful states. The result of sensitivity analysis unfolds a negative exponential distribution pattern of war occurrence over time, meanwhile, the overall global economy is thriving. The convergence behavior emerges following the substantial cooperative behaviors between states. Based on the results from both deduction and induction approaches, the possibility of initiating war is not the preferred option for most interconnected super states.

Keywords: Agent-based modeling · Power transition theory · Balance of power theory · Trade · Interdependence · War

1 Introduction

The cause of war has been the subject of intense scholarly debate in the field of international relations since World War II. War is always associated with death, exile, and poverty. The interdependent effect of war seems to be more detrimental in the perspective of economic growth and human capital development.

J. L. Wright et al. (Eds.): AHFE 2021, LNNS 264, pp. 103–110, 2021.
https://doi.org/10.1007/978-3-030-79763-8_12

From early human history to modern times, major wars have become less frequent. However, intra-state wars in developing countries and international disputes in general have increased simultaneously. The world is becoming less predictable than ever as the multi-scale interdependence and consistent interaction over time generate non-linear consequences that cannot be reliably predicted by traditional statistics.

The limitation of regression in the face of forecasting the endogenous relationship between theoretical and dependent variables is ineluctable. It is inefficient to assess theories through statistical analysis due to the regression's assumption of independency between variables. If the theoretical variable is simultaneously affected by the dependent variable, the estimates produced by OLS become biased and inconsistent. ABM offsets such limitation and particularly, it is aims to increase both theoretical and empirical verisimilitude, interdependence, and feedback loop throughout several disciplines across all scales of development.

2 Literature Review

The predominant theories of war are derived from two contrary schools: (neo)realism and others. The realism or neorealism theorists believe that anarchy characterizes the world structure, while some opponents assert that the world structure most resembles a hierarchy. From the perspective of neorealism, power balance theory addresses that parity in power reduce major wars [1, 2]. In contrast, the power transition theorists assert that parity leads to fatal conflicts if the challenger is unsatisfied with the status quo [3, 4]. Pollins et al., and Barbieri's related studies on the relationship between international trade and war suggest that international trade increases disputes [5, 6]. Axelrod and Keohane's research of cooperation under anarchy provides means of promoting the outcome of the Pareto optimal with conditions [7]. In the field of international relations, scholars' rigorous works point at the same direction of endeavoring to answer the question about how to prevent war with various perspectives.

Indeed, there is not a panacea that can provide a solid solution for preventing war. I propose that a more complete explanation of the causes of war should, at least, consider a state's status quo in terms of an integrated capacity, international activities in the matter of bilateral trade, the adaptability that is developed across sustainable interactions between states, and a probability of randomness [8].

3 Research Design

3.1 Model Entities and Variables

I design the simulation in Netlogo, where each "patch" represents a heterogenous individual state. There are 225 patches in the environment. Patches, or states, are the only entity in the simulation model. Each time step presents one year and the simulation starting point is set at the year of 1950. I use three colors to distinguish state's characteristics in terms of regime type. I label rational secular type or conservative in red, the self-express type or democracy in blue, and the neutral in black accordingly [9]. The major properties of each state are its own unique integrated capacity, presenting the overall power[1] a state obtains in the status quo, and the trade share between the state and its direct neighbors. All states are only programmed to have interactions with their direct neighbor states based on the assumption that contiguity is one of the most robust preconditions for the increase probability of war [10]. The capacity is calculated by multiplying the state's trade share with a scaler. The scaler is a random number that is drawn from a uniform distribution U (0, 225). The initial value of the bilateral-trade share is randomly drawn from a uniform distribution U (0, 99). Equation (1) below defines the capacity for a given patch or state i.

$$\text{Capacity}(i) = \text{U}\left(0,\ \left[\text{sqrt}\left[\text{U}(0,\ 225) \times \text{U}(0,\ 0.99)\right]\right]\right). \tag{1}$$

In Eq. (1), 255 stands for the maximum number of states and 0.99 represents the maximum percentage of trade share.

Three types of characteristics, rational secular, self-express and neutral, representing three different predominant states, are randomly assigned to the patches. At each tick time, each patch has choices, either selecting a target from his direct neighbor patch or doing nothing. The selecting action is determined by his preference, which is controlled by the slider named "prob-select-target" in interface. Following by selecting target, the agent also makes decision on whether or not to assimilate the target through a peaceful talk or war. The decision relies on the following certain situations: (1) If either the target's capacity is less than 50% of the initiator's capacity or the interdependence of bilateral trade between target and the initiator is equal to or greater than the threshold of not having war, then, it is assumed there will no need for the initiator to wage a war to against the target. The initiator can easily assimilate the target through various cooperation in terms of trade and strategic alliance; (2) If the target's capacity is equal to or greater than the initiator's capacity, the interdependence between the two states is less than the threshold and the two states do not share the same characteristic, then there will be an 80% chance that war occurs between the two states. The follow pseudo code depicts the logic of the procedure.

```
To make decision
if capacity of target < power-threshold-assimilation or
   (sqrt (my bilateral-trade ratio)×(target's bilateral-
    trade ratio) >= interdependence-threshold-for-peace,
    then cooperate.
if capacity of target >= power-threshold-assimilation and
   (sqrt (my bilateral-trade ratio)×(target's bilateral-
    trade ratio) < interdependence-threshold-for-peace and
    my regime type ≠ target regime type, and
    when random-float 1.0 < 0.8, then wage war.
end
```

[1] Power is an integrated indicator in the matters of population, political capacity, GDP per capita, and social stability.

The threshold of interdependence is calculated based on the result of Barberie's regression analysis on militarized disputes (Barbieri, 1996). I speculate that if the superpower state wages a war, it must at least have 80% chance to win it. A simplified flowchart, reflecting major behaviors across three dimensions, is shown in Fig. 1 below.

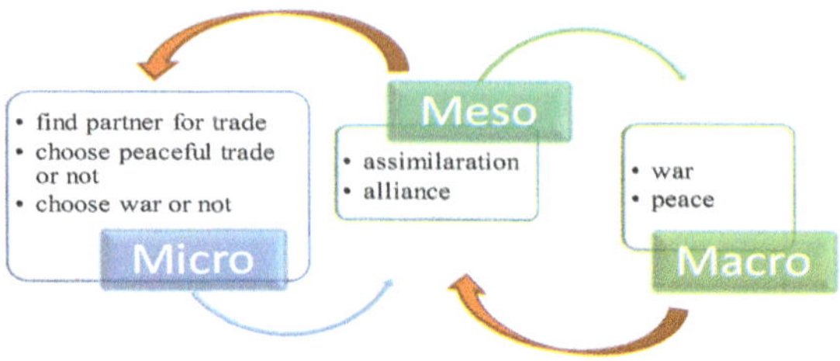

Fig. 1. A simplified flowchart consists of multi-scale behaviors within the endogeneity and feedback loops.

In Fig. 1, the interaction between states is either trade in a peaceful manner, trade made by war, or no interaction at all. Which event occurs depends on the powerful state's preferred strategy and the probability of its preference for war.

At meso level, when the powerful state chooses its target, there are two emerged consequences, which are led by the interaction at micro level. If it is a peaceful relationship between the two states, the color of the target states changes into the same color as the powerful state. In contrast, if war occurs between the two states, the lost state will change its color to white temporally and then change the color to the same color as the wining state.

In the macro system, the model portrays a picture of the dynamic interactions between states over time. The reciprocal actions between war and peace are captured by observing color changes across all iterations.

3.2 Parameterization and Simulation Experience

The baseline setup for the simulation is as follows: the interdependence (threshold) is set to 0.6. According to Barbieri, interdependence in trade can be measured by a reciprocal trade share for the two states multiplies the square root of total trade between two states (Barbieri, 1996). The probability of selecting a target is 50% and the probability of initiating a war is set to 50%, respectively. The model is terminated when tick time reaches 200 since I am interested in a long-term effect. Two snapshots of a sample run at $t = 100$ and 200 are shown in Fig. 2 below.

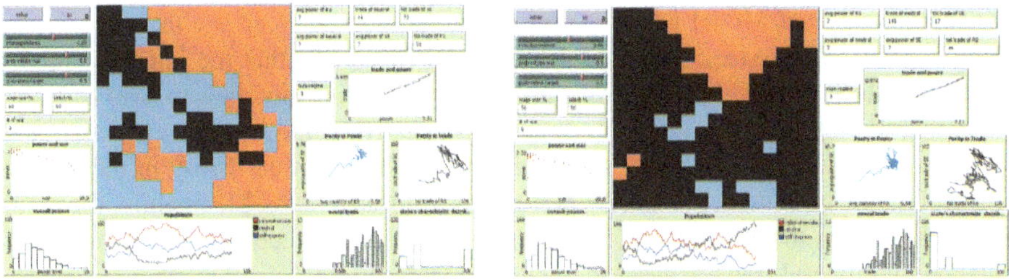

Fig. 2. The NetLogo interface. The snapshots of a sample run with baseline setup at time tick = 100 (*left*) and time tick = 200 (*right*). Three types of characteristics are colored by three different colors respectively.

3.3 Sensitivity Analysis

Sensitivity analysis not only helps examine the robustness of emerged behaviors, but also quantifying the variability in ABM outcomes resulting from model parameters. I instantiate it here to generate simulated data for OLS regression analysis in order to identify the correlation between theoretical variables and the dependent variable.

Three inputs of theoretical variables are adjusted in their feasible ranges simultaneously. For probability of initiating war, I set its value to 0.1, 0.5, 0.8, and 1 accordingly. For the probability of selecting a target to trade or fight, I adjust its value to 0.1, 0.5 and 1 accordingly. For the interdependence (threshold), I allow it to be 0.1, 0.6, and 0.95 correspondingly. The model generates 5,286 simulated data across 36 iterations. The pooled OLS regression is used and the detail of the regression is depicted in Table 1 below.

Table 1. OLS regression result based on simulated data. Confident level at 95%; significance: 0 '***' 0.001 '**' 0.01 '*' 0.05 'a'; all estimates are standardized. Time step as dummy variable is included in regression.

Pooled OLS	Model I: war		Model II: avg. capacity		Model III: avg. trade	
	Std. Coef.	Std. E	Std. Coef.	Std. E	Std. Coef.	Std. E
(Intercept)	0	0.423	0	0.157***	0	0.01***
Interdependence	0.124	0.096***	0.947	0.036***	0.955	0.004***
prob.initiate.war	−0.067	0.099***	−0.040	0.037***	0.003	0.004
prob.select.target	−0.005	0.091	0.018	0.034	0.019	0.004***
DF	5286		5286		5286	
RMSE	2.470		0.859		0.067	
Adj R-squared	0.484		0.93		0.94	
P(F)	0		0		0	

The objective of Model I is to assess the correlation between the theoretical variables and war occurrence. The result from model I suggests that the interdependence between states is positively and significantly associated with superpower-level wars, but the probability of initiating war with its target is negatively and significantly associated with war. The result is indirectly supported by power transition theory in the matters of the dissatisfaction which is led by non-reciprocal trade share. Meanwhile, the result indirectly reflects the interlinked consequences, which are directly generated and accumulated by bilateral trade, restrict the powerful states to fight against other states. The result also reveals that powerful states are less likely to initiate fatal disputes, but they are able to if war is their preferred strategy which is the case when states are at the onset of development. In addition, the standardized estimate of interdependence is the highest (β(est.) = 0.124) of all explanatory variables, thus, I suspect it is the most prominent cause that affects the occurrence of war.

The goal of Model II is to evaluate the association between the explanatory variables and the state's integrated capacity. The result indicates that the overall state's capacity is positively and significantly associated with the interdependence and the probability of waging war. It makes sense since that parity in power is correlated to war. Especially it is the case when two states, holding contradict opinions and misunderstanding, cannot settle down a shared agreement even though most states would not consider war was their optimal choice.

Model III is for examining the relationship between the interest variables and trade. The result from model III infers that the interdependence and the probability of selecting target for an intended relationship are positively and significantly associated with trade in general. Indeed, both trade and interdependence are associated with state's overall capacity. The probability of initiating a war does not show an association with trade. Thus, I postulate that trade is not the cause of war, but the interdependence which leads to dissatisfactions and unsolvable conflicts is.

4 Scenario Analysis

In this section, I particularly identify the impact of interdependence on war by implementing a scenario analysis.

The setup is as follows: Based on initialization, first, I set the range of the value of interdependence from 0.5 and 1.0 with an increment of 0.1 for each iteration. Holding all other inputs constant, the model generates 600 iterations. The frequency plot across iterations projects a negatively exponential distribution, indicating most wars occur in the beginning of the simulation while the probability of war is decreased significantly over time. Countries' characteristics, integrated capacities, and wars are positively correlated to each other. The frequency plot for the distribution of the types of characteristics exhibits the trend for convergence over time. This adaptive behavior implies that states tend to be more cooperative or prefer peaceful means to solve disputes (Fig. 3).

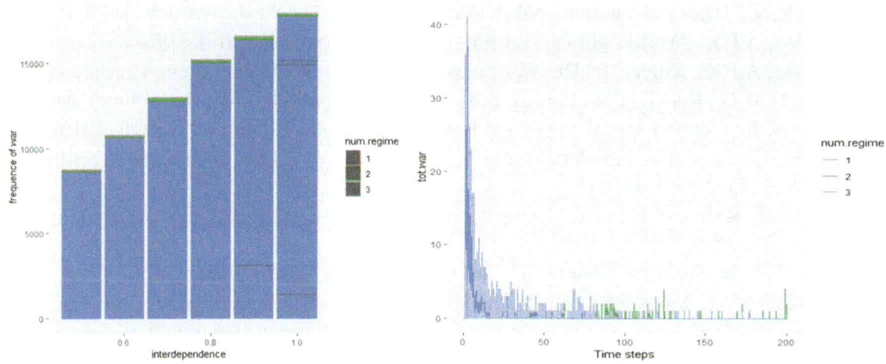

Fig. 3. War frequency plot of interdependence and war (*left*) and time series plot of war (*right*).

5 Conclusion

Agent-based modeling, combined with statistical analysis, is a promising approach to understanding the complexity. Simulation modeling uses assumptions about interactions between individual entities to simulate dynamic changes. This deductive conclusion is then evaluated by an inductive regression approach. In this frugal study, I merely focus on the impact of interdependence of trade and power distribution on war. Apparently, powerful state(s) would initiate war to any target state which is much weaker in terms of power. However, since a state's power is also positively associated with economic gains in terms of international trade and cooperation in various dimensions, the interlinked consequences that result in unpredicted cost constrains significantly restrict the powerful states from using their power over other states. Even though the interdependence is the inevitable trend of international development, the artistry of using efficient communication for mutual agreements rather than violence will be widely adapted and popularized by individual players. In consequence, the mutual interests and shared costs effectively prevent fatal conflicts and further prevent global wars.

6 Limitation

The limitations of this research include but not limited to: (1) The simulated result may not be generalizable; (2) It may not be an appropriate inference for those conflicts that are happened between non-powerful states; (3) Since the three characteristics as properties are simply designed to be randomly assigned to the states, without other conditions, the model may not be efficient in forecasting the most superior or longevity characteristic. In other words, I assume that three predominant characteristics are equivalent.

References

1. Waltz, K.N.: Man, the State, and War: A Theoretical Analysis. Columbia University Press, New York (1959)

2. Waltz, K.N.: Theory of International Politics, pp. 118–121. Addison-Wesley (1979)
3. Organski, A.F.K.: World Politics, 2nd edn. Alfred A. Knopf, New York (1968)
4. Organski, A.F.K., Kugler, J.: The War Ledger. University of Chicago Press, Chicago (1980)
5. Keshk, O.M.G., Reuveny, R., Pollins, B.M.: Trade and conflict: proximity, country size, and measures. In: Conflict Management and Peace Science, vol. 27, pp. 3–27. Sage (2010)
6. Barbieri, K.: Economic interdependence: a path to peace or a source of interstate conflict? J. Peace Res. **33**, 29–49 (1996)
7. Axelrod, R., Keohane, R.O.: Achieving cooperation under anarchy: strategies and institutions. World Polit. **38**, 226–254 (1985)
8. Gartzke, E.: War is in the error term. J. Int. Organ. **53**, 567–587 (1999)
9. Abdollahian, M., Yang, Z., Coan, T., Yesilada, B.: Human development dynamics: an agent based simulation of macro social systems and individual heterogeneous evolutionary games. Complex Adapt. Syst. Model. **1**(1), 1–17 (2013)
10. Bremer, S.A.: Dangerous dyads: conditions affecting the likelihood of interstate war, 1816–1965. J. Confl. Resolut. **36**(2), 309–341 (1992)

Modeling and Simulation of Disaster Medicine Processes for Resilience Assessment of Hospital BCPs

Kadono Shunsuke[1(✉)], Kanno Taro[1], Chen Chia-Hsin[2], and Chuang Sheuwen[2]

[1] The University of Tokyo, Tokyo, Japan
kadono-shunsuke038@g.ecc.u-tokyo.ac.jp
[2] Graduate Institute of Data Science, Taipei Medical University, Taipei, Taiwan

Abstract. Business continuity planning and management (BCP/BCM) is crucial for enterprises such as hospitals and healthcare agencies. However, there is no established method to evaluate the effectiveness and resilience of BCP. In this study, we develop a model for the in-hospital medical processes and resource usage required in mass casualty events. The process model is based on records from two hospitals that accepted several people injured in the Formosa Fun Coast Dust Explosion that occurred in Taiwan in 2015. We conducted a simulation to replicate the situation in the two hospitals for model verification and validation and confirmed that the simulation results were consistent with the actual data to some extent. The model developed in this study can be used to observe and compare the details of the disaster medicine processes under different settings and scenarios; thus, the model may be utilized for BCP evaluations.

Keywords: Disaster medicine · Mass casualty event · Process simulation · BCP · BCM

1 Introduction

Business continuity planning and management (BCP/BCM) is used to ensure the quick recovery and continuity of critical business processes during disasters. The International Organization for Standardization (ISO) has compiled a standard for business continuity in ISO 22301:2019 titled "Security and resilience — Business continuity management systems — Requirements." ISO 22301:2019 provides guidelines for managing the risks of disruption faced by organizations [1]. BCP is particularly crucial for hospitals and healthcare agencies to save the lives of the injured and reduce preventable mortalities during disasters, and it substantially affects the disaster resilience of the community and society. Moreover, it is necessary to consider how the scarce medical resources should be allocated and used to treat as many patients as possible after a disaster [2]. The importance of BCP/BCM has been widely acknowledged; however, there is no established method to evaluate their effectiveness and resilience [3, 4]. The purpose of this study is to develop a simulation-based method to evaluate the performance of hospital BCPs with a focus on resource consumption in consideration of various disaster conditions and scenarios.

J. L. Wright et al. (Eds.): AHFE 2021, LNNS 264, pp. 111–116, 2021.
https://doi.org/10.1007/978-3-030-79763-8_13

First, we develop a model of the in-hospital medical processes and resource usage in mass casualty events.

2 Modeling of In-Hospital Disaster Medicine

We developed a simulation model based on data from hospitals involved in actual mass casualty events. While modeling, we referred to the records of two hospitals that accepted several people injured in the Formosa Fun Coast Dust Explosion that occurred in Taiwan in 2015 [5, 6]. This is one of the worst accidents in the history of Taiwan, with 15 deaths and 497 injured survivors. We extracted the important disaster medicine processes from the records, including triage, emergency treatment, burn surgery, transfer of patients to ICU or general ward, and discharge of the patients, and modeled them using an input–process–output format. We also extracted the types and quantities of critical resources, such as ointment, gauze, and normal saline, as well as the number of physicians and nurses required for the processes and designated them as the variables for each medical process in the model. We developed a prototype of the process simulation by integrating these processes and variables as well as the configurations of the target hospitals.

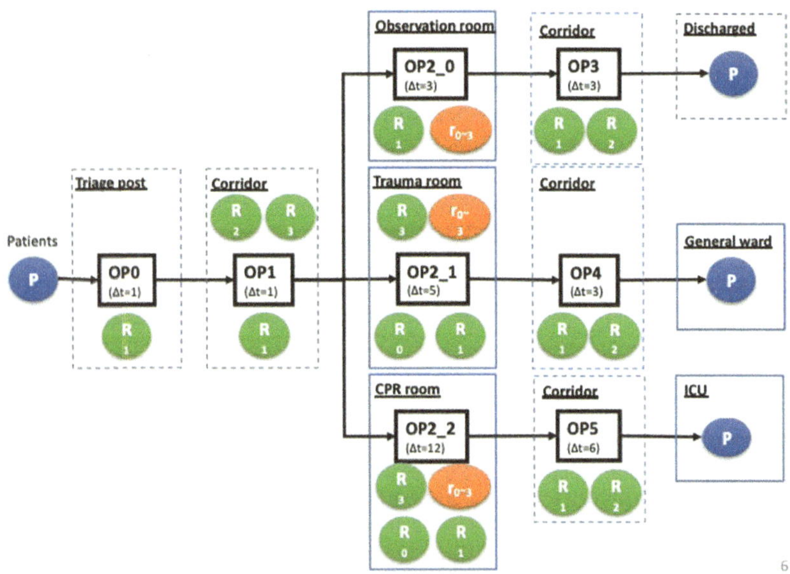

Fig. 1. Process model of disaster medicine in the target hospital

2.1 Process Model

We classified the processes in a hospital into six different operations (Fig. 1) represented as rectangles enclosed by a bold black line. As soon as burn patients reach a hospital, they are triaged at the triage post (OP0). Then, they are transported to appropriate rooms based on the triage results (OP1). Subsequently, they are given first aid in each room as required (OP2_0, OP2_1, and OP2_2). Patients who have received first aid are either discharged (OP3) or transported to the general ward (GW: OP4) or intensive care unit (ICU: OP5) as required. The locations and rooms in a hospital are represented as rectangles enclosed by blue solid or dotted lines.

The resources required for performing and completing each operation are represented using green or orange ellipses in Fig. 1. The details of each resource are described in the Table 1 and Table 2. Non-recoverable resources are the resources consumed or used up in an operation. Unconsumable resources are the resources that remain in use or occupied during an operation and become available again after the operation is over.

Table 1. Non-recoverable resources.

Resource number	Content
r_0	Gauze
r_1	Ointment
r_2	Insertion tube
r_3	Saline

Table 2. Unconsumable resources.

Resource number	Content
R_0	Physician
R_1	Nurse
R_2	Stretcher
R_3	Bed

2.2 Algorithm

The process model describes the flow of patients and resources in a hospital in disaster medicine. Figure 2 shows the basic operations in a unit process. The input to each process is the quantity of available resources and the number of patients entering or waiting at each time step. A process starts when there are patients waiting for the process and there is an adequate quantity of resources, including a vacant room/area to perform the process. The starting condition for each process is checked at each timestep of the simulation. Once the process starts, the non-recoverable resources are consumed and unconsumable resources become occupied. The process concludes after the time required for completion, which is estimated based on actual data. When the process concludes, the patients are moved on to the next process, the unconsumable resources are released, and their status is returned to unoccupied.

These processes are performed in parallel and repeated from the start to the end of the simulation. As the output of the simulation, we obtain the number of resources and patients in each room/place at each timestep.

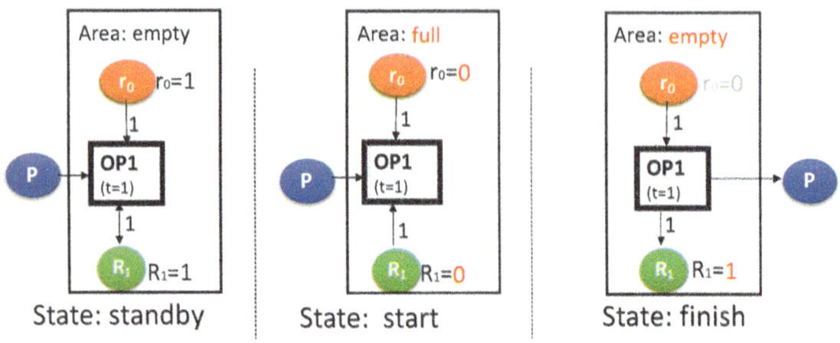

Fig. 2. Process diagram

2.3 Parameter Setting

The parameters for each process were determined based on actual data from hospitals. Triage (OP0) was estimated to take 5 min for each patient at the triage post. The capacity of the triage post was 24. If a room or area has already reached its capacity, a new process cannot be started there. The transportation of patients to each treatment room (OP1) through a corridor is estimated to take 5 min. The capacity of the corridor from the triage post to each room was set to 24. Treatment in the emergency department (OP2) was estimated to take 15 min, 25 min, or 60 min depending on the condition of the patients. The capacity of the treatment rooms was 6 or 11, depending on the case. Discharge (OP3) was estimated to take 5 min; this process can be performed regardless of the capacity. Transportation to the general ward (OP4) was assumed to take 15 min, including the waiting time, while transportation to the ICU (OP5) was set to take 30 min. The capacities of the corridors leading from each room to the ICU, general ward, and discharge were determined to be 36. This model can be applied to other hospitals by editing these parameters accordingly.

2.4 Input Scenario

The input to the simulation at each timestep includes the number of patients who arrive with different levels of injury and the resources supplied externally. In the simulations, we created an input scenario based on hospital records available in literature [5, 6], and we could simulate different disaster situations by changing the parameters of the input scenario.

3 Results

We conducted a simulation to replicate the situation in hospitals during disasters, and the simulation model was verified and validated. The input to the simulation is the actual number of injured people who arrive at the hospital, timing of their arrival, and quantity and timing of the external resource supply. We compared the results from the simulated disaster medicine processes, such as the consumption of resources and the time requirements for treatment, transfers, and discharges, with actual hospital records.

A comparison of the real and simulated numbers of burn patients according to their treatment duration is shown in Fig. 3. The results of the simulation matched well with the actual data. Figure 4 shows a comparison of the real and simulated quantities of available critical medical materials at each instant, which are also in good agreement with each other.

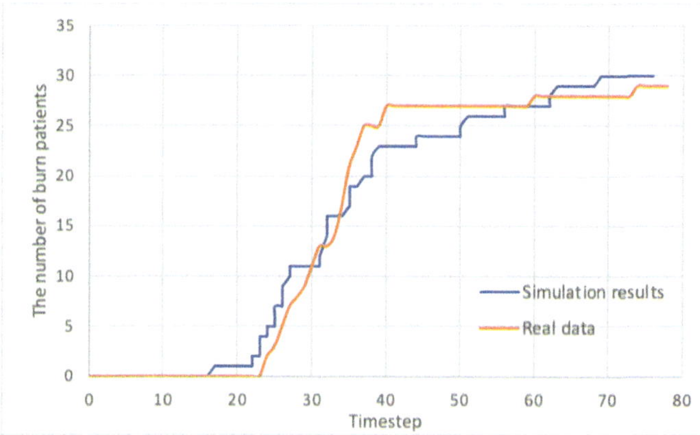

Fig. 3. Number of burn patients treated in the emergency department

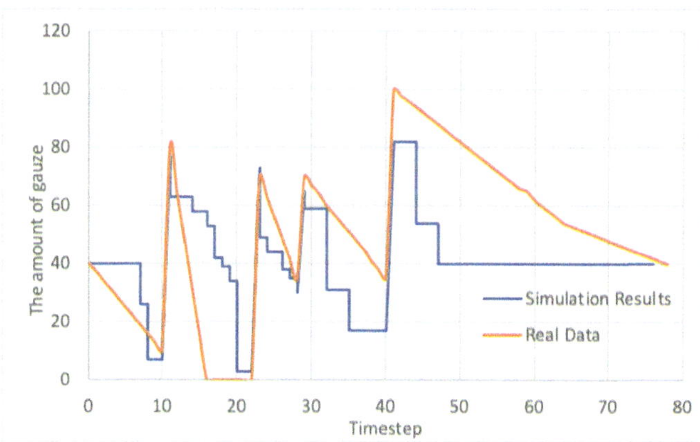

Fig. 4. Quantity of gauze available in the emergency department

4 Conclusion

We developed a process model and a simulation to represent disaster medicine procedures and successfully replicated the actual situation in the hospitals that responded to the Formosa Fun Coast Dust Explosion. In further studies, we hope to extend and refine the model to consider various processes that take place inside and outside the hospitals, along with the factors that influence the performance of the BCPs.

Acknowledgments. This work was supported by the JST-Mirai Program (Grant Number JPMJMI20B5), Japan.

References

1. ISO 22301:2019: International Standard — Security and resilience — Business continuity management systems — Requirements (2019)
2. Kanno, T., et al.: Resource-centric business continuity and resiliency planning. In: Proceedings of 8th REA Symposium on Resilience Engineering (2019)
3. Adi, G.W., et al.: Evaluation of the fulfillment of business continuity management system elements: a case study of clean water provision and services PT. XYZ in Indonesia. In: ICOHS (2017)
4. Daisuke, K., et al.: Resources for business continuity in disaster-based hospitals in the great east Japan earthquake: survey of Miyagi prefecture disaster base hospitals and the prefectural disaster medicine headquarters. Disaster Med. Public Health Prep. **7**(5), 461–466 (2013)
5. Chuang, S., et al.: Beyond surge: coping with mass burn casualty in the closest hospital to Formosa Fun Coast dust explosion. Burns **45**, 964–973 (2019)
6. Chuang, S., et al.: Coping with a mass casualty: insights into a hospital's emergency response and adaptations after the Formosa Fun Coast dust explosion. Disaster Med. Public Health Prep. **4**(4), 467–476 (2020)

Virtual and Augmented Reality

Information Ergonomics: Expediting Maintenance Workflow Using Mixed Reality

Chad Weiss(✉), Mary Freiman, Sylvain Bruni, Danielle Ward, Spencer Lynch, and Kristy Kay

Aptima, Inc., 12 Gill Street Suite 1400, Woburn, MA 01801, USA
{cweiss,mfreiman,sbruni,dward,slynch,kkay}@aptima.com

Abstract. Vehicle maintenance places high physical and cognitive demands on its practitioners, requiring technicians to access documentation of multiple media types that describe both the physical space and the actions within it. Technicians rely on documentation, collaboration with local and remote technicians, and their own research to address the challenges they face. While job aids such as manuals, build sheets, online tutorials, and contact with more expert or specialized technicians provide vital information, they also complicate the information landscape in which technicians work. We examine three everyday scenarios wherein vehicle maintenance technicians are impeded by poor information ergonomics. We explore the causes of unnecessary overhead, overload, and errors and their impacts on workflow. Taking lessons learned from the artifacts and coping strategies employed by seasoned maintenance technicians, we recommend uses of augmented reality to increase information fit to facilitate better workflow, increased efficiency, and reduced lifetime maintenance costs.

Keywords: Human factors · Human-systems integration · Systems engineering · Information ergonomics · Design · Human-centered engineering · Augment reality · Mixed reality

1 Introduction

Vehicle maintenance is challenging work that demands a lot of physical and cognitive strength of its technicians. In this paper we will focus primarily on maintenance work for the V-22 Osprey aircraft, but the findings are broadly applicable. The V-22 Osprey is a tiltrotor aircraft, which enables the vertical take-off and landing capabilities of a helicopter with the speed and power of an airplane. The systems and subsystems that make this amphibious assault vehicle possible were designed and manufactured by a web of subcontractors, and each has its own maintenance requirements. We conducted interviews with V-22 Maintenance technicians, examined artifacts they use on the job, and conducted a cognitive task analysis with subject matter experts to explore the pain points of maintaining increasingly complex machinery like the V-22.

The number and scale of procedures for a single aircraft make it impossible for even a seasoned technician to rely on memory alone. Maintenance technicians depend

J. L. Wright et al. (Eds.): AHFE 2021, LNNS 264, pp. 119–126, 2021.
https://doi.org/10.1007/978-3-030-79763-8_14

on job aids including procedure manuals (also referred to throughout as manuals and maintenance guides), build sheets, online tutorials, and contact with more experienced technicians to perform their job. Using these job aids is a struggle. Information systems are siloed, component identification schemas are inconsistent, and paper technical manuals make data aggregation and dynamic linking next to impossible. Each of these shortcomings costs time and frustration in the form of checks and re-checks, mistakes, and workflow friction.

To get a feel for the economic impact, consider the flight hour cost reduction in the V-22 Osprey Aircraft attributable to reduced maintenance needs. The V-22's current cost per flight hour is around $8,300 compared to $12,000 per hour in earlier years [1]. This decline is largely attributed to more reliable parts and different flying protocols which reduced the maintenance needs. With a ratio of up to 20 direct maintenance man hours per flight hour (DMMH/FH) [2] and total flight hours for all V-22s in excess of 500,000 [3], even a modest reduction in this ratio means significant annual cost savings.

We theorized that better information hygiene could make a serious impact on time and cost savings in vehicle maintenance. In order to test our theory, we designed a mixed reality prototype backed by a 7-dimensional data framework [4] to address the common challenges associated with information bottlenecks in vehicle maintenance. This framework affords contextual insights to enable just-in-time information delivery, automated intelligent procedure stacking, and frictionless guidance to reduce the time and cognitive energy demands on vehicle maintenance workers. We identified 10 emergent design principles centered around flexibly fitting the information to the technician, what we term information ergonomics.

In this paper, we will examine three work phases that highlight everyday information-centered inefficiencies. We will seek to explain the causes of frustration as well as the impacts on workflow. Lastly, we will explore the principles of information ergonomics that will facilitate a faster, more cost-effective maintenance force.

2 Maintenance Workflow

Vehicle Maintenance workflows require critical analysis and creative problem-solving skills in equal measure. We studied the work of vehicle maintenance technicians for both airborne and ground vehicles through interviews, artifact analysis, and cognitive task analysis [5], all supported by subject matter experts (SMEs). In both domains, technicians rely on manuals as an external knowledge base. Here, we combine the workflows from both domains and generalize across them to create a generic overview of maintenance workflows to illustrate (Fig. 1).

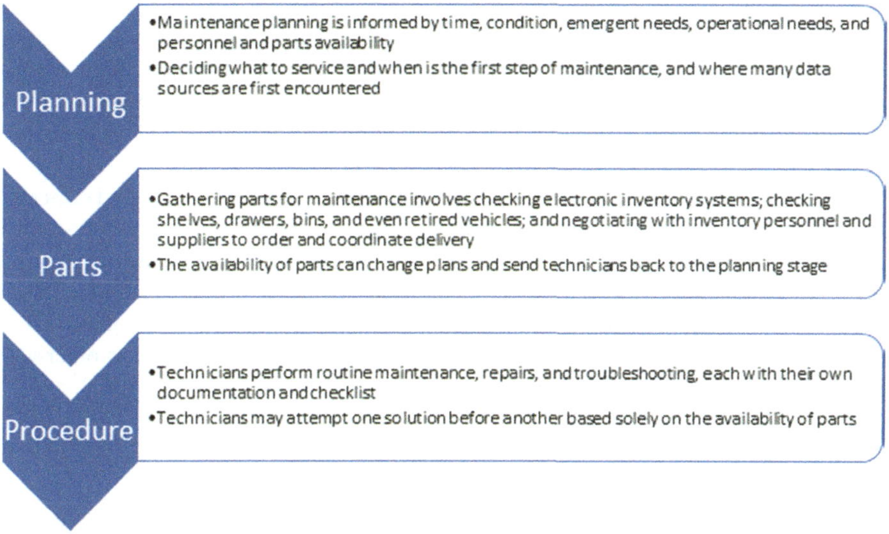

Fig. 1. Illustrates the stages of a maintenance technician workflow.

2.1 Planning

Maintenance planning has seen a revolution in the last twenty or so years. Maintenance approaches can be broken down into two broad categories, preventative and corrective. Preventive maintenance encompasses several prominent approaches, each roughly correlating to a strategy for maximizing up-time and avoiding failures [6]. Corrective maintenance, in contrast, happens after a failure has occurred, which is more expensive to perform and can result in tragedy as well as inconvenience. Condition-based maintenance (CBM) has risen in prominence as cheap and effective sensors and methods for data fusion have become available. CBM promises to shift maintenance from relying on time since last maintenance and emergent issues to using the actual condition of the vehicle to determine what maintenance procedures to perform [7].

Ideally, CBM-governed planning would rely on sensors from the vehicle to inform the decisions of maintenance managers. However, not all vehicles have been equipped with effective sensors on all components and analyzing data from sensors along with factors outside of the vehicle is still a challenge. Therefore, maintenance technicians rely on inspections, maintenance logs, scheduling and ticketing software, and vehicle manuals' time- and use-based recommendations to plan maintenance procedures. In our interviews with subject matter experts, we heard that these data sources are not correlated within a single data system, and problems in the data sources propagate throughout to cause inefficient scheduling. For example, missing or incorrect maintenance logs mean that the same procedures can be performed multiple times. Vehicle operators avoiding user-hostile scheduling systems are more likely to ask a teammate to perform maintenance than put a request into the system. Because the systems are siloed, manual updates are required to move information from one system to the other.

2.2 Parts

Prior to executing a maintenance procedure, the technician must make a list of all the necessary parts and order them through the inventory system. Part identification numbers found in procedure manuals do not match corresponding part ID numbers in the inventory system.

An inventory manager receives the order and collects the parts. When parts run out, the inventory manager locates yet another identification number and orders a resupply. If there are insufficient parts to cover the day's procedure, the maintenance shop has a part shortage. Part shortages can lead to maintenance delays and significantly increased maintenance times [2]. The inventory system is not a perfect representation of the inventory in stock at any given time, and parts that have been lost, damaged, or otherwise miscounted will further delay resupplies.

2.3 Procedure

The procedure includes diagnostics and execution of the repair. The procedure manual is the primary job aid used during the execution phase. Maintenance technicians noted challenges that arise from using the procedure manuals during procedure execution. The manual is cumbersome. Aircraft maintenance is hands-on work. It is easier to contort the body than it easier to bend the frame of an aircraft, and from an inopportune position it can be difficult to reference the procedure manual. Information in the procedure manuals is slow to update. Sometimes information is unintentionally omitted from early versions of a procedure manual. Sometimes an aircraft model is updated, and these changes often take months to work their way into the manual material. To fill gaps in the procedure manual, maintainers use other cognitive aids like notes from other technicians, handwritten checklists, and more experienced colleagues.

The technician is at heightened risk for physical injury during procedure execution. Inhalation of dangerous fumes can cause illness, contact with toxic chemical compounds can cause anything from skin irritation to severe illness to long term organ damage and central nervous system health issues [8]. Many dangers are documented in maintenance guides. For their health and safety, it is critical for a maintenance technician to be aware when an action they are about to perform will result in unsafe conditions (and whether that is by design or due to improper approach).

2.4 Summary

Currently, the best remedy for the challenges of maintenance planning, inventory management, and procedural knowledge is expertise. Technicians rely on their knowledge of vehicles to decide what and when procedures should be performed. Knowing of an extended or rough use of the vehicle, they know what to inspect and what issues to anticipate. They keep mental tabs on parts available to them in their workshops or on scrap components, and know who to call to find obscure parts, knowing that inventory management software and personnel are only part of the picture. And their knowledge and experience with procedures fill in the gaps in maintenance manuals with rare occurrences and quirks of the systems.

However, expertise is its own challenge across maintenance workforces. For example, the Air Force is losing its most experienced maintenance personnel each year, according to a 2019 Government Accountability Office report (Fig. 2). Maintainers are categorized by three levels of expertise, training, and experiences. The 7-level maintainers are the most experienced and 3-level maintainers are those with the least experience. As of 2017, there were documented shortages of 2,000 5-level and 400 7-level maintainers, and a surplus of 1,700 3-level maintainers. As 7-level maintainers are lost, so too is their experience-based knowledge and efficient performance on the job. The surplus of 3-level maintainers exacerbates the shortage of 5- and 7-level maintainers because maintainers at the 5-level are required to train 3-level maintainers, and 7-level maintainers train and supervise both 5-level and 3-level maintainers. The Air Force maintenance job role employs over 100,000 maintainers, making it the largest enlisted career field in the Air Force [9].

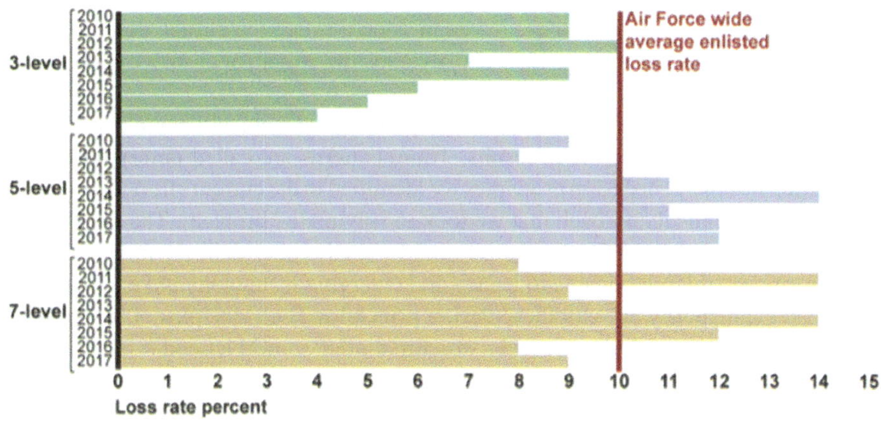

Fig. 2. Source: GAO analysis of Air Force Data, GAO-19-160. Air Force Aircraft Maintainer Loss Rates by Skill Level, Fiscal Years 2010–2017. (U.S. GAO, 2019)

In the 2015 and 2017 Air Force 2017 Aircraft Maintenance Retention Surveys [10], maintainers cited workload as one of the factors affecting their reenlisting decisions. They specifically noted increased stress and workload due to the lack of experienced maintainers, and the burden on experienced maintainers that is the result of training large numbers of 3-level maintainers. The scale of maintenance operations requires the precise coordination of people, systems, and data in order to maximize efficiency and effectiveness for the ultimate purpose of ensuring readiness and reducing downtime of systems. With the growing complexity of maintenance operations comes an increase in the number of human roles, in the variety of computerized software and hardware systems, and in the volume and intricacy of the information maintainers must manage and utilize. Expertise loss is seen across the phases of the maintenance workflow, making reliance on expertise as a mitigation strategy for the issues raised above doomed to fail.

3 Principles of Information Ergonomics

Our goal was to accelerate the maintenance process in order to decrease downtime and maintenance costs over time without introducing safety risk to individuals. We theorized that a cognitive agent built in accordance with these principles could help to mitigate many daily pain points of vehicle maintenance. We saw this as a natural fit for a mixed reality solution mediated by an Augmented Reality (AR) head-mounted device (HMD) to present information in-place, given the reliance on cognitive aids throughout the maintenance workflow. This would enable our cognitive agent to overlay relevant information over the physical environment, resulting in just-in-time information delivery while eliminating the latency faced when interacting with physical media. Mixed reality experiences can be mediated through both virtual and augmented reality technology. The key differentiator being that mixed reality facilitates a bi-directional information flow – interactions with objects in the physical world inform the virtual world and vice-versa. While AR facilitates contextual information placement, MR not only provides convenient naturalistic interactions (e.g. progression through the virtual world is driven by progression on the physical task), it also extends the cognitive agent's opportunities to learn through real-time observations.

During the execution of an iterative design process, the team will begin to recognize certain patterns that yield positive results. These patterns are called **emergent design principles.** The emergent design principles we identified in the current effort were shaped by the unique technical approach that we took to solving the problem. The intersection of our cognitive agent, our 7D Framework, and mixed reality delivery produced a set of emergent principles applicable to legacy and emerging technology alike. Because of their broad applicability, we call these design principles The Principles of Information Ergonomics. The principles are loosely structured after Dan MacLeod's 10 principles of physical ergonomics [11]. Our principles of information ergonomics are as follows:

Fit the Information to the Time and Task. Present information when it is needed and ensure that it in a format fit for the task it facilitates. For example, if the information at hand is a recipe that is intended to help an intermediate chef prepare a dish, adherence to this principle would favor a clear, specific ingredients list and step-by-step instructions over a recipe featuring a lengthy narrative lead-in outlining the history of the dish.

Help Make Sense of the Information to Offload Some of the Cognitive Work. Reduce the cognitive demands on your users by taking a mixed-initiative approach. Summarize, synthesize, disambiguate, and deconflict information where it is helpful to do so.

Minimize the Burden of Information Foraging and Search Tasks. Keep the relevant information within arm's reach. Aggregate data sources and bridge information siloes. Make it quick and easy to find the information a user needs.

Keep Information at the Appropriate Abstraction Level. Keeping focus on the users' tasks, ensure that the information is presented at the right level of detail. Facilitate user-driven exploration to traverse levels of granularity.

Present Information Close to the Locus of Action or Attention. Positioning is important. Position information where it is needed to increase the ease of application.

Provide Relevant Information in Digestible Chunks, Manage Information Density and Overcrowding. Make sure that the information presented to the user is not competing with itself for attention. Too much information density can be overwhelming.

Avoid Alert Fatigue. When busy workers become desensitized to important alerts, they ignore them, potentially missing critical or even life-saving information [12, 13]. Use alerts judiciously, alerting users only when the consequence warrants the user's attention.

Provide Clearance for Goal Time. Don't let the user get bogged down in the minutiae. Minimize tool time [14], automate repetitive information-related tasks, and facilitate smooth transition between tasks.

Allow the Mind to Stretch. Provide interesting, engaging content to keep the user stimulated.

Maintain a Comfortable Information Environment. This is a catchall for using good visualization design practices. The need to facilitate a comfortable information environment persists regardless of the mediating technology, but the implementation will differ from technology to technology. For example, designing for an AR environment introduces design considerations about focal plane and typography that don't apply when designing for the screen. Know the constraints of your technology and design around them.

4 Conclusion

The information ergonomics principles laid out above are not specific to the vehicle maintenance domain. They're not specific to AR, VR, MR, or any recombination thereof. They don't require a 7-dimensional framework (though it doesn't hurt). We expect that these information ergonomics principles will provide value to amateur chefs and soldiers alike. They are principles that will help practitioners ingest, interpret, and operationalize data with less friction.

They are also by no means final. We intend to operationalize these information ergonomics principles in a variety of other domains. The results of those efforts may uncover compelling reasons to adjust the principles as they stand today.

The information principles were formed as emergent design principles of a particular project, but they are informed by work from many great designers, design thinkers, human factors researchers, and ergonomists alike. Without these influences this work would not have been possible.

Acknowledgments. Thank you to our partners at Aviation Systems Engineering Company (ASEC) and Apex Analytics Group, Inc. for their significant contributions to this work.

References

1. Shalal-Esa, A.: U.S. eyes V-22 aircraft sales to Israel, Canada, UAE. Reuters, Thomson Reuters, February 2012. www.reuters.com/article/us-bell-boeing-v22-international/u-s-eyes-v-22-aircraft-sales-to-israel-canada-uae-idUSTRE81P0ND20120226
2. U.S. Government Accountability Office: Defense acquisitions: assessments needed to address V-22 aircraft operational and cost concerns to define future investments. GAO-09-482, May 2009. Publicly Released: June 23, 2009
3. Boeing MediaRoom: Bell boeing V-22 osprey fleet surpasses 500,000 flight hours. https://boeing.mediaroom.com/2019-10-07-Bell-Boeing-V-22-Osprey-Fleet-Surpasses-500-000-Flight-Hours
4. Bruni, S., Freiman, M., Weiss, C., Ward, D., Lynch, S., Kay, K.: A 7-dimensional framework for technical data in high-intensity vital environments and its application to aircraft maintenance. In: 2020 IEEE Conference on Cognitive and Computational Aspects of Situation Management (CogSIMA), pp. 211–215 (2020)
5. Hoffman, R., Crandall, B., Shadbolt, N.: Use of the critical decision method to elicit expert knowledge: a case study in the methodology of cognitive task analysis. Hum. Factors **40**(2), 254–276 (1998)
6. Barlow, R., Hunter, L.: Optimum preventive maintenance policies. Oper. Res. **8**(1), 90–100 (1960)
7. Prajapati, A., Bechtel, J., Ganesan, S.: Condition based maintenance: a survey. J. Qual. Maint. Eng. (2012)
8. Mehta, A.: How aviation injuries led to 8,670 lost work days across the military. Military Times (2018)
9. U.S. Government Accountability Office: Military personnel: strategy needed to improve retention of experienced air force aircraft maintainers. GAO-19-160, February 2019. Publicly Released: Feb 5, 2019
10. Air Force: Air force 2017 aircraft maintenance retention survey analysis (2018)
11. MacLeod, D.: Rules of Work: A Practical Engineering Guide to Ergonomics, 2nd edn. CRC Press, Boca Raton (2017)
12. Henneman, E.: Alert fatigue (2019). https://psnet.ahrq.gov/primer/alert-fatigue
13. Moacdieh, M., Sarter, N.: The effects of data density, display organization, and stress on search performance: an eye tracking study of clutter. IEEE Trans. Hum. Mach. Syst. **47**(6), 886–895 (2017)
14. Spool, J.: Dividing user time between goal and tool (2021). https://articles.uie.com/dividing-user-time-between-goal-and-tool/

Cybersickness and Its Implications for Using Virtual Reality Head Mounted Displays in Transport Psychology Research

Sára Klečková(✉), Petr Děcký, and Petr Zámečník

CDV - Transport Research Centre, Lisenska 33a, Brno, Czech Republic
{sara.kleckova,petr.decky,petr.zamecnik}@cdv.cz

Abstract. Cybersickness (CS), described mainly as unpleasant symptoms of nausea, eye strain, and headache, often occurs during usage of virtual reality (VR) head mounted displays (HMD). CS possesses a barrier for broader usage of VR not just in entertainment industry but also in research. This paper aims to review important findings about the prevalence and severity of CS, and main factors affecting CS. Further, opportunities and effectivity of VR usage in traffic psychology research are assessed.

Keywords: Cybersickness · Virtual reality (VR) · Head mounted display (HMD) · Theoretical review · Individual differences · Traffic psychology research · Simulated driving

1 Introduction

The development of new interface technologies, mainly Head Mounted Displays (HMD) such as the Oculus Rift have reestablished interest in virtual reality (VR) [1]. VR environment simulations became more accessible not just for private entertainment use but also for more serious research, education or training purposes [1–3].

However, there are long standing problems with unwanted side effects called cybersickness (CS). To the date, there is only limited understanding of this phenomenon and its causes. Also, there is also only limited knowledge about minimizing CS during VR experience. CS stands as the main barrier to wide and comfortable using of VR [1, 3, 4].

Traffic psychology research often uses simulated driving conditions to study driving behaviors. Especially research of hazardous or rarely observed behaviors could benefit from simulations, e.g. effects of fatigue, alcohol use, or distraction on driving. This paper introduces a unique connection of VR, cybersickness and traffic psychology, as the issue is very new to this field of study. For effective usage of VR in traffic psychology, practical implications from CS research should be implemented in order to create optimal conditions for VR simulated driving experiments.

J. L. Wright et al. (Eds.): AHFE 2021, LNNS 264, pp. 127–134, 2021.
https://doi.org/10.1007/978-3-030-79763-8_15

1.1 Description of Cybersickness Phenomenon

Cybersickness (or "visually induced motion sickness" or "virtual reality-induced symptoms and effects" [4]) is an unpleasant health condition occurring while people use VR. CS is understood as another kind of motion sickness when people experience visually induced movement from the VR while remain more or less static in real world [5]. CS usually includes symptoms such as nausea, eye strain and dizziness [1, 3–5]. Other symptoms could be – increased salivation, sweating, stomach awareness, burping, fatigue, headache, difficulty focusing, vertigo, eye dizziness, and blurred vision [6].

Similar symptoms share also other motion related phenomena – motion sickness and simulator sickness – although they occur under different situations [6]. Also, the main clusters of symptoms seem to have different severity. While nausea cluster is most severe during motion sickness, in simulator sickness and CS oculomotor cluster prevails [5–7]. In comparison to simulator sickness, CS poses more prevalent and approximately three times more severe condition [3].

In CS research, CS is usually assessed during relatively short time of VR experience. Most studies included immersion time up to 5 min [8, 9], 10 min [2, 10, 11], or 20 min [12–15]. Only a few studies included longer periods than 30 min [4, 7, 16]. Most CS symptoms occur relatively soon after beginning of the VR exposure – in first 10 min – and often increase with longer duration of VR exposure [17].

1.2 Prevalence and Severity of Cybersickness

According to most studies, some symptoms of CS experience overall majority of people. It seems that prevalence of CS symptoms and its severity among people strongly correlate with difficulty level of VR environment. When difficulty level was high, majority of participants did not hold in VR for the full immersion time, which was usually around 15 min. For example, in one study [12] examining the difficult condition of roller-coaster VR simulation, up to 66% of participants withdrew and from the rest, 41% of participants experienced severe symptoms of CS. In other study [10] where participants freely walked and explored VR environment, 55% of them did not finished VR session. In a study [15] where participants experienced relatively passive but horror-like VR simulation, 12% of sample withdrew. Another study with a larger sample [11] reported 7% of people with most severe symptoms and 79% of participants experiencing some symptoms of CS. In a study [2] using VR simulation with an action shooting game including walking and sometimes fast rotation movements, 12% of participants withdrew shortly after beginning of the immersion time and most of others experienced mild CS symptoms. In another study [18] with large sample which simulated driving conditions, approximately 40% of people experienced some CS.

Sometimes, the withdrawn part of participants was large due to the instruction to finish VR exposure when experiencing any symptoms of CS even minor [18].

Famous study [7] which is often cited for 80% prevalence CS, also confirms a mild to moderate symptoms for most of participants. Interestingly, this is one of the very few studies with larger sample size. A total of 148 participants were divided to 8 groups with mean size of 20 participants and maximum size of 40 participants. Unfortunately, we do not know much about the differences between groups who underwent different condition in VR. This study [7] reported 5% of people who must withdrew due the severe symptoms. Interestingly, some people underwent quiet long immersion time – up to one hour and more. While most of symptoms occurred during first 15 min, main worsening of symptoms happened between first 35 and 50 min and then slowly got better. When people came for repeated VR session, the second and third session included significant decrease of symptoms relative to the first session.

To summarize findings about the prevalence and severity of CS, there is always a minor part of people (about 5%) who withdrew shortly after beginning of the immersion time in VR [7, 11]; this minority represents group of people who clearly cannot handle even short and easy level of VR exposure. Also, there is a part of people who did not finish the whole VR session due to severe CS symptoms. Size of this group varies across the studies – from 10% to almost half of the participants. Severe symptoms are more prevalent in higher difficulty levels of VR simulations. Contrary, there is also minority of people who seems to be resistant to CS. Although majority of people experience some symptoms of CS, most of them suffer only from mild or moderate symptoms [7, 12]. In addition, symptoms usually do not last long, and their severity decreases to pre-immersion levels or they fully disappear shortly after exiting VR. If the severity of symptoms is mild, most people recover after 10 min [7, 19], or little longer [20].

1.3 Factors Influencing Prevalence of Cybersickness

Some theories offer description of causes of CS. However, there is still lack of complex understanding of the phenomenon. The issue gets even complicated with great individual variability [3]. The most popular explanation offers sensory conflict theory [1, 3, 21]. Theory describes conflicts of two key sensory systems engaged in virtual environments – visual and vestibular senses. For example, the vestibular system may be telling the individual that his body is stationary while the visual system is telling them that their body is moving, causing a sensory mismatch [21]. Unfortunately, like other theories, the sensory conflict theory does not explain the great variability of prevalence and severity of symptoms between people under same conditions [3].

Contrarily, important factors influencing CS are little more understood, as the most studies focus on assessment of these various possible causes and its control [3]. There is a great number of studied factors from psychological, physiological, demographic, or technical field. Studies of factors often include findings of optimal conditions that reduce CS, at least to some extent.

Technical Factors. Large number of studies draw attention to various technological characteristic of apparatus. Theoretical reviews [3, 22, 23] name a list of technical factors which should be controlled to decrease severity and prevalence of CS. These include display lags [9], flicker, calibration, general ergonomics, speed, acceleration, [24, 25], blur level, latency, refocus time, field of view, depth, persistence etc. More detailed environment with high resolution graphic or with quickly changing environment produces more severe symptoms [12, 23]. Monocular viewing appears to protect against CS relative to binocular viewing [9]. Also blurred scene during rotation movements significantly mitigated CS [2]. It seems that awareness of these device-related factors is essential in designing optimal conditions in VR [3, 22], as they probably have main influence to CS. Optimally, the simulation should not include sudden and unpredictable movements, accelerations and decelerations, rotations, as well as high speed [23, 25]. Changes of movements should be slow and smooth.

The position of the body during the VR simulation also plays the role. More severe symptoms occur when peoplae stand or walk, as the demands for postural control are greater. In the contrary, seating position shows less prevalence and severity of symptoms [21]. Interestingly, some studies observed compensatory head movements when participants experienced changes in direction of movement. These compensatory movements mitigate the mismatch between simulated and real movement and therefore help to decrease CS [8, 13, 26].

Probably the second most important factor are different kinds of tasks which participants perform during VR exposure. More control of the movement, such as driving simulations or simulations allowing free walking and manipulation with objects, produce less symptoms. If people control and can predict the movements, they can better prepare and adapt to them, so they are more protected to CS [2, 4, 12, 15, 21, 22, 24]. Contrary, uncontrollable, and unpredictable movement during VR session can produce severe symptoms. For example, when experiencing the fast ride on roller-coaster, most people could not finish the session [12].

Duration time also plays an important role for prevalence and severity of symptoms. Some studies recommend periodical breaks between short VR simulation episodes to mitigate the CS [23, 27].

Demographic Factors. Largely studied are differences between men and women. Some studies did not find significant differences between sexes [18]. This study found that from people who did not complete the full immersion time, women reported shorter time of immersion. Other studies revealed women more prone to CS than men [14]. The possible explanation is that women have wider field of view which increases the likelihood of flicker perception and this in turn increases their susceptibility to CS [1]. Another explanation could be that women are more willing to admit symptoms. Changes in levels of female hormones during the menstruation cycle also seems to have some impact on CS [21, 28].

Another largely studied factor is age. Symptoms get more severe with young children with significant decrease of symptoms in children older than 12 years and further decreasing with adults [29]. Contrary, older people (over 60 years) could be more prone to CS [14].

Psychological Factors. Familiarity with the content of VR environment helps people to mitigate CS [14]. Similarly, gamers are sometimes less prone to CS [3]. Importantly repetitive exposure to VR environment significantly mitigates symptoms. People who came to VR session for the second or third time had almost none symptoms of CS [7].

Physiological Factors. As the VR experience seems to be difficult for our body, probably any negative health condition such as acute or chronic disease could worsen CS. The same applies for hangover, fatigue, or alcohol use [1, 30].

History of motion sickness seems to be important predictor of current CS. In one study authors found significant correlation between history of motion sickness on carnival rides and current CS [16].

As the mismatch between simulated and real movement probably accounts for CS, some studies controlled postural factors. They found individual differences in spontaneous postural instability which predicted the likelihood of subsequently experiencing CS [8]. Other study [13] employed special moving platform during VR session, which could mimic the induced movements from VR simulation which resulted in significant mitigating CS.

Interestingly, one study [31] focused on mitigating CS through a specific eye exercise performed prior to VR experience. This 5-min-long exercise includes several head movements to various directions with focusing and fixating one constant spot with both eyes.

1.4 Measurement of Cybersickness

Traditionally, CS symptoms are measured by subjective measures, such as Simulator Sickness Questionnaire (SSQ) or verbal evaluation CS. More and more current studies also use physiological measures which often authors compare to SSQ scores.

Verbal self-evaluation of general discomfort during VR session was used in many studies [2, 10–13]. Participants reported severity of CS on a given scale, for example 0 to 4, where 0 refers to no CS and 4 refers to sever CS and intention to leave VR. Such verbal evaluation was recorded several times during VR session, for example every 2 min to observe development of symptoms.

SSQ [6] as a widely known and popular measure of CS [3, 23] was used in vast majority of studies mentioned in this paper [2, 4, 10, 14, 16, 18, 32]. SSQ includes 16 symptoms of CS categorized to three dimensions – oculomotor, disorientation, and nausea. Evaluation of SSQ results in calculation of Nausea-related subscore, an Oculomotor-related subscore, a Disorientation-related subscore, and the Total Score, representing the overall severity of CS experienced by the users of virtual reality systems [3, 6, 23, 33, 34].

Large amount of studies include objective measurement of physiological measures [7, 10, 11, 32, 35]. They allow a real time evaluation of changes related to CS and also cannot be biased by subjective evaluation of symptoms [3]. On the other hand, the equipment for physiological measures can be quite intrusive and analysis include great amount of data which can be difficult to analyze due also great intrapersonal and interpersonal variability of values [3]. Researchers found increase in heart rate, breathing, blinking, skin conductivity sweating, electroencephalography and stomach upset [7, 10, 11, 35]. Some also recorded significant changes of performance during VR session, such as reaction time [32, 35].

2 Discussion

Theoretical review of current CS research reveals some limitations and research challenges which should be addressed for more complex understanding. First, a lot of studies include only small number of participants. From above mentioned studies, a vast majority of them included a total number of participant smaller than 20 which underwent same VR conditions [2, 4, 8–10, 12–14, 16, 25, 28, 31, 32]. Only minority of studies found worked with bigger sample size than 20 participants [7, 11, 15, 18]. Therefore, it is difficult to generalized findings form these studies.

Majority of studies is based on subjective measure of SSQ or subjective verbal self-evaluation. These subjective measures could be biased due to differences in personal perceiving and evaluating of symptoms [3]. More precise objective measures of CS need to be developed.

Sometimes studies lack the detailed description of procedure, such as duration time of VR session, description of simulated environment and tasks, numbers of withdrawal, distribution of mild, moderate, or severe CS or reported values of SSQ for different groups. Also, understanding of outlasting effects of CS remain unknown, since only a few studies measured CS not just right after VR session, but after longer period.

The conditions of VR session greatly varied across studies – some included highly active environment with sudden changes movements, acceleration, and deceleration – some included much easier environment with free walking and exploring the environment. For the effective use of VR simulations in traffic psychology research, particularly studies implementing simulated driving software are most interesting. Unfortunately, only a few studies included simulated driving environments [7, 18, 24]. Finally, more complex control of various factors influencing CS is needed – e.g. evaluation of technical parameters and focus on optimal conditions of VR simulation.

3 Conclusion

This paper aims for theoretical review of CS with specific focus on practical implications of research findings for creating optimal conditions of VR simulation for its effective use in traffic psychology research. From studies can be deduced, that smooth, controlled driving without sudden changes of direction should refer to relatively easy conditions. Main challenge relates to sudden acceleration of car movement or sudden deceleration through braking. Ordinary traffic situations VR simulations which are close to real driving behavior should be less affected by CS than critical traffic events such as various mistakes or violating traffic rules resulting in near collision situations. However, particularly research of these critical events which are unreal or dangerous to perform in real traffic could benefit the most from simulated VR environment.

Acknowledgments. This article was produced with the financial support of the Ministry of Transport within the programme of long-term conceptual development of research institutions on the research infrastructure acquired from the Operation Programme Research and Development for Innovations (CZ.1.05/2.1.00/03.0064).

References

1. LaViola, J.J., Jr.: A discussion of cybersickness in virtual environments. ACM SIGCHI Bull. **32**, 47–56 (2000)
2. Budhiraja, P., Miller, M.R., Modi, A.K., Forsyth, D.: Rotation blurring: use of artificial blurring to reduce cybersickness in virtual reality first person shooters. arXiv preprint arXiv:1710.02599 (2017)

3. Davis, S., Nesbitt, K., Nalivaiko, E.: A systematic review of cybersickness. In: Proceedings of the 2014 Conference on Interactive Entertainment, pp. 1–9 (2014)
4. Sharples, S., Cobb, S., Moody, A., Wilson, J.R.: Virtual Reality Induced Symptoms and Effects (VRISE): comparison of Head Mounted Display (HMD), desktop and projection display systems. Displays **29**, 58–69 (2008)
5. Stanney, K.M., Kennedy, R.S., Drexler, J.M.: Cybersickness is not simulator sickness. In: Proceedings of the Human Factors and Ergonomics Society annual meeting, pp. 1138–1142. SAGE, Los Angeles (1997)
6. Kennedy, R.S., Lane, N.E., Berbaum, K.S., Lilienthal, M.G.: Simulator sickness questionnaire: an enhanced method for quantifying simulator sickness. Int. J. Aviat. Psychol. **3**, 203–220 (1993)
7. Cobb, S.V.G., Nichols, S., Ramsey, A., Wilson, J.R.: Virtual Reality-Induced Symptoms and Effects (VRISE). Presence Teleop. Virt. Environ. **8**, 169–186 (1999)
8. Arcioni, B., Palmisano, S., Apthorp, D., Kim, J.: Postural stability predicts the likelihood of cybersickness in active HMD-based virtual reality. Displays **58**, 3–11 (2019)
9. Palmisano, S., Szalla, L., Kim, J.: Monocular viewing protects against cybersickness produced by head movements in the Oculus Rift. In: 25th ACM Symposium on Virtual Reality Software and Technology, pp. 1–2 (2019)
10. Dennison, M.S., Wisti, A.Z., D'Zmura, M.: Use of physiological signals to predict cybersickness. Displays **44**, 42–52 (2016)
11. Kim, Y.Y., Kim, H.J., Kim, E.N., Ko, H.D., Kim, H.T.: Characteristic changes in the physiological components of cybersickness. Psychophysiology **42**, 616–625 (2005)
12. Davis, S., Nesbitt, K., Nalivaiko, E.: Comparing the onset of cybersickness using the Oculus Rift and two virtual roller coasters. In: Proceedings of the 11th Australasian Conference on Interactive Entertainment (IE 2015), p. 30 (2015)
13. Ng, A.K., Chan, L.K., Lau, H.Y.: A study of cybersickness and sensory conflict theory using a motion-coupled virtual reality system. Displays **61**, 101922 (2020)
14. Petri, K., Feuerstein, K., Folster, S., Bariszlovich, F., Witte, K.: Effects of age, gender, familiarity with the content, and exposure time on cybersickness in immersive head-mounted display based virtual reality. Am. J. Biomed. Sci. **12**, 107–121 (2020)
15. Sevinc, V., Berkman, M.I.: Psychometric evaluation of Simulator Sickness Questionnaire and its variants as a measure of cybersickness in consumer virtual environments. Appl. Ergon. **82**, 102958 (2020)
16. Rebenitsch, L., Owen, C.: Individual variation in susceptibility to cybersickness. In: Proceedings of the 27th Annual ACM Symposium on User Interface Software and Technology, pp. 309–317 (2014)
17. Min, B.-C., Chung, S.-C., Min, Y.-K., Sakamoto, K.: Psychophysiological evaluation of simulator sickness evoked by a graphic simulator. Appl. Ergon. **35**, 549–556 (2004)
18. Curry, C., Li, R., Peterson, N., Stoffregen, T.A.: Cybersickness in virtual reality head-mounted displays: examining the influence of sex differences and vehicle control. Int. J. Hum. Comput. Interact. **36**, 1161–1167 (2020)
19. Hildebrandt, J., Schmitz, P., Calero Valdez, A., Kobbelt, L., Ziefle, M.: Get well soon! Human factors' influence on cybersickness after redirected walking exposure in virtual reality. In: Chen, J.Y.C., Fragomeni, G. (eds.) VAMR 2018. LNCS, vol. 10909, pp. 82–101. Springer, Cham (2018). https://doi.org/10.1007/978-3-319-91581-4_7
20. Dziuda, Ł, Biernacki, M.P., Baran, P.M., Truszczyński, O.E.: The effects of simulated fog and motion on simulator sickness in a driving simulator and the duration of after-effects. Appl. Ergon. **45**, 406–412 (2014)
21. Kolasinski, E.M.: Simulator sickness in virtual environments. US Army Research Institute for the Behavioral and Social Sciences (1995)

22. orcino, T., Trevisan, D., Clua, E.: Minimizing cybersickness in head-mounted display systems: causes and strategies review. In: 2020 22nd Symposium on Virtual and Augmented Reality (SVR), pp. 154–163 (2020)
23. Porcino, T.M., Clua, E., Trevisan, D., Vasconcelos, C.N., Valente, L.: Minimizing cyber sickness in head mounted display systems: design guidelines and applications. In: 2017 IEEE 5th International Conference on Serious Games and Applications for Health (SeGAH), pp. 1–6. IEEE (2017)
24. Dong, X., Yoshida, K., Stoffregen, T.A.: Control of a virtual vehicle influences postural activity and motion sickness. J. Exp. Psychol. Appl. **17**, 128 (2011)
25. Kemeny, A., George, P., Mérienne, F., Colombet, F.: New VR navigation techniques to reduce cybersickness. Electron. Imaging **2017**, 48–53 (2017)
26. Palmisano, S., Mursic, R., Kim, J.: Vection and cybersickness generated by head-and-display motion in the Oculus Rift. Displays **46**, 1–8 (2017)
27. Melo, M., Vasconcelos-Raposo, J., Bessa, M.: Presence and cybersickness in immersive content: effects of content type, exposure time and gender. Comput. Graph. **71**, 159–165 (2018)
28. Clemes, S.A., Howarth, P.A.: The menstrual cycle and susceptibility to virtual simulation sickness. J. Biol. Rhythms **20**, 71–82 (2005)
29. Golding, J.F.: Motion sickness susceptibility questionnaire revised and its relationship to other forms of sickness. Brain Res. Bull. **47**, 507–516 (1998)
30. Kruk, R.: Simulator sickness experience in simulators equipped with fiber optic helmet mounted display systems. In: Flight Simulation Technologies Conference, p. 4135 (1992)
31. Park, W.D., Jang, S.W., Kim, Y.H., Kim, G.A., Son, W., Kim, Y.S.: A study on cyber sickness reduction by oculo-motor exercise performed immediately prior to viewing Virtual Reality (VR) content on Head Mounted Display (HMD). Vibroeng. Proc. **14**, 260–264 (2017)
32. Nesbitt, K., Davis, S., Blackmore, K., Nalivaiko, E.: Correlating reaction time and nausea measures with traditional measures of cybersickness. Displays **48**, 1–8 (2017)
33. Balk, S.A., Bertola, M.A., Inman, V.W.: Simulator sickness questionnaire: twenty years later (2013)
34. Kennedy, R.S., Drexler, J.M., Compton, D.E., Stanney, K.M., Lanham, D.S., Harm, D.L.: Configural scoring of simulator sickness, cybersickness and space adaptation syndrome: similarities and differences. In: Virtual and Adaptive Environments: Applications, Implications, and Human Performance Issues, p. 247 (2003)
35. Nalivaiko, E., Davis, S.L., Blackmore, K.L., Vakulin, A., Nesbitt, K.V.: Cybersickness provoked by head-mounted display affects cutaneous vascular tone, heart rate and reaction time. Physiol. Behav. **151**, 583–590 (2015)

Learning-Forgetting-Fatigue-Recovery Simulation Model

Vitor de Oliveira Vargas(✉) and Jung Hyup Kim

Department of Industrial and Manufacturing Systems Engineering, University of Missouri, Columbia, MO, USA
vdvx55@umsystem.edu, kijung@missouri.edu

Abstract. The present research aims to correlate the performance loss of a worker and fatigue during a day-work journey of a maintenance routine. The learning-forgetting-fatigue-recovery model seeks to answer why a worker cannot carry out such a maintenance routine for three or four hours long with a steady performance. Learning happens if someone performs a task multiple times. However, under fatigue, the worker will make more errors and lead to a longer completion time. On the other hand, forgetting occurs after a break, but due to recovery, s/he will make fewer mistakes, so the completion time tends to decrease. The current study results determine the best work and break configuration to minimize the whole maintenance routine duration based on the learning-forgetting-fatigue-recovery model.

Keywords: Maintenance routine · Work performance · Fatigue · Learning/forgetting/fatigue/recovery simulation model · Steepest descent direction · Response Surface Method

1 Introduction

The present research aims to correlate the performance loss of a worker and fatigue during a day-work journey. Usually, workers are submitted to the work routine with 8 h per day and 1 h for lunch & rest, regardless of their specialization and activity features. However, workers cannot work for 8 h uninterrupted, and they need multiple breaks while maximizing work performance. Hence, it is crucial to find the optimal configuration of work time/rest time in a workday to achieve the minimal routine completion time, including the best every work/rest period and the minimum worker fatigue level. The goal of this research is to develop a simulation model to compare the optimal work/rest configuration with the recorded fatigue level data.

In this study, the maintenance routine of performing borescope inspection of a gas turbine was selected to implement a simulation experiment. For that maintenance routine, the LM2500 General Electrics (GE) gas turbine was selected. That type of engine can be found in many ship propulsions all over the world. The reason for that specific choice is that this equipment is installed in enclosures inside ship machine rooms. Those rooms are usually noisy and very warm. Also, the work posture is awkward most of the time

J. L. Wright et al. (Eds.): AHFE 2021, LNNS 264, pp. 135–142, 2021.
https://doi.org/10.1007/978-3-030-79763-8_16

due to the poor accessibility. Plus, the maintenance routine is divided into several smaller sub-routines. For that reason, several actions are characterized as physically intensive motions.

This research will contribute to a systematic work scheduling method that will be applied to any industrial activity as a function of its work routine. The present study will also prove that the work performance is negatively related to worker's fatigue, and break times might be added to a workday schedule to achieve optimal work performance. Companies will then be able to reschedule their work routines by implementing the outcome of this simulation model, or simply by benchmarking, adopting the minimum acceptable fatigue level of similar industrial maintenance routines.

2 Methodology

In terms of fatigue, human factors modeling is concerned with muscular fatigue accumulation and recovery [0]. Although different aspects of fatigue can be included in human reliability analysis to identify potential risks, we considered the muscle fatigue accumulation and recovery curve behaviors as the basis of worker fatigue. Multipliers are related to the risk of a prolonged task performed without a break [0]. Similarly, in the absence of how the body accumulates fatigue and recoveries under a prolonged task performance followed by a break, it is reasonable that the fatigue accumulation is proportional to the associate risk level under that prolonged task performance.

The learning-forgetting-fatigue-recovery model calculates the completion time for a task by the Learn Forget Curve Model (LFCM), catching the interactions of learning–forgetting and fatigue–recovery of the worker [0, 0]. When a worker starts a task, the initial error rate is high and cumulative experience is low. As time passes on the job, the error rate decreases when the workers move down on the learning curve. However, with time, individual's fatigue level increases, leading to a rise in the error rate. Hence, in this study, we assume that the error rate can be translated to a longer task completion time.

3 Project Design and Routine

The simulation experiment will vary the factors of sufficient work and rest periods to minimize the routine completion time. It is essential to understand that factors, such as gender, age, expertise, and initial fatigue condition, are possible variability sources. In this study, we assume that all task durations are modeled using the normal distribution with a standard deviation representing 2% of its mean. The project design is detailed by a hierarchical task analysis (HTA) chart [1], which begins with selecting the model, experiment pattern, followed by the initial work/rest configuration setup. To calculate a minimal maintenance cycle time, Response Surface Methods and Design (RSMD) and the method of steepest descent were used [2, 3]. These methods consisted of fitting models based on the simulated data from Micro Saint Sharp software. The initial conditions for the experiment are, for work period, 60 min at a low level, 90 min as a center point, and 120 min at a high level, and for rest periods, 5 min for low level, 10 min as a center point, and 15 min at a high level. It is important to mention that the initial operation condition is randomly selected. Simultaneously, the worker fatigue level is recorded to identify a correlation between the optimal configuration work time/rest time and fatigue level. In this way, even if it is not possible to carry out the same procedure to other maintenance routines, by benchmarking a minimum fatigue allowance can be defined.

The experiment consists of describing the response according to the characteristic equation. Then, it should test for the relevance of the main effect interaction coefficient and the quadratic effect to reduce the model to a first-order design. Next, it should move sequentially toward the steepest descent direction to achieve a minimum region. Since there is evidence that the response is at a minimum region, it is time for building a second-order model, using the Central Composite Design (CCD), the most popular class of designs for fitting the second-order model [2]. So, a variable transformation will be carried out to a canonical form to identify the stationary point (minimum cycle time) and the controllable variable settings to achieve that value. Finally, the maintenance routine will be carried out using the optimal setting to record the fatigue level and define the maximum acceptable fatigue level.

4 Simulation Model Description

This study runs the Learning-forgetting-fatigue-recovery model using the Micro Saint Sharp software as described by its simulation flow chart (see Fig. 1).

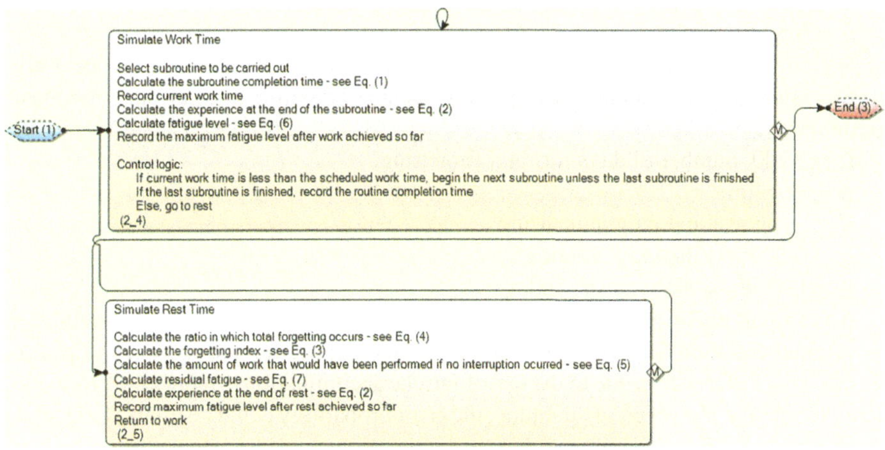

Fig. 1. Simulation flow chart.

This study uses Eqs. (1), (2), (3), (4), and (5) according to a learning curve that enables three types of learning rate: slow, moderate, and fast [9, 11]. This study uses the slow one because a few long tasks characterize a borescope inspection routine. So, the few repetitiveness implies a slow learning rate.

$$t_{x_i} = T_{x_1}(1 + u_i)^{-b} e^{\alpha F(t_i)}. \tag{1}$$

$$u_{i+1} = (1 + u_i)^{1+f_i/b} S_i^{-f_i/b}. \tag{2}$$

$$f_i = \frac{b(1-b)log(1+u_i)}{log(1+B)}. \quad (3)$$

$$B = \frac{D}{\frac{T_{x1}}{1-b}(1+u_i)^{1-b}}. \quad (4)$$

$$S_i = \left[\frac{1-b}{T_{x1}}\tau_i + (1+u_i)^{1-b}\right]^{1/(1-b)}. \quad (5)$$

$$F(t_i) = R(t_{i-1}) + \left[1 - R(t_{i-1})\right]\left(1 - e^{-\lambda t_i}\right). \quad (6)$$

$$R(t_i) = F(t_{i-1})e^{-\mu\tau_i}. \quad (7)$$

Where:

t_{x_i} = time to carry out task x at cycle i.

T_{x_1} = time to carry out task x at the first time.

u_i = experience at the beginning of the cycle i in the maintenance routine.

b = is the learning rate.

$\alpha = 1.6$, is the fatigue index.

$F(t_i)$ = fatigue accumulation at cycle i.

$u_1 = 0$, the initial experience in the first cycle is null.

f_i = forgetting index.

S_i = amount of work that would have been performed if no interruption occurred.

B = is the ratio in which total forgetting occurs to the time to carry out $1 + u_i$ tasks continuously.

$D = 1200$, number of days for total forgetting.

τ_i = rest time.

$R(t_{i-1})$ = residual fatigue measure.

$\lambda = 0.0096$, is the fatigue index.

$\mu = 0.0096$, is the recovery index.

It is important to mention that the fatigue accumulation effect [9] was added to the original equation to compound the learning-forgetting-fatigue-recovery impact on Eq. (1). This difference is due to the use of fatigue accumulation to assess the production error probability. However, in this study, the error probability could result in a longer task completion time. For example, workers are more prone to move borescope probe inappropriately. It means that they would have to find the correct inspection probe position, or they would have to go backward to re-check the damage. The fatigue index (α) implies that the maximum fatigue level is 100%, and the necessary time to complete a task will be five times greater. Due to the uncertainty of the relative contribution of learning and fatigue to the error rate, it is necessary to decide which one is more important [9]. In this study, fatigue is considered the most crucial factor because few and exceptionally long tasks characterize the chosen maintenance routine. For example, after the first task is finished, if it took 10 min, based on Eq. (4), the expected completion time for the next task is 10.43 min. It is 10 min times the learning contribution, 0.9, times the fatigue contribution, 1.15.

Although there are numerical examples under various forgetting times [10], 1,200 days were selected for the total forgetting time in this project. This simulation model treats the fatigue accumulation as in Eq. (9) and recovery function as in Eq. (10) based on the Learning-forgetting-fatigue-recovery model [12].

Both the fatigue accumulation index and the recovery speed index were determined using a test [9]. We also assumed that the workers are exhausted and are recovered at the end of 8 h. It means when the work starts with a zero-fatigued state, and as time passes by, fatigue converges asymptotically to 1. Moreover, the borescope inspection routine is simulated as established on the manufacturer manual [13] using Micro Saint Sharp software.

5 Results

5.1 First-Order Model

To build the first-order model, an experiment with two variables ($2^2 = 4$ runs) and five center points with ten replicates was used (total 90 runs). Equation (8) represents the built first-order model.

$$\hat{y} = 264 + 8.8325x_1 + 4.6375x_2. \quad (8)$$

Where:

$\hat{y}$ = estimated completion time (min).

x_1 = work time duration in coded level $[-1, 1]$.

x_2 = rest time duration in coded level $[-1, 1]$.

It is important to test the model for the relevance of the interaction and the lack of fit without the pure quadratic effect. The interaction effect (x_1x_2) is not significant since the p-value, 0.4945, so that effect can be neglected (see Table 1). There is enough evidence that the first-order model is not adequate.

Since it is not acceptable, the approach is to find another starting point. However, by comparing the main effects of F-ratios and the lack of fit F-ratio, it is possible to conclude that the main effects are more significant than the pure quadratic effect, which means that despite the significance of the pure quadratic effect, by following the steepest descent vector $(-\widehat{\beta}_1, -\widehat{\beta}_2)$ from Eq. (8), the objective function would decrease in any way.

By applying sequential experiments on the steepest descent vector, it is possible to reach the minimum output of 60 min for work time and 7.375 min for rest time.

Table 1. Analysis of variance for the first-order model (significance level $\alpha = 0.05$).

Effect tests					
Source	DF[a]	Sum of squares	Mean square	F ratio	P-value
x_1	1	3,120.5223	3,120.5223	304.1148	<0.001
x_2	1	860.2562	860.2562	83.8375	<0.001
x_1x_2	1	4.8302	4.8302	0.4707	0.4945
Errors					
Lack of fit	1	256.8133	256.8133	34.8912	<0.001
Pure error	85	625.6318	7.3604		
Total error	86	882.4451	10.2610		
Total					
Total	89	4,868.0538			

[a]DF are the initials for Degrees of Freedom

5.2 Building a Second First-Order Model

Since the model is near a minimum point, it is necessary to build another first-order model and check both main effect interaction and pure quadratic effect again. The experiment with two variables ($2^2 = 4$ runs) and five center points with ten replicates was also used. Equation (9) represents the built first-order model.

$$\hat{y} = 252.7 - 4.96x_1. \tag{9}$$

The interaction effect (x_1x_2) is not significant (p-value $= 0.3361$). In fact, from Eq. (9), the interaction effect is not significant and the main effect x_2, with a p-value of 0.1251. On the other hand, the lack of fit presented a p-value less than 0.001, which means that the quadratic effect is significant, so the experiment must be designed as a second-order model.

5.3 Central Composite Design for the Second-Order Model

Using JMP software, the experiment was designed as rotatable, with the axial value of 1.414 [2], five center points, and ten replicates, turning out 130 runs in total. The model is characterized by Eq. (10) and by a variable transformation, Eq. (11) in a canonical form, which tells that the solution to this experiment is a minimum point. Moreover, the curve is very reasonable to w_1 and almost insensible to w_2, that is, the response is acceptable to variations in work time, while it is almost insensible to variations in rest time. Plus, JMP software also turns out that the solution is $(x_1, x_2) = (0.1141, -1.2111)$. Transforming coded variables to natural ones turns out 61.141 min for work time and 6.32 min for rest time.

$$\hat{y} = 249.5 - 0.391x_1 + 0.8291x_2 + 0.7075x_1x_2 + 5.4656x_1^2 + 0.3756x_2^2. \tag{10}$$

$$\hat{y} = 248.9 + 0.9976w_1^2 + 0.069w_2^2. \tag{11}$$

The first first-order model turned out the average routine cycle time of 264 min and 83% of fatigue after work. After applying the steepest descent vector theory, a second first-order model was built with the outcome of the average routine cycle time of 252.7 min (−4.3%) and 79.3% of fatigue after work. Finally, applying the RSMD, the optimal solution is 61.141 min of work periods and 6.32 min of breaks. Then, running the model using the reached solution with ten replicates, it turned out an average completion time of 245.3 min (−7.1% over the first model), with an average maximum fatigue level after work of 79.3%.

6 Discussion and Conclusion

The results show that the response is extremely sensitive to work time variations, while it is almost insensible to rest time. It does not mean that rest time is not important, but it is essential that the worker carry out the maintenance routine with a work time near 61.141 min with a flexible break time, within its second-order model range [6.14 min, 8.61 min], for decreasing the worker fatigue levels and hence reducing the maintenance routine completion time.

Those fatigue levels are useful if companies do not intend to spend time and resources for an experiment to identify the maximum fatigue level of the tested conditions (i.e., working without a break and after resting). They might apply our findings to a worker who resumes the maintenance routine.

Moreover, it is worth mentioning that this experimental design does possibly fit any maintenance routines. In this aspect, the determination of the maximum acceptable fatigue level might be an important tool to pinpoint when a worker should take a break. Based on our study, when a worker reaches 79.3% of fatigue, it is time to rest, returning to work not before reaching 69.1% of fatigue.

The most important conclusion in this study is that fatigue decreases work performance. In other words, if the objective function is to minimize the cycle time for a maintenance routine, it is mandatory scheduling breaks after working periods. Moreover, the model solution cannot be considered a precise outcome; however, it can be viewed as a starting point. This study recommends a real experiment to ratify our simulated results in future works. But still, if the borescope routine is carried out using the same starting point, this study results will be validated, and the work performance will be improved.

The fatigue (α) and recovery (μ) indexes are adaptations from the original model that correlates fatigue to error rate [9]. One limitation of the current study is that those indexes must be defined using real experiment results to link fatigue level, task completion time, and the fatigue and recovery indexes. In this way, the simulation model might be more precise.

For future studies, applying this simulation model to more complex maintenance routines with multiple workers is recommended. Also, by including different break time variables in the model, it would be possible to predict the best resting periods in various routine stages to minimize the whole maintenance completion time.

References

1. Staton, N.A.: Hierarchical task analysis: developments, applications and extensions. Appl. Ergon. **37**(1), 55–79 (2006)
2. Montgomery, D.C.: Design and Analysis of Experiments, 9th edn. Wiley, Hoboken (2017)
3. Bradley, N.: The response surface methodology. Department of Mathematical Sciences, Indiana University of South Bend. MS in Applied Mathematics and Computer Science thesis (2007)
4. Socha, V., Hanakova, L., Freigang, M., Kraus, J., Stojic, S.: Impact of pilots' tiredness on the outcome of psychological testing. Transp. Probl. **14**(2) (2019)
5. Tsuchiya, M., Takahashi, M., Miki, K., Kubo, T., Izawa, S.: Cross-sectional associations between daily rest periods during weekdays and psychological distress, non-restorative sleep, fatigue, and work performance among information technology workers. Ind. Health **55**, 173–179 (2017)
6. Sagherian, K., Clinton, M.E., Abu-Saad Huijer, H., Geiger-Brown, J.: Fatigue, work schedules, and perceived performance in bedside care nurses. Workplace Health Saf. **65**(7), 304–312 (2017)
7. Dode, P., Greig, M., Zolfaghari, S., Neumann, W.P.: Integrating human factors into discrete event simulation: a proactive approach to simultaneously design for system performance and employees' well being. Int. J. Prod. Res. **54**(10), 3105–3117 (2016)
8. Rasmussen, M., Laumann, K.: The evaluation of fatigue as a performance shaping factor in the Petro-HRA method. Reliab. Eng. Syst. Saf. **194**, 106187 (2020)
9. Givi, Z.S., Jaber, M.Y., Neumann, W.P.: Modelling worker reliability with learning and fatigue. Appl. Math. Model. **39**(17), 5186–5199 (2015)
10. Jaber, M.Y., Bonney, M.: Production breaks and the learning curve: the forgetting phenomenon. Appl. Math. Model. **20**(2), 162–169 (1996)
11. Wright, T.P.: Factors affecting the cost of airplanes. J. Aeronaut. Sci. **3**(4), 122–128 (1936)
12. Jaber, M.Y., Givi, Z.S., Neumann, W.P.: Incorporating human fatigue and recovery into the learning–forgetting process. Appl. Math. Model. **37**(12–13), 7287–7299 (2013)
13. GEK 50501 - GE Marine Gas Turbines. GE Marine Engines, Cincinnati, Ohio

Human Body Posture Recognition in Virtual Reality System for Astronauts Training

Junpeng Guo(✉) and Jin Yang

National Key Laboratory of Human Factors Engineering, China Astronaut Research and Training Center, Beijing, China

Abstract. Virtual reality provided a new way for training before flight in the weightless environment. In this study, we established a system that could recognize the human body's posture in real-time and the recognition results could be used for VR training system. An experiment was conducted to testify the effectiveness of posture recognition in VR training, and the results showed that the recognition results could help promote the sense of immersion and reality in the virtual environment.

Keywords: Virtual reality · Human body posture recognition · Training simulation · Deep convolution neural network

1 Introduction

Preflight training is crucial to guarantee astronauts safety and efficiency during the on-orbital flight [1, 2]. With its rapid development, virtual reality now has provided a new method of preflight training for astronauts. In the virtual environment, astronauts can experience the feeling of weightlessness via vision or even proprioception as this simulated environment can show abundant details from the perspectives which frequently appear in the weightlessness environment and difficult to replicate by other means on earth due the gravity, e.g., from the perspective with the body upside-down and so on [3, 4]. However, up to now, the behavior of the trainee's whole body is not easy to be captured or reproduced in real-time at a low cost during the virtual training, which leads to the behavior of the avatar in the virtual environment always be inconsistent with the trainee [5]. This inconsistency could undermine the realistic and immersive experience during the virtual training process and the information provided by the trainee's body is also not integrated into the training system.

In this paper, we proposed a new method to recognize the human body posture via an ordinary RGB camera and integrate the human body posture information into the virtual reality system for astronaut training, which could effectively improve the experience during the virtual training. Our method was based on deep convolution neural network, and was developed using the DensePose [6], which was an open-source human body posture recognition tool that could be used to recognize the whole human body parts and establish the correspondence from every pixel of the picture captured from human by

J. L. Wright et al. (Eds.): AHFE 2021, LNNS 264, pp. 143–150, 2021.
https://doi.org/10.1007/978-3-030-79763-8_17

camera to the human body. And we had modified some steps in the establishment of the correspondence and made it more suitable for the virtual reality application. With this algorithm, the trainee's behavior in the virtual training system could be recognized and reproduced in real-time efficiently. Then we integrated the recognition results obtained above into the virtual reality training system to make the behavior of the avatar in the virtual environment consistent with the trainee. With some optimization, the avatar could behaved consistently with the trainee in real-time which made it practicable for improving the virtual training experience. And an experiment was conducted to testify the effectiveness of this improved method in virtual training.

2 Modeling of Human Body

To recognize the human posture, the first step was to establish an appropriate model that could describe the human body. There were two kinds of human body model that generally used for posture recognition [7, 8]. The first one was named skeleton model, which described the human body by several joints and the connections between these joints. With these joints and connections, all the parts of human body could be defined and described. For example, some vital joints such as head, neck, shoulder and hand were usually chosen to be used in the model. Figure 1 was a schematic diagram that describing the skeleton model, including 18 joints, namely head, neck, left shoulder, right shoulder, left elbow, right elbow, left wrist, right wrist, left hand, right hand, left hip, right hip, left knee, right knee, left ankle, right ankle, left foot and right foot.

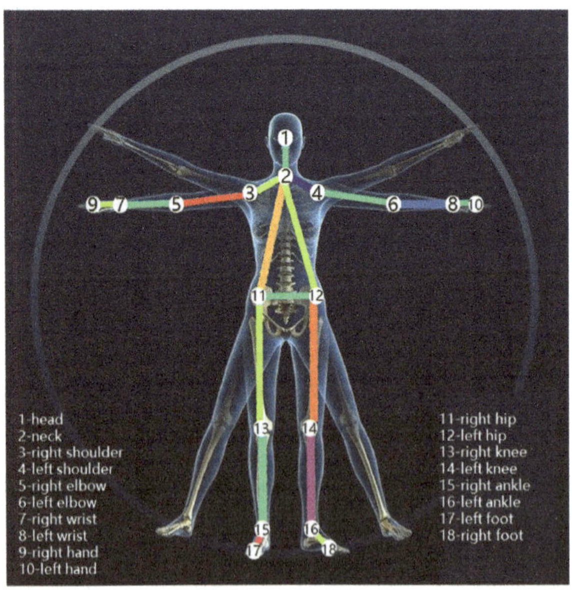

Fig. 1. The schematic diagram of skeleton model. The entire human body could be described by the eighteen key joints and the connections between them.

Another model was named dense model, this kind of model described the human body via dense points and mapped the two-dimensional coordinate of the picture to the three-dimensional surface of human body. There were two stages in building the dense model. The first stage was the partition of the human body and the second stage was finding the key points in the different parts. The way to partitioning the human body was alike the method in choosing the vital joints in the skeleton model. And for the simplification of the UV parameters, the different parts of the human body were designed to be isomorphic to the two-dimensional panels, as shown in Fig. 2.

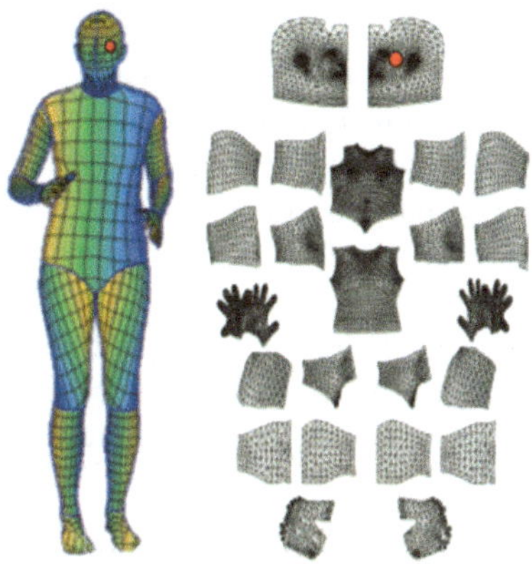

Fig. 2. Partitioning and UV parametrization of the body surface in dense pose model. This model could provide a more accurate way to describe the human body.

In the second stage, the key points were usually obtained by the methods of K-means clustering algorithm, which could generating a group of equidistant points in each parts of the human body, with the maximum of 14 points. For example, in the part of the torso, the result of the equidistant points obtained was shown as Fig. 3.

Fig. 3. The schematic diagram of equidistant key points obtained in torso from K-means clustering algorithm.

In this paper, we adpoted the second kind of human body model, namely the dense model for posture recognition. As we use the method of deep convolution neural network, training data in large amount was needed. We the Dense Coco dataset to achieve this. The dataset of Dense Coco was developed from the dataset of MSCoco [9], and included about 50000 human body pictures with annotations. These annotations were marked with UV coordination and were convenient for data training.

3 CNN-Based Human Body Posture Recognition Algorithm

Deep learning has developed rapidly in recent years and has made lots of achievements in various areas, especially in computer vision [10, 11]. Convolution neural network (CNN) is one of the most popular implementations for deep learning. The main idea of CNN is establishing an artificial neural network that imitates the human brain and the connections between the neural nodes determine the structure of the network. CNN usually consists of input layer, convolution layer, active layer, pooling layer, fully-connected layer and output layer. Through the use of convolution kernel, the CNN could abstract more local and meaningful information more efficiently. And with the back-propagation algorithm, the parameters, namely the weights between the different nodes, could be calculated and evaluated. After training process, the parameters of the CNN could be identified and used for future predicting [12].

In this paper, our goal was to segment the instance of every single person from pictures and mark the different parts of the human bodies. In order to achieve this, firstly, we needed to scan the picture to obtain features and produce proposals which represented the potential areas that contained the aimed regions. And secondly, we needed to predict the classification of the proposals and make regression of the target box and the mask. The whole structure was shown as Fig. 4.

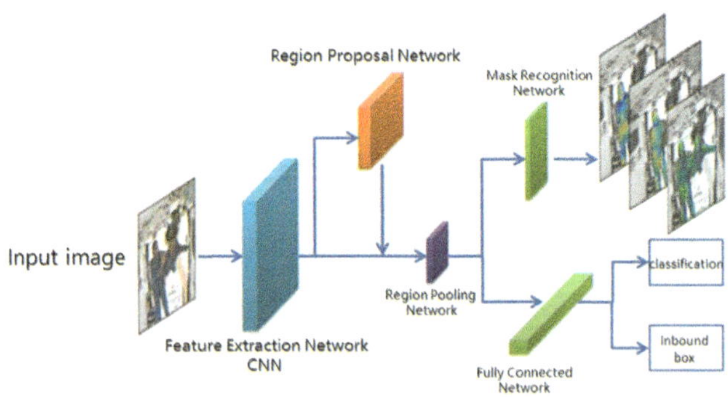

Fig. 4. The whole structure and workflow of the recognition network. In application, the input information could be images or videos (as in the situation of driving avatars in virtual reality) obtained from an ordinary RGB camera.

4 Results of the Posture Recognition

To testify the validation of the human body posture recognition algorithm, we captured some pictures containing single or multiple persons using an ordinary RGB camera. And with these pictures, we testified the correctness of the recognition results and the extent of the dense grids' coverage. And we checked the recognition results in various way, including the semantic results of the human body's parts, the U picture and V picture in the UV coordination, the frame picture and the grid picture. And the frame picture included the probability that represent whether certain object belonged to the human body.

4.1 Results of the Picture with Single Person

The result of recognition in the picture with single person in various postures, was shown as Fig. 5, which successively included the semantic results, the U picture, the V picture and the frame & grid picture. In quantitative analysis, the human body object recognition result of the single person picture was about 99%, which was quite high. And as the semantic results showed, the different parts of the human body were marked as different colors correctly. Lastly, according to the frame and grid picture, the recognition accuracy was also very high and the representation was neither excessive nor deficient.

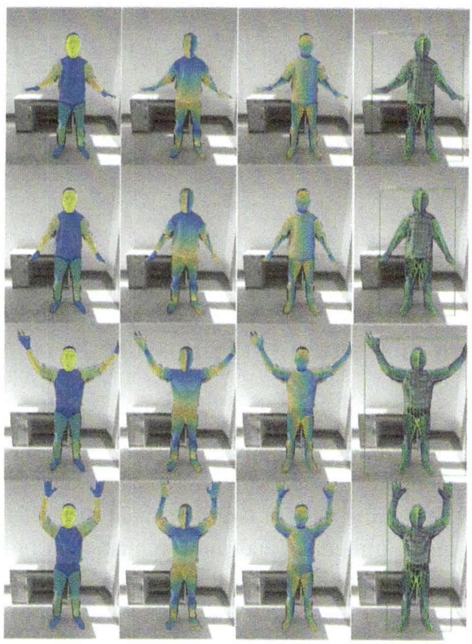

Fig. 5. The result of recognition in the picture with single person in various postures. The semantic results, the U picture, the V picture and the frame & grid picture was shown in each posture.

4.2 Results in the Picture with Multiple Persons

The result of the human body posture recognition in the picture of multiple persons, was shown in Fig. 6, which also successively included the semantic results, the U picture, the V picture in UV coordination and the frame & grid picture. For the picture of multiple persons, the recognition results were not reduced due to the additional person. And the semantic results and recognition accuracy were also as good as the situation of single person.

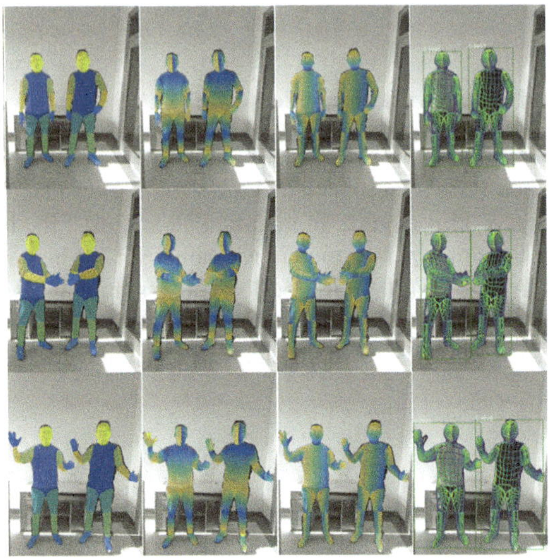

Fig. 6. The result of recognition in the picture with multiple persons in various postures. The semantic results, the U picture, the V picture and the frame & grid picture was shown in each posture.

5 The Application of Human Body Posture Recognition in Virtual Reality

Virtual reality is a useful training equipment for some special situations, such as the task in dangerous environment or in the weightless space. With the help of human body posture recognition, VR training system could provide more abundant information to make the trainee feel more immersive and more realistic. In order to do this, we had developed a data interface that could drive the avatars in VR environment by the results of human body posture recognition so that the different parts of the avatars' body were behaved in accordance with the trainee. The results showed that our recognition results could provide effective driving data for the avatar especially in the action of walking and delivering goods.

In order to testify the effectiveness of this improved method in virtual training, we conducted an experiment. The experiment was designed to fulfill an on-orbital task in the virtual environment that simulating the weightless space and was finished with thirty-two subjects (mean age = 28.12, SD = 2.41, ranging from 25 to 33). All the subjects were divided into two groups with equal numbers and finished the on-orbital tasks with\without the help of human body posture recognition results. And the related data of the trainees were collected, including the performance data such as the time used and the results of accuracy in the task, and the questionnaire data that recorded the sense of immersion and reality. Analysis of these data collected was shown in Table 1. From the analysis we could see that although there were no significant difference between the groups with\without the help of recognition results, the participants in the group using the recognition results felt more immersive and realistic than the other group.

Table 1. The contrast between the two groups with\without recognition results. The higher score of sense of immersion and reality meant stronger sense.

Item	Averaged performance data in the group with recognition results (s)	Averaged performance data in the group without recognition results
Time consumed in the task	325	412
The score of sense of immersion and reality	8.7	6.5

6 Conclusion and Future Study

In this study, we proposed a new method of VR training for the simulated weightless environment with the help of human body posture recognition algorithm. The recognition algorithm was based on deep convolution neural network and could effectively provide the driving data for the corresponding parts' behavior of the avatar in VR. The results showed that this method could significantly improve the sense of immersion and reality in VR training.

In the future, in order to provide more full-scale recognition results in VR training system, study in making use of the gesture and feet posture data should be focused to improve the sense of immersion and reality in the VR training system.

Acknowledgments. This study was supported by the foundation of National Key Laboratory of Human Factors Engineering (No. SYFD06180051805).

References

1. Oman, C.: Spatial orientation and navigation in microgravity. In: Mast, F., Jäncke, L. (eds.) Spatial processing in navigation Imagery and Perception. Springer, Boston (2007). https://doi.org/10.1007/978-0-387-71978-8_13

2. Aoki, H., Oman, C.M., Natapoff, A.: Virtual-reality-based 3D navigation training for emergency egress from spacecraft. Aviat. Space Environ. Med. **78**(8), 774–783 (2007)
3. Clément, G., Reschke, M.F.: Neuroscience in Space, 1st edn. Springer, New York (2010). https://doi.org/10.1007/978-0-387-78950-7
4. Zhu, X., et al.: Design of a virtual training system and application of an evaluation scheme for orientation in a spacecraft cabin. In: Long, S., Dhillon, B.S. (eds.) MMESE 2015. LNEE, vol. 356, pp. 653–661. Springer, Heidelberg (2015). https://doi.org/10.1007/978-3-662-48224-7_78
5. Jiang, F., Yang, X.: Real-time full-body motion reconstruction and recognition based on rare sensors. Doctoral dissertation, Shanghai Jiao Tong University, China (2017)
6. Güler, R.A., Neverova, N., Kokkinos, I.: DensePose: dense human pose estimation in the wild. In: Proceedings of the IEEE Conference on Computer Vision and Pattern Recognition (2018)
7. Martinez, J., Hossain, R., Romero, J., et al.: A simple yet effective baseline for 3D human pose estimation. In: IEEE Computer Society (2017)
8. Wei, S.E.: Convolutional pose machines: a deep architecture for estimating articulated poses. In: IEEE Conference on Computer Vision and Pattern Recognition (2016)
9. Lin, T.-Y., et al.: Microsoft COCO: common objects in context. In: Fleet, D., Pajdla, T., Schiele, B., Tuytelaars, T. (eds.) ECCV 2014. LNCS, vol. 8693, pp. 740–755. Springer, Cham (2014). https://doi.org/10.1007/978-3-319-10602-1_48
10. LeCun, Y., Bengio, Y., Hinton, G.: Deep learning. Nature **521**(7553), 436 (2015)
11. Yoo, H.J.: Deep convolution neural networks in computer vision: a review. IEIE Trans. Smart Process. Comput. **4**(1), 35–43 (2015)
12. LeCun, Y., et al.: Backpropagation applied to handwritten zip code recognition. Neural Comput. **1**(4), 541–551 (1989)

Hospital Escape VR: A Virtual Reality Simulation for Hospital Fire Evacuation Training

Victor Saint Martin[1(✉)], Eduardo Palmeira[1], Victor Gonçalves[1], Alexandre Siqueira[2], Edgard Lamounier[1], and Alexandre Cardoso[1]

[1] Faculty of Electrical Engineering, Federal University of Uberlândia (UFU), Uberlândia 38400-902, Brazil
{lamounier,alexandre}@ufu.br

[2] Department of Computer and Information Science and Engineering, University of Florida, Gainesville 32611, USA
agomesdesiqueira@ufl.edu

Abstract. Every establishment is vulnerable to imminent threatening situations. Hence, traditional hands-on evacuation training is paramount to assure the establishments' safety. However, typical evacuation training has limitations in its ability to reproduce threatening situations. In Health Care Establishments (HCE) such as a hospital, traditional practical training may affect the daily operation. Also, usually, only a small number of healthcare workers participate in regular training. Moreover, the inclusion of features that can threaten patients' safety — such as fire and smoke — is not allowed. Consequently, the healthcare workers' engagement and retention of knowledge acquired through training are impaired. Thus, Virtual Reality (VR) emerges as an alternative to traditional evacuation training. Therefore, we designed a VR simulator specifically focused on hospital evacuation training in the face of a fire situation. This VR evacuation training has shown the potential to aid healthcare workers to acquire abilities concerning evacuation in the face of fire.

Keywords: Virtual reality · Evacuation · Hospitals · Training

1 Introduction

According to reports published by the Instituto Sprinkler Brasil [1], there was an increase of 96% in the number of hospital fires in 2020, compared to 2019. Between January and October of 2020, 45 cases were recorded — one hospital fire every six days. Despite the high number of records, the actual number of cases is even greater since there is a lack of statistical data in Brazil, and casualties of this nature are not limited to that country; between 2011 and 2015, 5750 structure fires in healthcare properties were recorded annually in the USA. Among these structure fires, approximately 1110 occurred in hospitals [2].

J. L. Wright et al. (Eds.): AHFE 2021, LNNS 264, pp. 151–158, 2021.
https://doi.org/10.1007/978-3-030-79763-8_18

Regardless of the leading cause of hospital fire, healthcare workers need to deal with the emergency quickly and correctly. According to the Joint Commission [3], hospital leaders are responsible for managing hospital emergency plans. Leaders need to review hospital evacuation plans and promote hands-on evacuation training for healthcare workers. Although hands-on evacuation training is crucial for healthcare workers, introducing fire and smoke in real-life simulations is not a viable approach. Fire and smoke are essential to simulate a real life-threatening situation; however, the presence of these elements go against hospital environments' safety regulations. In this context, Virtual Reality (VR) emerges as an evolution of traditional training [4]. Through VR, users' immersion in the training simulation can be enhanced, for example by developing Virtual Environments (VE) with high levels of realism [5–9]. Besides, healthcare workers' engagement and knowledge retention can be increased by using simulated virtual fires and smoke [9, 10]. Also, the VR training can be practiced asynchronously [5, 10–12]. Finally, traditional training has a high cost to perform; consequently, it is accomplished only a few times throughout the year. Differently, virtual training only has a high cost to be implemented, becoming a cheaper option in the long term [11].

2 Related Work

Serious games are activities that entertain users in an environment where they can learn and train well-defined areas and tasks [13]. As pointed out in some studies [4, 9], serious games offer considerable advantages compared to traditional learning approaches. Currently, serious games linked with VE is a trend. Hence, users are immersed in realistic environments and deal with dangerous situations without threatening their lives. In this context, serious games associated with VR simulators can be used for training actions in a hospital fire situation [9, 10, 12]. For instance, VR training would not disrupt the hospital's activity, and the user could repeat the exercise as much as they want. Also, features as fire and smoke could be simulated; thus, neither the hospital nor the users would be exposed to imminent risks.

Rahouti et al. [10] designed a serious game for healthcare workers in a VE that simulates a hospital. The authors emphasize actions required in fire emergencies such as firefighting and the triage of non-ambulant patients. In contrast, Lu et al. [9] also designed a hospital's VE but focused on a different audience. Their serious game was designed for firefighters to train rescue actions under a post-earthquake fire scenario. Unlike the other studies, the serious game by Silva et al. [12] focused exclusively on hospital fire evacuation.

Unlike the mentioned related works [9, 10, 12], we propose an immersive VR serious game to teach healthcare workers how to evacuate from a virtual hospital in the face of a fire situation. The serious game simulates a hospital fire evacuation during the night shift. According to Hunt et al. [14], night scenarios increase evacuation time considerably. The increase in evacuation time can be explained by the reduced staff typically working night shifts and, commonly, weary compared to the day staff. We chose the night shift to increase the adverse conditions — there is no sunlight illuminating the VE. In addition to fire and smoke, hospital lights are turned off due to a short circuit. We intended to train healthcare workers to reach the nearest emergency exit following just the photoluminescent emergency signaling.

3 Hospital Escape VR

3.1 Software and Hardware

The 3D hospital environment was built using as a reference a computer-aided design (CAD) plan and a building information modeling (BIM) file from a Brazilian hospital. However, we applied modifications in the hospital project to reach the simulation objectives. First, virtual objects were designed using Blender[1]. Next, textures for virtual objects were acquired from Quixel Megascans[2]. In addition, Epic Games Store's[3] free products — already made objects and textures — were used to increase the development speed. Then, all virtual objects and textures were imported into the Unreal Engine 4[4] (UE4) game engine. Finally, all functionalities of this serious game were coded using the programming languages C++ and Blueprints.

The avatar was designed using MakeHuman[5], and then its mesh was exported to Blender for small adjustments. Mixamo[6] was used to obtain animations for the avatar. Afterward, adjustments inside UE4 were made to convert the avatar into a first-person character. We choose a Head-Mounted Display (HMD) to increase the realism of the experience, as suggested by Rahouti et al. [10]. An Oculus Rift S and its remote controllers were used to visualize and interact with the VE.

3.2 Virtual Environment

We chose the hospital environment in a fire situation as the scenario for our virtual simulation. The hospital environment was chosen because patients with different mobility restrictions need to be evacuated correctly and safely from the Health Care Establishment (HCE). Moreover, the fire situation was chosen due to its recurrence in hospitals and the danger associated with fire smoke. To focus specifically on fire evacuation training of healthcare workers, only a hospital section was built. Figure 1 portrays a hospital room present in the VE.

Equipment and furniture belonging to the hospital — stretcher, serum support, bed head panel, armchair, and reception bench — were included in the scenario to increase the user's immersion. Besides, emergency system components directly related to evacuation procedures were designed to assist users during the HCE evacuation. Emergency lights, photoluminescent emergency signaling, emergency exit, and alarms are some emergency system components included. Through the emergency system, we aimed to induce distinct emotions in the users, highlight the danger associated with a fire emergency, and assist them to find the nearest emergency exit.

The primary purpose of our simulation is to provide realistic and reliable training. To that end, we developed two features based on the fire propagation system designed by

[1] https://www.blender.org.
[2] https://quixel.com/megascans.
[3] https://www.unrealengine.com/marketplace/en-US/store.
[4] https://www.unrealengine.com/en-US.
[5] http://www.makehumancommunity.org.
[6] https://www.mixamo.com.

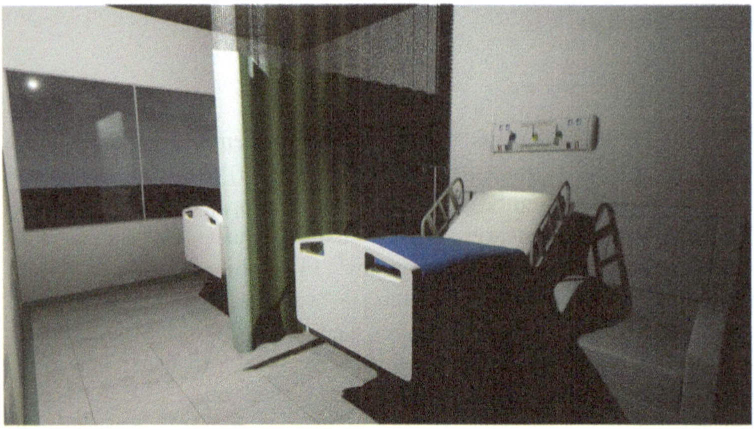

Fig. 1. Hospital room number 101. Textures and lighting effects were carefully designed to simulate a real hospital room.

Jean-Francois Lévesque in Far-Cry [16]. The first feature simulates fire propagation and destruction of virtual objects, while the second feature simulates fire smoke propagation and visibility reduction. Both features occur dynamically and automatically. The fire propagation is based on the use of a Grid-Box (Fig. 2). Each cube from Grid-Box identifies, individually, if there is a collision with a virtual object. If there is no collision, the cube is removed from Grid-Box. After verifying all the cubes, only those colliding with virtual objects remain in Grid-Box, creating an object outline in the tridimensional space. In each cube, there are a fire particle, a smoke particle, a decal — to simulate the fire damage on the object —, and a sound emission — to reproduce the object's burning sound. Then, the algorithm randomly chooses one cube from the Grid-Box to start the fire. Finally, the cube on fire recognizes neighbors cubes and activates them after some seconds, propagating the fire and restarting the cycle.

Smoke propagation is also based on Grid-Box. However, there is just a smoke particle without collision detection with other objects. After the fire begins, the smoke gradually fills the cubes from the Grid-Box near the ceiling and ends filling the cubes near the floor. In other words, when the cubes near the ceiling reach the maximum opacity, the cubes below begin to fill with smoke. As a result, the smoke propagation impairs the avatar's visibility. Note that both features — fire and smoke propagation — can be used anywhere and with any object without significant algorithm changes. Moreover, an object's position or room's size can be changed even after introducing these features in the VE.

Through the emergency system, smoke, and fire propagation, this immersive VR simulator has the potential to reproduce an emergency in a VE. In this context, healthcare workers could use the simulator to enhance the hands-on evacuation training.

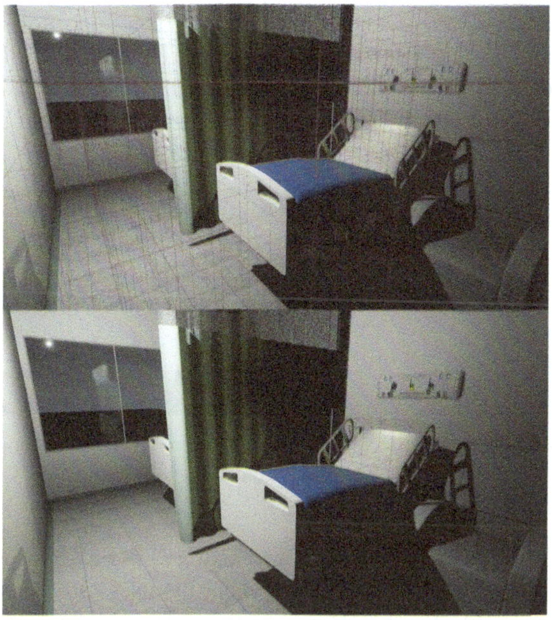

Fig. 2. Initially Grid-Box spread all over the hospital room (top) and cubes remained in Grid-Box after collision detection with hospital stretcher (bottom).

3.3 Prototype

In an HCE, there are patients with mobility restrictions and patients who cannot stand up and walk, considering they may be connected to life-saving technologies. In this context, the importance of healthcare workers to assist these patients to evacuate the building stands out as a primary action. Although the simulator's main objective is the HCE evacuation, we expect to observe altruism in the users of our serious game. Since we expect the sense of presence to be high in the simulation, we hypothesize that we will observe altruism in the users of our serious game, and this will be expressed by the users visiting multiple rooms looking for other patients. Even though there are no virtual agents in the hospital — the users will not be informed of this fact —, we intend to observe if the users will immediately evacuate the HCE when the fire starts or if they will search for patients in the rooms before evacuate. For this purpose, a virtual fire starts in a hospital wing with several rooms. Afterward, the user may check each room of this wing, searching for patients needing assistance to evacuate — in each room, there is a collision box to identify if the user has investigated the room. However, if the user realizes that his permanence in the environment may threaten his avatar's health, he must choose to evacuate the building.

To ensure the user's familiarity with the VE and interaction devices, the serious game begins with an acclimatization phase where instructions are delivered to them through screens with objectives. In this phase, the user's avatar is initially placed in the Intensive Care Unit (ICU) (location A in Fig. 3). The first task is to reach room number 108 (location B in Fig. 3) to check if the patient discharged from the hospital has already left the room. Following, ten seconds after the user reaches location B, the fire begins in room number 101 (location C in Fig. 3). As the fire begins, the acclimatization phase ends. Sounds of distant screams and emergency alarms start, after which the user can navigate the environment freely to evacuate that hospital wing.

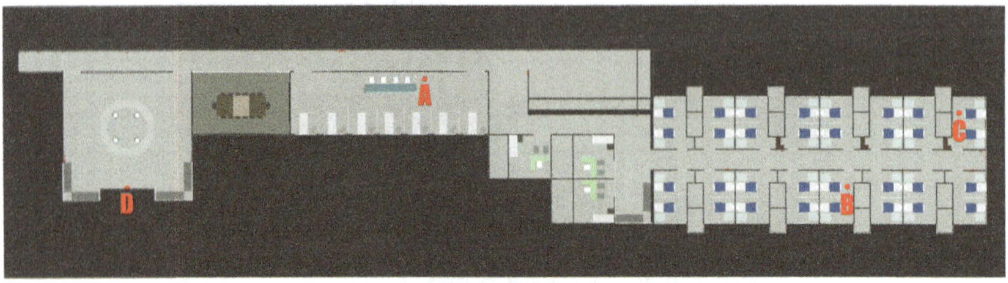

Fig. 3. Hospital floor top view: (A) avatar's initial position; (B) hospital room number 108; (C) hospital room number 101; (D) hospital entrance/emergency exit.

Initially, smoke begins to fill room 101 (location C in Fig. 3). Because the door is open, smoke spreads to the corridor. Meanwhile, the user can keep investigating the hospital wing looking for patients. However, the user needs to keep crouch to avoid inhaling smoke. As immediate feedback, if the avatar stands up, the user will hear the avatar coughing. Also, the smoke in the environment interferes with the avatar's visibility. Aiming to collect simulation data, a collision box was placed near the hospital ceiling to identify if the user is crouched or inhaling smoke.

Simultaneously to the smoke propagation, the fire propagates in the room, reaching the bed head panel. Consequently, a short circuit occurs in the hospital, turning off all the lights. Instantly, emergency lights are activated, and photoluminescent emergency signs shine in the absence of light (Fig. 4). Hence, the user must reach the nearest emergency exit assisted by the emergency signaling.

Fig. 4. Emergency signaling in a normal condition (left) and emergency signaling in the absence of light (right).

The simulation finishes when the user reaches and crosses the emergency exit (location D in Fig. 3). The evacuation time starts at the end of the acclimatization phase. At the end of the simulation, the total time to evacuate the HCE, the quantity of investigated rooms, and the total time of inhaled smoke are displayed to the user. Also, we created a function (Eq. 1) to calculate an overall user score. Therefore, we expect to encourage the user to keep training and improve his score.

$$overall\ user\ score = ((investigated\ rooms * 1000) - (evacuation\ time) - (time\ inhaling\ smoke * 100)) \quad (1)$$

4 Conclusion and Future Work

In this paper, we designed an evacuation training simulator for healthcare workers in the face of a hospital fire. Aiming to reach the simulation objectives, serious games concepts and immersive VR were used. Through immersive VR serious games, dangerous actions and tasks — paramount in the face of an emergency — can be taught and trained without threatening users' lives or endangering the environment's safety. Also, through immersive VR, the sense of presence in VE perceived by the users can be increased. For these reasons, Hospital Escape VR has the potential to be an evolution of hands-on training currently drilled in hospitals. In addition to simulating a virtual emergency with main adverse conditions — fire, smoke, alarms, and light absence —, the training can be practiced at various times throughout the year without affecting the hospital routine.

Although the designed serious game has the potential to be an evolution of traditional training, it can be classified as a pilot project. In future works, we intend to carry out a case study with healthcare workers, analyzing the viability and efficacy of Hospital Escape VR in the hospital evacuation training. Note that this pilot project focused exclusively on hospital evacuation. In the future, we intend to implement user interaction with virtual characters, either to have users assist them to evacuate the hospital or to have users coordinate other characters of the hospital staff during the evacuation. Finally, we hypothesize that a multiplayer level of Hospital Escape VR could increase the user experience in the evacuation training.

Acknowledgments. This study was financed in part by the Coordenação de Aperfeiçoamento de Pessoal de Nível Superior - Brasil (CAPES) - Finance Code 001.

References

1. Notificações de Incêndios em Hospitais Aumentam 96% no País [Fire Hospital Notifications Increase 96% in the Country]. https://veja.abril.com.br/brasil/notificacoes-de-incendios-em-hospitais-aumentam-96-no-pais
2. Campbell, R.: Structure Fires in Health Care Facilities. National Fire Protection Association (2017). https://www.nfpa.org//-/media/Files/News-and-Research/Fire-statistics-and-reports/Building-and-life-safety/oshealthcarefacilities.pdf
3. The Joint Commission: Hospitals Emergency Management (EM) (2019). https://store.jcrinc.com/assets/1/7/cc_hap_em.pdf
4. Saint Martin, V., et al.: Virtual reality simulations for hospital fire evacuation: a systematic literature review. In: 2020 22nd Symposium on Virtual and Augmented Reality (SVR), pp. 313–320. IEEE (2020)

5. Farra, S., et al.: Effects of virtual reality simulation on worker emergency evacuation of neonates. Disaster Med. Publ. Health Preparedness **13**(2), 301–308 (2019)
6. Feng, Z., et al.: How people make decisions during earthquakes and post-earthquake evacuation: using verbal protocol analysis in immersive virtual reality. Saf. Sci. **129**, 104837 (2020)
7. Feng, Z., et al.: An immersive virtual reality serious game to enhance earthquake behavioral responses and post-earthquake evacuation preparedness in buildings. Adv. Eng. Inform. **45**, 101118 (2020)
8. Lovreglio, R., et al.: Prototyping virtual reality serious games for building earthquake preparedness: the auckland city hospital case study. Adv. Eng. Inform. **38**, 670–682 (2018)
9. Lu, X., Yang, Z., Xu, Z., Xiong, C.: Scenario simulation of indoor post-earthquake fire rescue based on building information model and virtual reality. Adv. Eng. Softw. **143**, 102792 (2020)
10. Rahouti, A., Salze, G., Lovreglio, R., Datoussaïd, S.: An immersive serious game for firefighting and evacuation training in healthcare facilities. Int. J. Comput. Electr. Autom. Control Inform. Eng. **11**(9), 1088–1094 (2017)
11. Farra, S.L., et al.: Comparative cost of virtual reality training and live exercises for training hospital workers for evacuation. CIN: Comput. Inform. Nurs. **37**(9), 446–454 (2019)
12. Silva, J.F., Almeida, J.E., Rossetti, R.J., Coelho, A.L.: A serious game for EVAcuation training. In: 2013 IEEE 2nd International Conference on Serious Games and Applications for Health (SeGAH), pp. 1–6. IEEE (2013)
13. Checa, D., Bustillo, A.: A review of immersive virtual reality serious games to enhance learning and training. Multimedia Tools Appl. **79**(9), 5501–5527 (2020)
14. Hunt, A., Galea, E.R., Lawrence, P.J.: An analysis and numerical simulation of the performance of trained hospital staff using movement assist devices to evacuate people with reduced mobility. Fire Mater. **39**(4), 407–429 (2015)
15. Lévesque, J.F.: Far Cry: How the Fire Burns and Spreads (2012). http://jflevesque.com/2012/12/06/far-cry-how-the-fire-burns-and-spreads/

Simulation and Human Factors for Nuclear Control Rooms

Comparing the Sensitivity of Workload Measures for Different Task Types Using Nuclear Power Plant Main Control Room Simulators

Jinchao Lin[1(✉)], Gerald Matthews[1], Daniel Barber[2], and Niav Hughes[3]

[1] University of Central Florida, Institute for Simulation and Training, Orlando, FL 32826, USA
{jlin,gmatthews}@ist.ucf.edu

[2] Soar Technology Inc., Orlando, FL 32817, USA
daniel.barber@soartech.com

[3] U.S. Nuclear Regulatory Commission, Rockville, MD 20852, USA
niav.hughes@nrc.gov

Abstract. This paper presents the analyses of data collected from four previous studies to compare the sensitivity of multiple physiological and subjective workload measures in detecting the workload changes induced by common nuclear power plant (NPP) main control room tasks in three types using three simulators. Analyses of effect sizes were used to quantify the magnitude of response or rating changes in the workload metrics. The results suggest that the majority of the workload measures utilized in the Human Performance Test Facility (HPTF) studies show practically relevant sensitivity to the workload changes induced by the experimental manipulations in the simulated NPP operations.

Keywords: Workload · Sensitivity · Nuclear power plant operations

1 Introduction

Nuclear power plants (NPPs) are controlled by a human-system interface (HSI) located in the main control room (MCR) [1]. According to 10 CFR 50.54 (m), a minimum of three operators, including one Senior Reactor Operator (SRO) and two Reactor Operators (ROs), are required to manage and maintain a single nuclear reactor. Operators in NPP MCRs often face varying workload from different task types, such as maintaining vigilance to monitor specific instrumentations, locating and checking status of other instrumentations using multiple attention resources, or physically manipulating controls. The U.S. Nuclear Regulatory Commission (NRC) is required to review proposals for NPPs to ensure their safety. According to the *Human Factors Engineering Program Review Model* (NUREG-0711 Rev.3), one element of the review is to determine appropriate function allocation. "Function allocation is the assignment of functions to (1) personnel, (2) automatic systems, and (3) combinations of both. Exploiting the strengths of personnel and system elements enhances the plant's safety and reliability, including

J. L. Wright et al. (Eds.): AHFE 2021, LNNS 264, pp. 161–170, 2021.
https://doi.org/10.1007/978-3-030-79763-8_19

improvements achievable through assigning control to these elements with overlapping and redundant responsibilities. Functions are allocated to human and system resources and are separated into tasks. The subsequent analysis of personnel tasks identifies the alarms, displays, controls and task support needs required for performing the task. Tasks are arranged into jobs and assigned to staff positions. Each position is evaluated to verify the workload is acceptable" [2]. Therefore, profiling operators' workload is one basis for evaluating NPP designs in compliance with NRC's human engineering principles and guidelines. However, the guidance does not specify what method should be used to assess workload in the NPP domain in different tasks and scenarios. Several studies of the NRC Human Performance Test Facility (HPTF) project performed in collaboration with the University of Central Florida have utilized a multivariate assessment strategy that includes psychophysiological, subjective, and performance-based measures to evaluate the impact of task type and simulator type on workload. Results from assessment of workload using multivariate strategies suggested that the measures of workload may differ in their sensitivity to task types.

1.1 Types of NPP Simulated Environments

As NPP reactor technology and control room design has modernized and evolved, so too has the NPP simulator technology and capability. In the HPTF studies, we characterize the types of NPP simulators with five main features, summarized in Table 1.

Table 1. Summary of types of NPP simulated environments.

Features	NPP simulator types
Scope	a. Full scope simulator – has the capability to simulate all of the physical and underlying thermodynamics occurring in the would-be plant b. Part task simulator – has the capability to simulate only part of plant behavior
Layout	a. Spatially dedicated – all I&Cs are available and continuously in view to the operator and presented in a fixed location b. Hierarchical – all I&Cs are available but not continuously in view; the I&Cs can be displayed in a hierarchical manner embedded within the workstation displays
Interface types	a. Analog – conventional hard panels or bench boards with hard wired analog I&Cs b. Digital – computer-based workstations with digital I&Cs c. Hybrid – analog hard panels and computer-based workstations d. Simulated Analog – digital representation of emulating analog I&C hard panels

(continued)

Table 1. *(continued)*

Features	NPP simulator types
Workstation design	a. Sit-down workstations b. Stand-up workstations
Control interaction techniques	a. Mouse click input (for digital and hybrid interfaces) b. Touch-screen input (for digital and hybrid interfaces) c. Manual manipulations of hard-wired controls (for conventional analog interfaces)

Based on these definitions, the simulators used in the HPTF studies can be characterized into three types: 1. a full-scope simulator with hierarchical layout, simulated analog interface in sit-down desktop mouse click workstations; 2. a full-scope simulator with hierarchical layout, simulated analog interface in stand-up touchscreen workstations; 3. a full-scope simulator with spatially dedicated layout, analog interface in stand-up manual manipulation benchboard.

1.2 Workload Assessment

There has been a longstanding debate in human factors over the optimal methodology for workload assessment. A major challenge has been that different measures may dissociate [3]. That is, manipulations of task demands may have different impacts on subjective workload, psychophysiological indicators of brain response, and objective performance metrics. Thus, while the NASA Task Load Index (NASA-TLX) [4] is the most popular measure, it does not provide a comprehensive workload assessment.

Work conducted in the HPTF has supplemented the NASA-TLX with additional subjective measures including the Multiple Resource Questionnaire (MRQ) [5] which has greater diagnosticity for different sources of demand such as working memory and spatial attention. Stress response is assessed with the Dundee Stress State Questionnaire (DSSQ) [6]. Workload is also assessed with an integrated suite of psychophysiological sensors, including electrocardiogram (ECG), electroencephalogram (EEG), transcranial doppler (TCD), and functional near-infrared spectroscopy (fNIR). Performance measures include those capturing accuracy of execution of both three-way communication and task execution. Taken together, these multiple subjective and objective measures provide a comprehensive picture of operator response to changing task demands.

Results from assessment of workload using multivariate strategies suggested that the measures of workload may differ in their sensitivity to task types [7]. In addition, few studies have directly compared the various objective and subjective workload metrics for their sensitivity in the NPP domain. Therefore, we performed analyses based on the data collected in four previous studies to compare the sensitivity of the various workload indices to task type in terms of task distribution. Understanding which indices are more and less sensitive will help guide choice of workload measures for operational assessments.

2 Methods

2.1 Summary of Studies

Data collected in four previously completed NRC HPTF studies were reanalyzed to investigate the sensitivity of workload measures. These four studies evaluated workload responses of participants from different populations using simulators with different types of interface in three common NPP operation task types. Designs of the four studies are summarized in Table 2.

Table 2. Summary of study designs.

Study	Participant	Sample size	Simulator	Interface
Study 1	Student novice	81 ($M = 21$, $SD =$ 4.1)	GSE GPWR	Simulated analog/Mouse
Study 2	Student novice	71 ($M = 20.2$, $SD =$ 2.7)	GSE GPWR	Simulated analog/Touch
Study 3	Former operator	18 ($M = 45.9$, $SD =$ 10.6)	GSE GPWR	Simulated analog/Touch
Study 4	Former operator	30 ($M = 55.5$, $SD =$ 7.8)	TTC Westinghouse PWR	Analog/Manual

The experimental scenario developed for Studies 1–3 consisted of tasks reflecting common activities required when completing operating procedures: checking, detection and response implementation. Checking requires a one-time inspection of an I&C to verify that it is in the appropriate state. Detection requires continuous monitoring of a control parameter to identify a change in the state of the plant. Response implementation requires a fine motor response (mouse usage or finger touch) to change the state of the NPP by locating a control and subsequently manipulating the control in the required direction. There were twelve steps in the experimental scenario, grouped by task type (4 checking steps, 4 detection steps, and 4 response implementation steps). The order of task type block was partially counterbalanced across participants.

The experimental scenario for Study 4 was developed based on a generic version of an emergency operating procedure (EOP) for a "Loss of All Alternating Current Power (ECA-0.0)" scenario but modified for experimental use. The experimental procedure contained 69 steps supporting three different task types. In the experimental procedure, there were 30 steps (16 checking, 5 detection, and 9 response implementation) for RO 1, and 39 steps (27 checking, 1 detection, and 11 response implementation) for RO 2. The number of steps was not balanced nor was task type due to the nature of the original, realistic EOP, which requires steps to be taken in a prescribed sequence.

2.2 Quantifying Sensitivity of Workload Measures

The multivariate assessment of workload in the HPTF project included both subjective measures (self-assessment rating scales) and objective measures (psychophysiological and performance-based measures). This report focuses on investigating and comparing the sensitivity of the subjective measures and psychophysiological measures.

Effect sizes were used to quantify the sensitivity of metrics. Cohen's *d* was calculated to compare the sensitivity of the workload metrics in different task types. We took the responses in checking tasks as a reference and compared the means from response implementation and detection tasks to the means from checking task. According to Cohen [8], $d = 0.2$ is considered as a small effect size, 0.5 suggests a medium effect size, and 0.8 indicates a large effect size. Differences in workload across task types, especially the elevated workload of detection, have been robust across multiple studies [1, 7]. Thus, Cohen's *d* for task type comparisons would differentiate more and less sensitive measures, and the consistency of sensitivity differences across studies.

The sensitivity of workload metrics from the NASA-TLX, MRQ, ECG, EEG, fNIR, and TCD was analyzed for Studies 1–3, in which the subjective measures were administered after each task type block. In Study 4, only the psychophysiological metrics were available for task type based comparison.

3 Results

3.1 Physiological Metrics

Table 3 summarizes the effect sizes for workload differences between detection or response implementation tasks and checking tasks. Generally, the effect sizes were larger in the comparisons between detection tasks to checking tasks than in the comparisons between response implementation tasks to checking tasks. Oxygen saturation (especially in left hemisphere) measured by fNIR showed consistent small to moderate effect sizes across studies 1–4, with the only exception for response implementation to checking comparison in study 4 in which the effect size was less than a small effect. In addition, EEG metrics showed consistent effect sizes across studies 2–4. Similar EEG responses in Studies 2 and 3 suggest that, with a touchscreen interface, the metrics are highly sensitive to the demands of detection. The analog, full-scope/scale simulator may have some elements of manual or physical involvement from moving around to control the panels. EEG may also pick up the same physical engagement which we observed from the touchscreen interface in studies 2 and 3 as well, but not necessarily with the desktop interface (Study 1). EEG beta band and oxygen saturation in right hemisphere, it should be noted, seemed to be contradictory. Oxygen saturation measured by fNIR may indicate overall workload, consistent with the higher workload which can be attributed to the vigilance tasks. EEG beta band may pick up some other task-induced response, such as mindlessness. People usually find it hard to stay focused as they daydream and mind wander. EEG beta band may be picking up drift in focus.

Table 3. Physiological metrics: summary of effect sizes for task type differences.

	Study 1		Study 2		Study 3		Study 4	
	Cohen *d* (C – D)	Cohen *d* (C – RI)	Cohen *d* (C – D)	Cohen *d* (C – RI)	Cohen *d* (C – D)	Cohen *d* (C – RI)	Cohen *d* (C – D)	Cohen *d* (C – RI)
HR	.422	.016	−.218	.262	−.120	.385	.410	−.387
HRV	−.235	−.187	.505	−.479	.051	.052	−.255	−.252
IBI	−.384	−.101	.237	−.275	.125	−.359	−.521	.336
Alpha	−.278	−.177	.215	−.334	.355	.158	.536	−.394
Beta	−.020	−.061	.968	−.352	.589	.211	.682	.108
Theta	−.101	−.117	.483	−.226	.362	.238	.449	.774
SO_2-L	−.287	.228	−.435	.271	−.720	.121	.085	.165
SO_2-R	−.534	.095	−.423	.103	−1.112	−.025	−.398	−.155
CBFV-L	.268	.224	.081	.086	.039	−.224	.303	.532
CBFV-R	−.003	−.210	.194	.297	−.556	.179	−.259	−.515

C: checking; D: detection; RI: response implementation

3.2 NASA-TLX

Table 4 summarizes the effect sizes for comparisons between detection or response implementation tasks and checking tasks from participants' responses on the NASA-TLX in Studies 1–3. Due to the study design of using the original realistic EOP in Study 4, the NASA-TLX was not able to be administered after the task type manipulations. Therefore, the analysis by task type was not available for Study 4.

Table 4. NASA-TLX: summary of effect sizes for task type differences.

	Study 1		Study 2		Study 3	
	Cohen's *d* (C – D)	Cohen's *d* (C – RI)	Cohen's *d* (C – D)	Cohen's *d* (C – RI)	Cohen's *d* (C – D)	Cohen's *d* (C – RI)
Global workload	−.224	.069	−.266	.020	−.925	−.085
Mental demand	−.128	.104	.165	.102	−.421	−.077
Physical demand	−.109	−.120	−.386	−.482	−1.214	−.622
Temporal demand	.275	.130	.177	.097	−.012	.294
Effort	.047	.019	.049	−.041	−.475	.104
Frustration	−.711	.103	−.801	.081	−.705	−.274
Performance	.094	.079	−.065	.123	−.544	.091

C: checking; D: detection; RI: response implementation

In the task type analysis, the effect sizes were generally larger in the comparisons between detection tasks and checking tasks than in the comparisons between response implementation tasks and checking tasks. Analyses revealed small or (near) large effects in global workload and frustration subscale from the NASA-TLX across Studies 1–3. In addition, across studies 2 and 3 using simulators with a touchscreen interface, the effect sizes in physical demand subscale were in the medium effect size range in the comparisons between response implementation tasks and checking tasks. Comparable effect size was not revealed in study 1, which used a similar simulator with a desktop interface. The differences in sensitivity across simulator interface types may suggest that the NASA-TLX can be used to determine the dependence of task effects on the particular interface used by the operator, such as the touchscreen.

3.3 MRQ

Table 5 summarizes the effect sizes between detection or response implementation tasks and checking tasks from participants' responses on the MRQ in Studies 1–3. Again, due to the experimental design of using the original realistic EOP (i.e., full scenario) in Study 4, the analysis of Cohen's *d* was not available for Study 4. Several mental processes measured by MRQ were found to be sensitive in profiling the workload differences induced by task type manipulations across multiple studies. Analyses revealed consistent meaningful effect sizes in spatial quantitative process (small to large effect), visual temporal process (near-medium effect), and vocal process (small to medium effect) in the comparisons between detection tasks to checking tasks across Studies 1–3. By contrast, in Study 1 NASA-TLX data identified only a difference in frustration in these tasks and a consequent small global workload effect. Thus, the MRQ may perform better than the NASA-TLX in identifying specific sources of cognitive demand consistent across different interface types, i.e., it has higher diagnosticity.

Table 5. MRQ: summary of effect sizes for task type differences.

	Study 1		Study 2		Study 3	
	Cohen's *d* (C – D)	Cohen's *d* (C – RI)	Cohen's *d* (C – D)	Cohen's *d* (C – RI)	Cohen's *d* (C – D)	Cohen's *d* (C – RI)
Auditory emotional	−.013	−.208	−.206	−.518	.003	.226
Auditory linguistic	.081	.005	.171	.249	−.030	−.002
Manual	.143	−.077	−.236	−.347	−1.173	−.486
Short term memory	.100	.136	.069	.055	.069	.234
Spatial attentive	−.072	.242	−.034	.224	−.711	−.065
Spatial concentrative	−.316	−.062	−.069	−.079	−.710	−.150
Spatial categorical	−.070	.061	−.361	−.356	−.355	−.512

(continued)

Table 5. *(continued)*

	Study 1		Study 2		Study 3	
Spatial emergent	−.065	−.041	.171	−.018	.330	−.045
Spatial positional	.113	.340	.330	.092	−.226	.270
Spatial quantitative	−.348	.041	−.513	−.106	−1.428	−.113
Visual lexical	.241	.192	.132	.244	.186	.201
Visual phonetic	.132	−.035	−.040	−.356	.387	.128
Visual temporal	−.415	−.093	−.397	−.269	−.589	−.237
Vocal process	.262	−.018	.435	.138	.639	.408

C: checking; D: detection; RI: response implementation

In addition, visual lexical process showed a consistent small effect in the comparisons between response implementation tasks and checking tasks across Studies 1–3. Besides the consistency across three studies, the results also revealed some consistencies in the studies using touchscreen interface (Studies 2 and 3) and in the studies using college student sample (Studies 1 and 2). Specifically, spatial categorical process exhibited consistent small to medium effects and manual process showed consistent small effects in Studies 2 and 3. Consistent small effects were noticed in spatial attentive process in the comparisons between response implementation tasks checking tasks in Studies 1 and 2. Again, these findings add to NASA-TLX data in diagnostic identification of workload sources.

4 Discussion

The analyses compared effect sizes for the responses or ratings measured by different metrics. Results showed that workload responses induced by task type manipulations vary in magnitude, depending also on sample and interface type, suggesting sensitivity variation of the metrics in detecting workload changes in different circumstances. Psychophysiological workload measures are validated based on evidence of associations between variations in task demands and changes in physiological response. Psychophysiological measures are irreplaceable as they allow researchers to administer continuous measurement of participants' responses to experimental manipulations. In the HPTF studies, brain activity and cardiac activity were recorded and analyzed. Generally, the metrics of the psychophysiological workload measures are more sensitive in detecting the workload changes induced by the detection tasks compared to the response implementation tasks since the detection tasks are more demanding. In other words, from a resource theory point of view, detection tasks require more mental resource allocation, consistent with existing evidence [9]. The similarly consistent EEG results from Studies 2–4 indicate that the touchscreen interface may reproduce some comparable levels of human-system interaction which real operators experience in the MCRs with analog systems. EEG may pick up the similar physical engagement (e.g., moving around to manipulate I&Cs) involved in operations using analog and touchscreen simulators. ECG, as a means of unobtrusive and continuous cardiac monitoring, is one of the most common psychophysiological measurements of workload. Three metrics, heart rate, HRV, and IBI, derived from ECG were used in the studies. Unfortunately, no consistent trends were revealed to draw a solid conclusion in regard to ECG's sensitivity in profiling workload in nuclear domain. However, this does not suggest that ECG is never a good measure in practice. ECG is relatively easy to use and has been proved as a fairly reliable indicator of workload. Generally, an increase in mental workload leads to an increase in heart rate and a decrease in HRV [10, 11]. Divergence of cardiac data and other metrics is not uncommon. Literature shows that speech, respiration, muscle activity, body position, physical fitness, and many other factors can affect cardiac responses to workload manipulations in an experiment [12, 13]. In the HPTF studies, physical movement and frequent verbal communication were required. Such complexity makes the ECG data hard to explain. In practice, when ECG or similar cardiac measures are used for any evaluation in the nuclear domain, administrators must take all possible factors that may affect cardiac responses into consideration, if not able to control all the factors, and explain the results with caution.

Although psychophysiological measures may be more precise in some circumstances, subjective measures are more practical and easier to use. In addition, subjective measures reflect how participants feel about the experimental workload manipulations. Due to the limitation of the experimental design for Study 4, analyses for sensitivity of the subjective measures were only available among Studies 1–3. Similar to the general trend revealed by the psychophysiological measures, NASA-TLX and MRQ appear to be more sensitive in distinguishing the workload changes induced by detection tasks than the workload changes induced by response implementation tasks when comparing to the workload level induced by checking tasks. The MRQ may be more diagnostic than the NASA-TLX of specific sources of cognitive demand including the spatial quantitative, visual temporal, and vocal process demands common to both interfaces investigated here. The physical demand subscale of the NASA-TLX seems to be fairly sensitive in detecting the workload changes in both detection and response implementation tasks in the studies using simulators with a touchscreen interface. The difference in sensitivity in multiple studies using the same type of simulator may suggest that the NASA-TLX may be particularly useful in investigating human-system interaction involving touchscreen interfaces. The MRQ as a measure to characterize the nature of the mental processes used during a task was found to be sensitive in profiling the workload differences induced by task type manipulations across multiple studies. Multiple subscales showed a similar trend consistent with the trend revealed by the NASA-TLX. Such convergence of the subjective measures adds supportive evidence to the NASA-TLX as a reliable tool for practical use.

Generally, the reanalysis of current data from the HPTF studies suggests that the majority of the workload measures utilized in the studies show practically relevant sensitivity to the workload changes induced by the experimental manipulations in the simulated NPP operations. Some variation in sensitivity of the different measures was found, depending on sample and interface, but the higher workload associated with detection tasks relative to the other two task types was evident in multiple objective and subjective measures. Considering the complexity of NPP operations and workload variation involved, using multivariate assessment of workload facilitates a more comprehensive understanding of factors impacting workload.

References

1. Reinerman-Jones, L., Lin, J., Barber, D., Matthews, G., Hughes, N.: Human Performance Test Facility Experiment 4: Former Operator Workload and Performance Comparison between Two Simulated Environments, U.S. NRC, Washington, DC (2019)
2. O'Hara, J., Higgins, J., Fleger, S., Pieringer, P.: Human Factors Engineering Program Review Model (NUREG-0711, Rev.3). U.S. NRC, Washington, DC (2012)
3. Hancock, P., Matthews, G.: Workload and performance: associations, insensitivities, and dissociations. Hum. Factors **61**, 374–392 (2019)
4. Hart, S., Staveland, L.: Development of NASA-TLX: results of empirical and theoretical research. In: Hancock, P., Meshkati, N. (eds.) Advances in Psychology, pp. 139–183. North-Holland, Amsterdam (1988)
5. Boles, D., Adair, L.: The MRQ. HFES Ann. Meet. **45**(25), 1790–1794 (2001)
6. Matthews, G., et al.: Fundamental dimensions of subjective state in performance settings: task engagement, distress, and worry. Emotion **2**, 315–340 (2002)
7. Reinerman-Jones, L., Lin, J., Barber, D., Matthews, G., Hughes, N.: Multi-dimensional workload in two types of NPP simulators. HFES Ann. Meet. **63**(1), 1777–1781 (2019)
8. Cohen, J.: Statistical Power Analysis for the Behavioral Sciences, 2nd edn. Lawrence, Hillsdale (1988)
9. Warm, J., Parasuraman, R., Matthews, G.: Vigilance requires hard mental work and is stressful. Hum. Factors **50**, 433–441 (2008)
10. Veltman, J., Gaillard, A.: Physiological indices of workload in a simulated flight task. Biol. Psych. **42**(3), 323–342 (1996)
11. Hankins, T., Wilson, G.: A comparison of heart rate, eye activity, EEG and subjective measures of pilot mental workload during flight. Aviat. Space Environ. Med. **69**(4), 360–367 (1998)
12. Wilson, G.: Applied use of cardiac and respiration measures: practical considerations and precautions. Biol. Psych. **34**(2), 163–178 (1992)
13. Jorna, P.G.A.M.: Spectral analysis of heart rate and psychological state: a review of its validity as a workload index. Biol. Psych. **34**(2), 237–257 (1992)

Using Qualified On-Site Nuclear Power Plant Simulators in Human Factors Validations of Control Room Upgrades

Jeffrey C. Joe(✉) and Casey Kovesdi

Idaho National Laboratory, Idaho Falls, Idaho, USA
{Jeffrey.Joe,Casey.Kovesdi}@inl.gov

Abstract. Existing commercial nuclear power plants (NPPs) are valuable assets in the infrastructure portfolio of the United States because they safely and reliably generate about 1/5th of all the electricity used. The instrumentation and control (I&C) systems in commercial NPPs are the 'eyes and ears' of the operator, allowing operators to maintain situation awareness, thereby allowing the plant to operate safely and efficiently for all phases of operation. Because the I&C systems in NPPs are still mostly based on analog technologies, which are reliable but not especially cost-effective, upgrading the existing I&C in commercial NPPs to new digital I&C is very important. Furthermore, simulators and simulation are critical human factors engineering tools for this I&C modernization work. NPP control room simulators are essential test beds to simulate normal, abnormal, and emergency operations that allow human factors researchers to safely and realistically evaluate early design phase prototypes of the upgraded digital I&C and validate final as-built digital I&C systems that have been modernized and deployed in NPPs. This paper describes research performed using qualified on-site NPP simulators to perform human factors validations of digital I&C control room upgrades.

Keywords: Human factors engineering · Simulators · Simulation · Nuclear power plant modernization

1 Introduction

In January 2021, the United States (U.S.) Department of Energy (DOE) Office of Nuclear Energy (NE) published their strategic vision [1]. DOE-NE's strategic vision includes five goals, with supporting objectives to address "challenges in the nuclear energy sector, help realize the potential of advanced technology, and leverage the unique role of the government in spurring innovation" (pg. 3). As seen in Fig. 1, the first goal is to enable the continued operation of existing commercial nuclear power plants (NPPs) by conducting research and development (R&D) to create technologies that reduce operating costs and expand nuclear energy beyond the electricity market.

In support of this goal, DOE-NE has been sponsoring the Light Water Reactor Sustainability (LWRS) Program, which facilitates R&D performed in collaboration with utilities to enable the continued operation of existing commercial NPPs. The LWRS

J. L. Wright et al. (Eds.): AHFE 2021, LNNS 264, pp. 171–178, 2021.
https://doi.org/10.1007/978-3-030-79763-8_20

Fig. 1. Vision, Mission, Goals, Objectives, and Performance Indicators from the U.S. DOE-NE 2021 Strategic Vision [1].

Program's scope includes human factors R&D and human factors engineering (HFE) to support plant modernization and, in particular, the digital transformation of legacy control room instrumentation and control (I&C) systems.

I&C systems in commercial NPPs are the 'eyes and ears' of the operator, allowing operators to maintain situation awareness, thereby allowing the plant to operate safely and efficiently for all phases of operation. The I&C systems in U.S. commercial NPPs are still mostly based on analog technologies, which are reliable but not especially cost-effective. Upgrading the existing I&C systems in commercial NPPs to more cost-effective digital I&C is thus a primary way to achieve the goal of continued operations of existing U.S. nuclear reactors.

1.1 Nuclear Power Plant Simulators and Simulation

NPP simulators and simulation are critical HFE tools for this I&C modernization work. NPP control room simulators are essential test beds to simulate normal, abnormal, and emergency operations that allow human factors researchers to safely and realistically evaluate early design phase prototypes of the upgraded digital I&C and validate final as-built digital I&C systems that have been modernized and deployed in NPPs.

The simulators and simulation used in the nuclear industry can vary significantly in their size and complexity. While others have provided a more complete and thorough treatment of simulators used in human factors [3], for the purposes of this paper, a simplified categorization scheme to delineate the types of simulators used in the nuclear industry is provided here. At one end of the spectrum are part-task simulators, which are sometimes referred to as microworlds. Microworlds are a simplified representation of the system being simulated, which often means that nonexperts of that system can be trained and used as participants in controlled experiments [4].

Somewhere in the middle, but near the higher end of the spectrum, is the Human System Simulation Laboratory (HSSL) at Idaho National Laboratory (INL). INL's HSSL, as seen in Fig. 2, is a full-scope, full-scale NPP control room simulator. The HSSL simulator uses the same software that is run on the on-site simulators at NPP that are used to train, qualify, and license NPP operators. As such, the HSSL simulator model includes all functions found in a control room and is capable of modeling normal, abnormal, and emergency plant operations. In addition, the HSSL is reconfigurable, such that it can mimic multiple control room layouts and simulate both analog and digital I&C systems.

Fig. 2. The Human System Simulation Laboratory at Idaho National Laboratory.

At the top end of the spectrum are the qualified, on-site training simulators. These simulators are as close to the actual main control room (MCR) as is possible, using the same hardware and control systems that are in the MCR. Qualified on-site simulators are required to meet rigorous standards for simulator fidelity, as specified in [5], which establishes, "The functional requirements for full-scope nuclear power plant control room simulators for use in operator training and examination, as well as criteria for the scope of simulation, performance, and functional capabilities of simulators". (pg. 4) [6] The consensus standard [5] also includes information on, "Simulator scenario-based performance testing, new sections addressing simulator core performance testing and post-event simulator testing, and other important guidance required to ensure accurate simulation for use in operator training and examination" (pg. 4) [6]. It is because on-site training simulators meet these rigorous standards that they are assured to be very high fidelity.

There are a number of trade-offs with respect to the advantages and disadvantages of using microworlds, the HSSL, or qualified on-site simulators in human factors R&D. As mentioned previously, microworlds are simplified, low-fidelity simulations of the system, which allow experimental human factors researchers to use large pools of participants who are not licensed operators. This allows the researchers to collect large amounts of data for use in inferential statistical tests at a relatively low cost. Attempting to collect the same amount of data from licensed NPP operators on their qualified on-site simulator would be prohibitively time-consuming and expensive, but, if high simulator fidelity is an important consideration in the research being conducted (e.g., integrated system validations), there is no better option than the qualified on-site simulator[1]. This paper

[1] The HSSL is an excellent simulation tool because its flexibility allows human factors researchers to collect larger samples of data using a simulator that has at level of fidelity that is very close to the qualified on-site simulator.

describes research performed using qualified on-site NPP simulators to perform human factors validations of digital I&C control room upgrades.

2 Using the On-Site Simulator for Human Factors R&D

This research project involved conducting human factors operator-in-the-loop studies using the utility's qualified on-site control room simulators. The purpose of the studies was to conduct HFE validation assessments of the planned control room I&C upgrades for four NPP units and had the objective of identifying potential HFE issues with the upgrades prior to installation in the MCR.

One unique aspect of this research was the availability of two different qualified on-site control room simulators. One on-site simulator had been upgraded with the new digital I&C, and the second was at an identical NPP located a few hours away and had not been upgraded. This was a very rare opportunity to conduct comparative operator-in-the-loop research because not all commercial NPPs are alike with respect to the exact layout and design of their control rooms. In this case, however, the I&C upgrades were occurring at two different NPP stations that were built at nearly the same time and had virtually identical control rooms. The fact that this study was conducted right in the time window where one simulator had been upgraded and the other had not adds to the true uniqueness of the opportunity.

The operator-in-the-loop studies entailed direct observations and assessments of key operator interactions with the existing and new human system interfaces (HSIs) across a number of normal, abnormal, and emergency scenarios. The scenarios were designed to evaluate the functional ergonomic and human factors aspects of the existing and upgraded I&C systems, with a particular emphasis on the ability of the new I&C system to support operators' cognitive processes and their ability to facilitate the operators' ability to perform the correct control actions. The general workshop flow is depicted in Fig. 3.

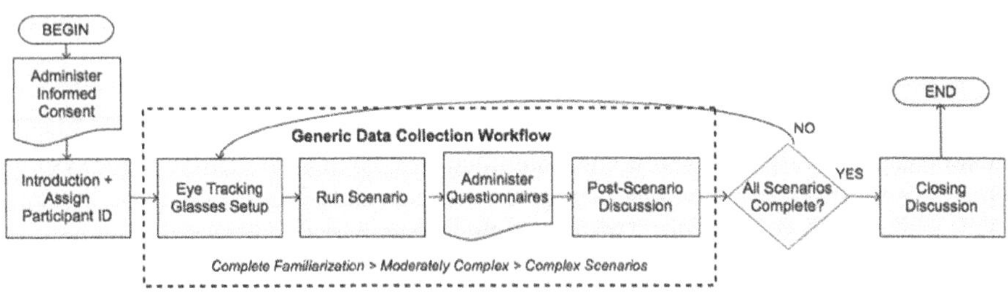

Fig. 3. Workshop execution workflow.

The assessment strategy focused on assessing plant safety and overall human-system performance. As such, measures of usability, workload, and situation awareness (SA) were collected during select scenarios to evaluate the suitability of the new HSIs in an operational context. During each scenario, the evaluation team also identified key information that determined the success or failure of the new HSIs as installed.

Human factors researchers and operationally experienced monitors observed the operators who performed actions in the scenarios using relevant procedures (see Fig. 4). Observations focused on operator interactions that were affected by the upgraded controls and indicators. Changes in time available for operator action, as well as information availability, were identified by observers during the scenarios.

Fig. 4. Human factors researchers observing NPP operators interacting with the new HSI.

In addition to collecting performance data, a comparison between the existing and upgraded configurations was achieved by collecting self-report measures of perceived usability, workload, and SA. Specifically, the single ease question [7], National Aeronautics and Space Administration Task Load Index [8], and Situation Awareness Rating Technique [9] were used to collect self-report measures of usability, workload, and SA, respectively. These questionnaires were provided to operators together after each scenario and operators were asked to complete each questionnaire independently. The ratings for each questionnaire type were aggregated across crew members for each scenario using the arithmetic mean to compare existing and upgraded configurations.

Notably, the extent to which perceived usability, workload, and SA were practically equivalent across the two configurations was of interest to the research team. That is, the extent to which any of the observed differences in ratings across these questionnaires had any meaningful difference, potentially reflecting a change in usability, workload, or SA, when introduced to the new upgrades, was evaluated through a series of two one-sided tests (TOSTs). TOSTs provide a means of testing statistical equivalence by setting up explicit composite null hypotheses based on creation of an upper (Δ_U) and lower ($-\Delta_L$) equivalence bound. The logic of TOST is highlighted below:

$$H01 : \Delta_{Observed} \leq -\Delta_L. \tag{1}$$

$$H02 : \Delta_{Observed} \geq \Delta_U. \tag{2}$$

$$H1 : -\Delta_L < \Delta_{Observed} < \Delta_U. \tag{3}$$

TOST provides a more accurate way of testing equivalence, as opposed to accepting the null hypothesis from a traditional two-sided statistical test of differences [10]. As such, rather than concluding equivalence by the absence of an observed effect, TOST requires more specific predictions of potential differences by comparing it to explicit equivalence bounds (i.e., $-\Delta_L$ and Δ_U).

This study set equivalence bounds for each questionnaire as two units of measurement. Since there were different operators used for the existing and upgraded configurations, the tests were set up as unpaired and assumed unequal variance. Type I error rates were set up as $\alpha = .20$ given the exploratory nature of these tests.

3 Workshop Results

Across each of the scenarios run for the two conditions, the human factors researchers and operationally experienced monitors observed no safety critical errors and no meaningful differences in completion times that were time critical in nature when introducing the new HSIs in the upgraded configuration. The TOSTs indicated statistical equivalence across perceived usability (ease of completion), workload, and SA. Figure 5 presents the results from these questionnaires.

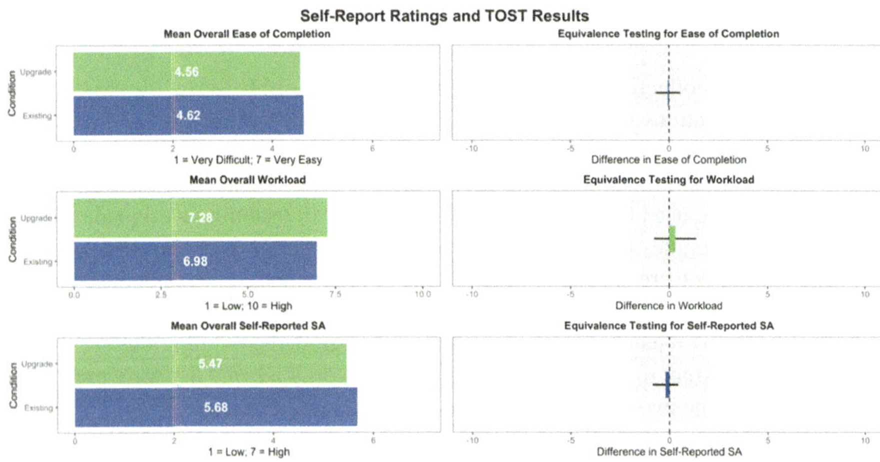

Fig. 5. Mean self-report ratings and TOST results for usability, workload, and SA.

In Fig. 5, each row represents perceived usability (ease of completion), workload, and SA. On the left side, the colored bar charts illustrate the mean ratings for each questionnaire by condition (upgraded HSIs in green versus existing HSIs in blue). On the right side, TOST results are illustrated; the gray band indicates the region of practical equivalence as defined by $-\Delta_L$ and Δ_U bounds. Where the upgraded HSI ratings were greater than existing HSIs, the difference is illustrated by a green bar in the positive direction. Where the existing HSIs configuration ratings were greater than the new HSI configuration, a blue bar is shown in the negative direction. There are 80% confidence intervals presented to illustrate whether the difference was practically equivalent; the intervals were all within the equivalence bound (gray region), indicating statistical equivalence. These results were further supported by the observation of no safety critical errors.

4 Conclusion

This paper summarized operator-in-the-loop studies that were performed using the utility's qualified on-site NPP control room simulators to assess the human factors aspects and overall system performance of upgraded digital I&C systems. Having access to two different qualified on-site simulators, one which had been upgraded with the new digital I&C and another at an identical NPP located a few hours away that had not been upgraded, provided a unique opportunity to evaluate and compare operator and overall system performance in two high-fidelity simulation environments.

The fact that this study observed no safety critical issues with the new HSIs in the upgraded simulator meets the industry-accepted standard of proof that the new I&C system had been validated. What was unique to this research was that it also demonstrated that there were no meaningful differences in completion times for operators using the old I&C versus the new I&C that were time critical in nature. This was additional comparative evidence supporting the validation conclusion that before now had not been empirically demonstrated and provides additional assurance that commercial NPPs in the U.S. are safely operated. This assurance, using the qualified on-site simulators to validate final as-built digital I&C systems that have been modernized and deployed, directly supports DOE's strategic vision to enable the continued operation of the existing fleet of commercial NPPs.

Acknowledgments. This manuscript has been authored by Battelle Energy Alliance, LLC under Contract No. DE-AC07-05ID14517 with the U.S. Department of Energy. The United States Government retains and the publisher, by accepting the article for publication, acknowledges that the United States Government retains a nonexclusive, paid-up, irrevocable, worldwide license to publish or reproduce the published form of this manuscript, or allow others to do so, for United States Government purposes. The INL issued document number for this paper is: INL/CON-21-61472.

References

1. United States Department of Energy: Office of Nuclear Energy: Strategic Vision. Washington, D.C (2021)
2. International Atomic Energy Agency: Modern Instrumentation and Control for Nuclear Power Plants: A Guidebook. IAEA Technical Report Series #387, Vienna, Austria (1999)
3. Hancock, P.A., Vincenzi, D.A., Wise, J.A., Mouloua, M. (eds.): Human Factors in Simulation and Training. CRC Press, Boca Raton (2008)
4. Lew, R., Ulrich, T.A., Boring, R.L., Werner, S.: Applications of the rancor microworld nuclear power plant simulator. In: 2017 Resilience Week, pp. 143–149. IEEE Press, New York (2017)
5. American National Standards Institute/American Nuclear Society: ANSI/ANS-3.5 Nuclear Power Plant Simulators for Use in Operator Training and Examination. American Nuclear Society, La Grange Park, IL (2018)
6. United States Nuclear Regulatory Commission: Regulatory Guide 1.149 Nuclear Power Plant Simulation Facilities for Use in Operator Training, License Examinations, and Applicant Experience Requirements (Reg Guide 1.149). Washington, DC (2011)
7. Sauro, J., Lewis, J.R.: Quantifying the User Experience: Practical Statistics for User Research. Morgan Kaufmann, Burlington (2016)

8. Hart, S.G., Staveland, L.E.: Development of NASA-TLX (Task Load Index): results of empirical and theoretical research. Adv. Psychol. **52**, 139–183 (1988)
9. Taylor, R.M.: Situational Awareness Rating Technique (SART): the development of a tool for aircrew systems design. In: Situational Awareness in Aerospace Operations (AGARD-CP-478), pp. 3/1–3/17. NATO, Neuilly Sur Seine (1990)
10. Lakens, D.: Equivalence tests: a practical primer for tests, correlations, and meta-analysis. Soc. Psychol. Personal. Sci. **8**(4), 355–362 (2017)

A Dual Full-Scope and Reduced-Scope Microworld Simulator Approach to Evaluate the Human Factors of a Coupled Hydrogen Production Concept of Operations

Thomas A. Ulrich[1(✉)], Roger Lew[2], Torrey Mortenson[1], Heather Medema[1], Ronald L. Boring[1], Steffen Werner[2], Lucas Terry[2], and Nathan Minard[2]

[1] Idaho National Laboratory, Human Factors and Reliability Department, 2525 N Fremont Ave, Idaho Falls, Idaho 83415, USA
{thomas.ulrich,torrey.mortenson,heather.medema, ronald.boring}@inl.gov
[2] University of Idaho, Psychology and Communications Studies, 875 Perimeter Drive, Moscow, Idaho 83844, USA
{rogerlew,swerner,nminard}@uidaho.edu, terr9729@vandals.uidaho.edu

Abstract. Simulators serve as a valuable human factors research tool for control room modernization as nuclear power plants continue to upgrade their analog technology with digital equivalents. Beyond upgrades, plants are beginning to explore alternative concepts of operation including using their thermal output for chemical processes. These upgrades represent more substantial plant modifications and mandate new human factors techniques to develop and evaluate the HSIs, procedures, and operations required to control the new systems. This paper describes the strategy and use of a full-scope and reduced-scope simulator in a comprehensive research approach to develop the concept of operations. Specifically, the full-scope simulator serves as a means to extract invaluable operational knowledge from operators and the reduced-scope simulator, namely the Rancor microworld, serves as a means to address targeted human factors issues and is sufficiently flexible to be more easily integrated with small scale physical test beds than the full-scope counterpart.

Keywords: Rancor · Microworld · Nuclear process control · Flexible plant operations and generation

1 Introduction

Over the last 10–15 years, the commercial nuclear power industry has sought ways to modernize their operations to yield enhanced efficiency while retaining high levels of safety. From a human factors perspective, the modernization effort has focused mainly on main control rooms, which represent the central command capability for each nuclear

J. L. Wright et al. (Eds.): AHFE 2021, LNNS 264, pp. 179–186, 2021.
https://doi.org/10.1007/978-3-030-79763-8_21

power plant. The drive to modernize main control rooms is fueled by license extensions allowing plants to operate well beyond their original planned lifespan into 60–80 years of operation [1]. The antiquated though impressively reliable technologies such as analog indication and controls are no longer economically viable to maintain.

As the analog equipment is replaced with a digital variant, human factors research is focused on finding ways to safely reduce staffing by providing advanced diagnostics and decision-making tools as a part of these modernization efforts. However, the swiftly evolving grid has placed additional economic pressure on nuclear utilities and driven a shift from simply finding efficiencies to exploring alternative technologies and concepts of operations that can restore economic competitiveness to nuclear power plants. Thermally and electrically coupling a nuclear power plant to a nearby hydrogen production facility is one such economically promising concept of operation under development. The requisite human factors activities to support the development of this new concept of operation represent a fundamental shift from small to large plant modification. As such, additional development and evaluation techniques are required to support the human factors activities for new system development and integration with existing light water reactor nuclear power plants. This paper describes the unique simulations, equipment and test facilities, and human-in-the-loop human factors activities undertaken to develop and evaluate this new flexible concept of operations. Findings and lessons learned are discussed in an attempt to advance human factors techniques for the commercial nuclear power domain.

2 Flexible Plant Operation and Generation

The existing fleet of U.S. nuclear power plants are facing increasing challenges to maintain economic competitiveness. The economic challenges stem from high costs to maintain their impeccable safety record, especially in light of new requirements in response to the Fukushima Daiichi accident, as well as the low cost of competing technologies such as fossil fuels and subsidies for renewable energy. Fortunately, nuclear power plants are unrivaled thermal energy producers. As such, leveraging this unique capability to provide large volumes of heat and electricity for chemical processes can improve the economic viability of existing large-scale nuclear power plants. Specifically, the coupling of a nuclear power plant to a hydrogen production plant is being pursued, since the high-temperature steam electrolysis process represents a desirable use case that nuclear power plants can uniquely supply.

The authors are part of a larger interdisciplinary team developing and demonstrating the integrated technologies required to safely and reliably couple an existing nuclear power plant to a nearby high temperature steam electrolysis hydrogen production plant. The central concept for the system entails diverting a portion of the thermal energy carried by the main steam line from its traditional use to turn a turbine connected to a generator to produce electricity. The thermal energy is extracted through a series of heat exchangers and travels a kilometer through insulated pipes where it can then be used by the High Temperature Steam Extraction plant to enhance its efficiency in producing hydrogen [2]. The modelled nuclear power plant can also be electrically coupled with the hydrogen plant to provide its electricity needs. Together the thermal and electrical coupling representant a substantial shift in the current plant physical configuration and operations. The systems must be developed and thoroughly evaluated to meet the strict regulatory compliance with the U.S. Nuclear Regulatory Commission before the economic benefits can be realized. Additional details on the project at large can be found in [3], while the remainder of this paper will detail the use of both full-scope and reduced-scope simulators to support human factors activities to develop a concept of operations for coupling a hydrogen plant to an existing nuclear power plant.

3 Human Factors Approach for the New Concept of Operations Development

The modifications required to support coupling a hydrogen plant to a nuclear power plant are substantial and require an interdisciplinary team to address all the respective elements of the design. The project is multiyear and spans developing an initial concept through simulations and then ultimately demonstrating the capability to industry through a full-scale physical implementation with a collaborating utility. The project followed a systems engineering "V" model process comprised of designing individual components based on requirements, integrating those elements at a systems level, and then evaluating and refining the design in integrated evaluations [4]. This approach is compatible with the U.S. Nuclear Regulatory Commission's prescribed guidance on plant modifications contained in NUREG-0711, *Human Factors Engineering Program Review Model* [5]. Additional and more specific guidance is needed for new and more extensive modifications, such as those pursed in this effort and in direct contrast to existing system replacement types of modifications. A discussion of the regulatory needs is outside the scope of this paper, which focuses on the use of simulators as a method to perform the human factors work (Fig. 1).

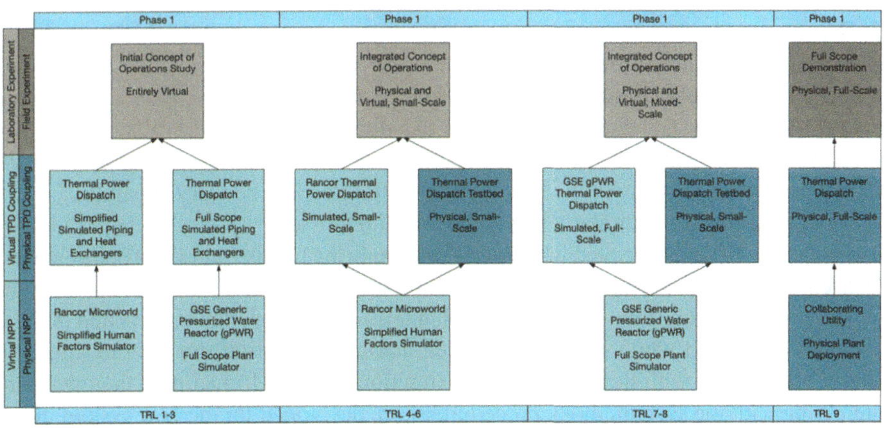

Fig. 1. Research timeline depicting the use of virtual simulations and physical testbeds to develop the concept of operation for coupling a hydrogen plant to a nuclear power plant.

The human factors team represents a subset of the overall team and lead the task to develop and evaluate the human systems interfaces (HSIs), sample procedures, and the concept of operations for the overall project. One unique aspect of this project was the early engagement of the human factors team during the design of the physical system and identifying an initial concept of operations. This early engagement proved fruitful as the human factors team was able to engage licensed operators and capture operational requirements for the design, which ultimately yielded a more compatible system design that matched operator expectations [6]. Once a somewhat matured system design concept was available, the human factors team, in tandem with support from the engineering team, used a combination of simulations, physical equipment and testbeds, and human-in-the-loop studies to develop and evaluate an HSI and mock procedures. The following sections describe the use of different types of simulators to support the research and highlight the use of a microworld simulator to fill a niche in the design and evaluation process.

4 Simulator-Based Approach to Support Concept of Operations Development

The human factors work was performed with full-scope and reduced-scope simulator platforms in a two-pronged approach to enhance the development of the concept of operations. The two different simulator platforms provide complementary capabilities that render them more suitable to address particular aspects of the concept of operations. This section describes the role of each simulator to support the concept of operations development and how the approach enables a comprehensive and rigorous evaluation.

Full-Scope Simulator. GSE System's Generic Pressurized Water Reactor (GPWR) full-scope simulator served as the main testbed to develop the concept of operations. The simulator was first modified to include the thermal extraction and thermal delivery systems to simulate physical processes beyond the existing overall GPWR simulation [7]. The modified simulator provides high fidelity responses in terms of feedback on the main steam system and the reactor temperature and reactivity. The simulator then served as a testbed to develop a prototype HSI to control the thermal extraction and delivery process. The HSI was then overlaid on the existing GPWR analog virtual panels displayed within the Human Systems Simulation Laboratory (HSSL) to create a full-scale control room simulation with the new thermal extraction and delivery capability.

Since the full-scope simulator is high fidelity and full-scale it is well suited to address human factors issues associated with operations requiring highly trained and experienced licensed or formerly licensed operators as participants. Therefore, it serves as conduit to extract information from expert operators in terms of their mental models and control strategies employed while using the existing equipment and inform how to integrate similar approaches with the new thermal extraction system. To facilitate the operators interacting with the system, mock procedures were developed for the new system in the same format and terminology used with the GPWR simulator.

The final evaluation for the initial year of the project was performed using the full-scope simulator. COVID-19 restrictions prevented the experiment from being performed in the physical HSSL. Instead, a remotely administered variant was conducted, focusing more on the prototype HSI itself as a sort of part-task simulator due to limitations of display real estate for remote testing. A web meeting format was adopted in which four expert operators connected and were given remote control of the interface as they performed basic operating scenarios to bring the system into service and remove it from service. The GPWR simulator was used to drive the simulation for the operators, albeit in a much-reduced format due to the limited display available within the remote format. Future efforts will resume using the full-scale control room simulation within the HSSL post-COVID restrictions as the project moves forward.

Specific results from the study are still being analyzed, but the preliminary results from the remote usability study indicate minor human factors issues associated with nomenclature, thresholds for valve indicators showing open and closed, and general modelling critiques of the thermal extraction and delivery loop in terms of the feedback it provided to some of the larger plant systems. Overall, operators were able to execute all procedures without any major human errors, and the demonstration was successful in identifying an initial concept of operations for thermal energy extraction and delivery to an offsite hydrogen production facility. In tandem with the full-scope simulator effort, a microworld was used to evaluate the HSI design with a greater emphasis on the visual human factors as opposed to validating the overall system design and concept of operations as was the focus with the full-scope simulator.

Reduced-Scope Simulator. The Rancor microworld served two primary roles to support the concept of operations. First, the microworld served as a testbed to refine and optimize the HSI design through human factors evaluations. The Rancor microworld is a reduced-order and reduced-scope nuclear process control simulator originally designed to specifically address human performance issues within the context of human factors and human reliability analysis [8]. The reduced-order and simplified nature of the microworld makes it well suited to address human factors and usability. The simplified code base is easier to modify and more flexible, which supports more flexibility in the prototyping effort and allows for more variants of the HSI to be evaluated. The interface for the thermal extraction and delivery systems developed on the full-scope simulator was ported to the microworld with minimal modifications. There was no compromise in the functionality of the interface; however, the larger context of the simulated plant within Rancor was reduced. Therefore, the translation of the prototype HSI into the microworld was not exact. Nonetheless, it was sufficiently similar to support performing the same four scenarios—start up, hot standby to online, online to hot standby, and shutdown—that were developed for the full-scope simulator. A central benefit of the microworld is the ability to use students as participants to run through the same four procedurally driven scenarios. Using students is inexpensive and affords greater sample sizes, which are difficult to obtain in any nuclear process control experiment using a full-scope simulator.

The results of the student microworld evaluation are still being analyzed, but some qualitative results are worth noting here. The students experienced and reported many of the same usability issues as the operators, including the nomenclature for the valves and the opening and closing thresholds for the valves. This provides promising evidence for using this type full-scope and reduced-scope simulator approach to vet HSI designs before evaluating and validating the HSIs with experienced operators, to yield substantial cost reductions in development effort. Students lack the expertise operators possess and therefore the generalizability of the results may be questioned. The replication of the HSI and the procedures for the thermal extraction and delivery in both the full-scope and microworld provide a common base for comparison and the preliminary results, at least qualitatively demonstrate similar patterns of usability feedback.

The simulator approach described above was employed during the first year of this multi-year project. The second and active year of the project, is using the same dual simulator approach. Fortunately, with the COVID-19 vaccine role-outs occurring at the time of writing this paper, it appears that the full-scope HSSL facility can be used in lieu of the remote format that was adopted due to COVID-19 issues last fiscal year. At the time of this writing, the GPWR is undergoing active development to incorporate electrical coupling with a simulated hydrogen plant in the form of a virtual switchyard. The switchyard inclusion supports an expanded concept of operations in which operators must engage the thermal extraction, thermal delivery, and electrical systems to support hydrogen production. The Rancor microworld continues its role also being used to refine the HSI and will continue to be used in this specific augmenting role as the project moves forward. The resulting HSI improvements will then be integrated into the full-scope GPWR HSI for human-in-the-loop testing in the HSSL.

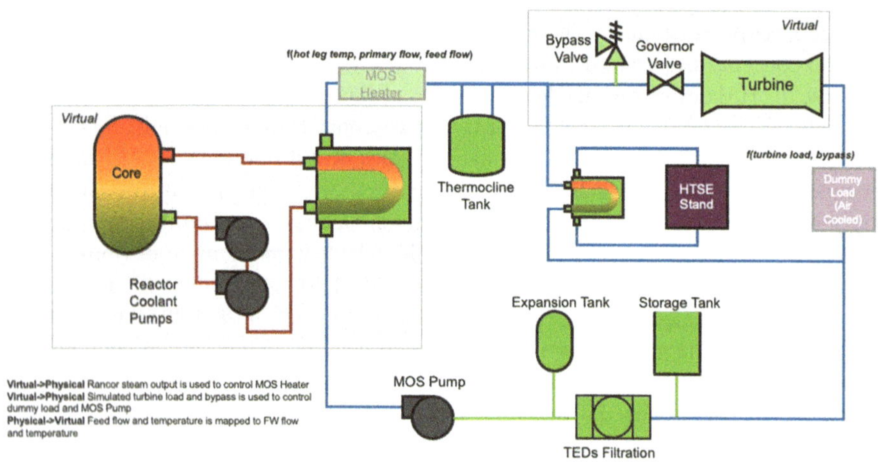

Fig. 2. Integration of the Rancor microworld with the physical Thermal Energy Delivery System testbed for human- and equipment-in-the-loop testing.

The Rancor microworld has taken on a second role within the larger project during the second year. The microworld serves as the nuclear process aspect of human- and equipment-in-the-loop testing for the physical Thermal Energy Delivery System (TEDS), which includes a small-scale hydrogen production unit. The scale of the microworld affords it the ability to serve as the virtual nuclear process for integration with the physical test equipment. The microworld's underlying thermohydraulic model is based on a small modular reactor and is scaled for a thermal output of ~160 MW thermal. This scale is a fraction of the thermal output of a commercial light water reactor that is modelled within the full-scope simulator. As such, its scale is more appropriate to support the initial integrated physical and virtual simulations. During the initial testing, the project uses a highly reduced scale in comparison to what would actually be deployed with an existing light water reactor and therefore a small scale simulator is ideal. Furthermore, the Rancor microworld is a much simpler piece of software with significantly smaller models which eases integration of the virtual simulation with the physical control systems of TEDS much easier and faster in comparison to a full-scale simulator such as GPWR. Rancor's original underlying physics model required modifications to increase its fidelity, which included refactoring the thermohydraulic mode based on a small modular reactor [9] and incorporating simple point kinetics reactor model. With these modifications, the microworld can be integrated with the physical TEDS test stand (as outlined in Fig. 2) to support human- and equipment-in-the-loop testing.

5 Conclusions

The human factors work described in this paper used a dual simulator approach to develop a concept of operation for joint electricity and hydrogen production with a nuclear power plant. The Rancor microworld was used as a means to augment the human-in-the-loop testing by focusing specifically on human factors issues to refine and optimize the HSI. The microworld can also be used where a full-scope plant model is not appropriate as is the case with the reduced-scale virtual and physical evaluations of the initial thermal extraction and delivery concept. Using multiple simulators of different scales and complexity offers a greater toolset to develop a concept of operations and provides a richer capability to evaluate the usability of new systems for adoption in nuclear process control. Future efforts can leverage this approach to augment traditional existing approach centered on full-scope engineering and training simulators.

Acknowledgments. This work of authorship was prepared as an account of work sponsored by Idaho National Laboratory (under Contract DE-AC07-05ID14517), an agency of the U.S. Government. Neither the U.S. Government, nor any agency thereof, nor any of their employees makes any warranty, express or implied, or assumes any legal liability or responsibility for the accuracy, completeness, or usefulness of any information, apparatus, product, or process disclosed, or represents that its use would not infringe privately owned rights.

References

1. Boring, R.L., Ulrich, T.A., Joe, J.C., Lew, R.T.: Guideline for operational nuclear usability and knowledge elicitation (GONUKE). Procedia Manuf. **3**, 1327–1334 (2015)
2. Frick, K., et al.: Evaluation of Hydrogen Production Feasibility for a Light Water Reactor in the Midwest, INL/EXT-19-55395. Idaho Falls: Idaho National Laboratory, September 2019
3. Ulrich, T., et al.: Preliminary Human Systems Evaluation of Thermal Power Dispatch Concept of Operations, INL/EXT-20-59898. Idaho Falls: Idaho National Laboratory, September 2020

4. Haskins, C.: International Council on Systems Engineering (INCOSE) Systems Engineering Handbook v. 3.2. 2. A guide for system life cycle processes and activities. INCOSE-TP-2003-002-03.2. 2 (2011)
5. NRC. Human factors program review model (NUREG-0711). Rev. 3 (2012). Washington, D. C.: U. S. Nuclear Regulatory Commission (2012)
6. Ulrich, T.A., Lew, R., Boring, R.L., Mortenson, T., Park, J., Medema, H.: Developing an integrated energy system interface for electricity-hydrogen hybrid nuclear operations. In: Proceedings of the Human Factors and Ergonomics Society Annual Meeting, vol. 64, no. 1, pp. 92–96. Sage CA: Los Angeles, CA: SAGE Publications, December 2020
7. Hancock, S., Shigrekar, A., Westover, T.: Incorporation of Thermal Hydraulic Models for Thermal Power Dispatch into a PWR Power Plant Simulator, INL/EXT-20-58766. Idaho Falls: Idaho National Laboratory, July 2020
8. Ulrich, T.A., Lew, R., Werner, S., Boring, R.L.: Rancor: a gamified microworld nuclear power plant simulation for engineering psychology research and process control applications. In: Proceedings of the Human Factors and Ergonomics Society Annual Meeting, vol. 61, no. 1, pp. 398–402. Sage CA: Los Angeles, CA: SAGE Publications, September 2017
9. Poudel, B., Joshi, K., Gokaraju, R.: A dynamic model of small modular reactor based nuclear plant for power system studies. IEEE Trans. Energ. Convers. **35**(2), 977–985 (2019)

Analysis and Validation of the Main Control Room Staffing of the Xe-100 SMR

Borja Hervás(✉) and Luis Rejas

TECNATOM S.A., Avda. Montes de Oca 1, San Sebastián de Los Reyes, Madrid, Spain
{bhervas,lrejas}@tecnatom.es

Abstract. X-Energy is currently designing an innovative concept of nuclear power reactor: the Xe-100. It is a Small Modular Reactor (SMR) based in High Temperature Gas Cooled Reactor (HTGR) technology and it is currently under licensing process in the United States and Canada. In addition to the generic goals, the Xe-100 Human Factors Engineering (HFE) Program is conducted to provide evidence that a reduced staffing can safely operate several reactor modules from one control room. Advanced reactor designs are changing tasks, jobs, roles, responsibilities, size and composition of the operation crews relying on the increased use of automation technologies, passive safety features and adapted control stations designs. This paper explores the details and insights of analysis and validation activities of the HFE Program, providing the key-aspects to succeed in the Staffing Validation.

Keywords: Human factors engineering · Validation · Staffing · V & V · SMR licensing · Safety · Control room

1 Introduction

The Xe-100 is a High-Temperature Gas-cooled Reactor (HTGR) cooled by helium and moderated by graphite, implementing features that make the reactor inherently safe. The plant configuration is optimized with a pack of four Xe-100 reactors, because it maximizes shared systems capabilities. The main purpose of a Xe-100 plant is to safely convert nuclear energy to electricity, although the plant design ensures multiple missions, such as process heat applications.

The initial concept of operations of the plant establishes a crew of three licensed operators who will oversee the operation of four operating reactors from a single control room. This control strategy goes beyond the traditional rules for control rooms, so the licensees shall probe the safe operation of the plant with that configuration.

The Xe-100 HFE Program covers the analysis and validation of the control room staff. In this case, the HFE Program represents a new challenge completely different from a conventional large Nuclear Power Plant staff analysis. Besides designing human-oriented Human-System Interfaces (HSIs) and reducing human error risk to the minimum, the Program's goals target to lower drastically operation and maintenance costs.

J. L. Wright et al. (Eds.): AHFE 2021, LNNS 264, pp. 187–193, 2021.
https://doi.org/10.1007/978-3-030-79763-8_22

New plant missions and different critical scenarios due to the HTGR technology, such as online refueling, and the multi-unit operation from one control room are facts that need to be addressed in the HFE Program.

The guidance to be followed is NUREG-0711 [3]. The Program basically addresses the technical elements described on that document. However, this guidance is complemented with NUREG-1791 [4] requirements in order to comply with the challenging plant staffing reduction goals. Among all the NUREG-0711 [3] technical elements, the following ones are essential to ensure a correct staffing analysis and validation:

- Operating Experience Review (OER)
- Functional Requirements Analysis and Function Allocation (FRA & FA)
- Task Analysis (TA)
- Staffing & Qualifications (S & Q)
- Human Factors Verification & Validation (HF V & V) (Fig. 1)

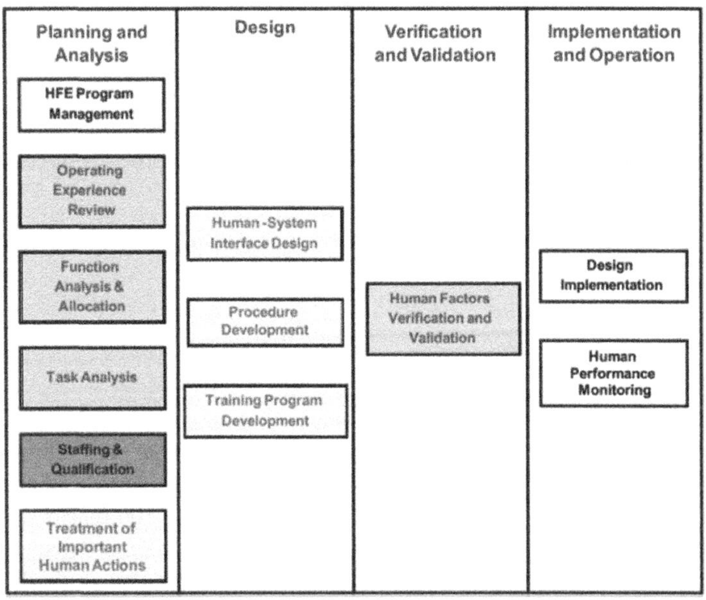

Fig. 1. NUREG-0711 [3] technical element's relative importance regarding the control room staff analysis and validation. From darkest grey, most important, to white, least important.

The following sections of this paper describe how the control room staffing is determined, evaluated, and tested, from the initial Concept of Operations until the last Integrated System Validation.

2 Concept of Operations

The Concept of Operations introduces the staffing, qualifications and operation assumptions that will be subsequently consolidated, modified or dismissed during the development of the HFE Program.

The Xe-100 plant control philosophy relies on a high level of automation and passive phenomena. Considering these passive features, high level of automation of the control system poses no risk to the plant safety, and therefore enables a much-simplified control that significantly reduces the workload of the operator and the control system.

The Xe-100 plant control concept facilitates continuous monitoring, plant maneuvering, and allows the operators to avoid repetitive tasks. This design objective aims to minimize operator human error. Automation performs functions associated with parameter and process monitoring, continuous process control, safety function monitoring, and safety actuations. Operators periodically check process parameters being continuously monitored by automation.

As previously mentioned, the concept operations set the main control room staff which is made up of three operators, one shift supervisor and two reactor operators. Each reactor operators monitors two out of the four modules.

3 Analysis

The initial iteration of the analysis relies on the assumptions made in the Concept of Operations and provides reasonable inputs to the S & Q evaluation.

Operating Experience Review. The starting activity of the analysis is the Operating Experience Review. which should identify two types of experiences: staffing issues to be avoided, and effective staffing practices to be incorporated into the operation environment. Research shall look into past HGTR designs, prototype reactors and recent modular designs. Multi-unit operation experiences can be found in other industries and plants giving valuable insights for the Xe-100 plant operation.

Functional Requirements Analysis and Function Allocation. The second activity of the analysis is the Functional Requirements Analysis and Function Allocation. The first one aims to identify the functions that must be performed in order to achieve Xe-100 plant goals. The different applications of the Xe-100 plant may add new functions, such as hydrogen production or district heating. The Function Allocation assigns the functions to personnel, system elements, or combination of both.

Task Analysis. The functions allocated to personnel require actions that are integrated into a hierarchical structure of tasks and activities, which is the subject of the next analysis. The task analysis provides information to identify task timing and workload issues, giving potential information to determine conflicts that would affect the staffing configuration.

The results of these analysis, specially from Task Analysis, are a powerful tool for the subsequent steps. The following activities consistently share the task requirements for different purposes: job definition, staffing & qualifications evaluation, HSI design, and procedures and training programs.

4 Evaluation

The Xe-100 staffing evaluation shall be done by using the tasks information. At this point, an evaluation is performed by assigning each activity to the adequate operator in several scenarios. In this case, the evaluation focuses in the more challenging scenarios, not typically seen in large nuclear plants analysis. For example, multi-unit emergencies, online refueling, and critical scenarios identified through the Probabilistic Risk Assessment (PRA) of the Xe-100 plant. The staffing evaluation is based in preliminary sequence and timing of task on these scenarios set by subject matter expert, showing the potential limitations of the staff. The conclusions lead to staffing modifications and design change decisions.

It is worth noting that operators are planned to be multi-skilled and versatile, so extra tasks related to plant maintenance can be assigned also to them. Also, maintenance personnel are going to be instructed in control room tasks, so the entire plant staff would be able to share its functions across the plant in a flexible and efficient way.

The results of the evaluations, after iterations of the analysis if required, confirm the size of the control room shift crew, which might vary from the initial established in the Concept of Operations.

5 Validation

The staffing validation refers to an assessment using performance-based tests to determine whether the staffing plan meets performance requirements and supports a safe operation of the plant.

The validation tests of the Xe-100 are planned to be performed during several phases of the design. In the beginning, by early partial validations with portions of the HSI design. Finally, when the complete simulation environment is ready, the Integrated System Validation (ISV) is performed. In this way, the HFE Program ensures operator's gradual adaptation to the HSI and control room environment from the beginning of the design until meeting the final one. Tools such as augmented reality, virtual reality and 3D models of the control room are planned to be used during partial validations at intermediate design phases.

The validations should demonstrate that the Xe-100 plant shift crew can satisfy the plant and human performance requirements identified in the analysis.

The fundamentals for conducting a successful validation are:

Sampling of Scenarios. The range of operational conditions identified as relevant during the analysis should be part of this sample. As previously commented, multi-unit emergencies and critical sequences from the PRA should be part of the sample.

Testbed Conditions. A description of the validation environment, providing: the scenarios to be tested, a representative number of trained crews and scenario repetition, the state of the simulator and simulation conditions, and the number and duties of the HFE observers.

In the early stages of the validation a wide range of profiles can be selected as operators. Due to it is a new design, people from different fields of engineering and industry might be candidates to receive the training and perform the simulations.

Human Performance Measures. Data shall be collected during the tests through observation, questionnaires, plant data extracted from the simulation console, and other methods.

The data should be analyzed and processed considering a set of performance measurements that are established prior to the validation:

- Time to complete actions, especially those identified as critical;
- Timeliness of actions;
- Accuracy and completeness of actions;
- Omitted actions;
- Percentage of automated actions verified by the operators;
- Workload metrics based in NASA-Task Load Index (NASA-TLX);
- Situation awareness metrics based in Behaviorally Anchored Rating Scale (BARS); and
- Teamwork metrics based in Situation Awareness Rating Technique (SART).

Validation Criteria. To determine the acceptability of the results, a set of criteria, related to the performance measures, is identified in advance of the validation. Examples of validation criteria are:

- Nominal task performance times have not been exceeded,
- No more than N seconds have been required to begin Task X after Event Y,
- No actions in the procedures have been be omitted,
- Critical actions have been performed on time,
- Percentage of verified automated actions is above 90%,
- Main plant parameters have been kept inside normal operational range,
- The scenario finished with the plant in a safe state, and/or
- Plant mode X was reached.

6 Conclusion

Concluding, the four facts below will ensure the Xe-100 plant success on its staff analysis and validation.

1. To perform a complete Task Analysis, which is key-important to ensure proper correspondence between staffing evaluation and validation, procedures, training program and HSI design. A multi-skilled and versatile operations crew aids to reduce operation and maintenance costs.
2. To perform a staggered validation through human-in-the-loop simulations from the initial partial validations until the final integrated system validation in order to ensure end-user and human-oriented HSI design. New technologies such as augmented reality, virtual reality and 3D models help to develop these tests without significantly increasing project costs.

3. To include simultaneous abnormal and emergency events on multiple modules in the control room simulator in the validations. Those scenarios likely produce challenging and high workload conditions which are needed to assess the proposed staffing reduction in order to operate the plant safely.
4. A wide range of profiles should be selected for the early validations, including technicians, engineers, and experienced operators who will be previously trained in the new technology. This variety of profiles will enhance the quality of the test results.

For the final goal from X-Energy of getting the design certification from the Nuclear Regulatory Commission, a topical report on this topic is elaborated and periodically updated during the process.

Acknowledgments. To my family and to the X-Energy and Tecnatom teams involved in the development of a new era of reactors, for a clean and reliable world's energy future.

References

1. Nuclear Regulatory Commission: NUREG-0800, Chapter 18 Human Factors Engineering (revision 3)
2. Nuclear Regulatory Commission: NUREG-0711, Human Factors Engineering Program Review Model (revision 3)
3. Nuclear Regulatory Commission: NUREG-0700, Human-System Interface Design Review Guidelines (revision 3)
4. Nuclear Regulatory Commission: NUREG/CR-7190 Workload, Situation Awareness and Teamwork (2015)
5. Canadian Nuclear Safety Commission: REGDOC-2.5.1 General Design Considerations Human Factors (2019)
6. Canadian Nuclear Safety Commission: REGDOC-2.2.5 Minimum Staff Complement (2019)
7. Canadian Nuclear Safety Commission: REGDOC-2.2.1 Human performance management Human Factors (2019)
8. Nuclear Regulatory Commission: NUREG/CR-7126 Human-Performance Issues Related to the Design and Operation of Small Modular Reactors (2012)
9. Nuclear Regulatory Commission: NUREG/CR-6838 Technical Basis for Regulatory Guidance for Assessing Exemption Requests from the Nuclear Power Plant Licensed Operator Staffing Requirements Specified in 10 CFR 50.54(m) (2004)
10. Nuclear Regulatory Commission: NUREG/CR-7202 NRC Reviewer Aid for Evaluating the Human Factors Engineering Aspects of Small Modular Reactors (2012)
11. Nuclear Regulatory Commission: NUREG/CR-6753 Review of Findings for Human Performance Contribution to Risk in Operating Events (2002)
12. Nuclear Regulatory Commission: NUREG/CR-6400 Human Factors Engineering (HFE) Insights for Advanced Reactors Based Upon Operating Experience (1997)
13. Nuclear Regulatory Commission: NUREG-1764 Guidance for the Review of Changes to Human Actions (revision 1)
14. Brookhaven National Laboratory: BNL-20918-1-2015, Methodology to Assess the Workload of Challenging Operational Conditions in Support of Minimum Staffing Level Reviews (2015)

15. Nuclear Regulatory Commission: SECY-14-0095, Status of the Office of New Reactors Readiness to Review Small Modular Reactor Applications (2014)
16. Nuclear Regulatory Commission: SECY-11-0098 Operator Staffing for Small or Multi-Module Nuclear Power Plant Facilities (2011)

Strategies for Human Factors Validation of Fire-Event Scenarios in a Nuclear Power Plant

Borja Hervás(✉) and Luis Rejas

TECNATOM S.A., Avda. Montes de Oca 1, San Sebastián de los Reyes, Madrid, Spain
{bhervas,lrejas}@tecnatom.es

Abstract. The Spanish Nuclear Power Plants are currently incorporating the National Fire Protection Association (NFPA) Standard 805, "Performance-Based Standard for Fire Protection for Light Water Reactor Electric Generating Plants". The main requirement from this standard is to provide a reasonable assurance that a fire initiated at any plant location with any plant configuration will not prevent the plant to achieve safe shutdown. This paper develops the process of Human Factors Engineering (HFE) validation of scenarios related to fire events. The document describes the preparation, presents the validation strategy, and shows the results and recommendations obtained in terms of human performance.

Keywords: Human factors engineering · Validation · NFPA · Simulation · Fire-event scenarios · Human Factors · V&V

1 Introduction

During the last decade the Nuclear Regulatory Commission (NRC) encouraged the U.S. plants to adopt NFPA Standard 805, "Performance-Based Standard for Fire Protection for Light Water Reactor Electric Generating Plants" instead of their existing fire protection requirements basis. Following them, the Spanish Nuclear Power Plants (NPPs) are currently enhancing their safety by incorporating the NFPA Standard 805. Tecnatom has been supporting the Spanish nuclear fleet in this process performing, among other activities, the Human Factors Engineering Validation.

Due to the requirements from this new standard, several fire-postulated scenarios have been simulated and validated in the NPP full scope simulators during the last few years. Spanish fleet has seven nuclear units, five Westinghouse PWR, one General Electric BWR and one Siemens KWU. The first validations were performed in the PWR simulators, being expected to be replicated in the future in the rest of simulators. The Spanish Nuclear Regulatory Body (CSN) is taking part in this task to assure the compliance of each plant with the NFPA 805 requirements. These experiences have allowed the national industry, including Tecnatom S.A., to gain specific knowledge in the strategies and insights of this type of scenarios, specially from the HFE perspective.

J. L. Wright et al. (Eds.): AHFE 2021, LNNS 264, pp. 194–200, 2021.
https://doi.org/10.1007/978-3-030-79763-8_23

2 Validation Plan

The validation is the activity that ensures that a system accomplishes its designed function and goals in a particular operational condition. This definition suits perfectly with the objective of the NFPA Standard 805: provide assurance of a safe shutdown in the case of a fire event started at any plant location. Due to this, several scenarios were conducted for validation in the simulators to ensure that the plants systems, control means, operation crews and procedures were adequate to face the fire events.

The following paragraphs describe the plan and strategy that Tecnatom is following to validate the fire-related scenarios from the human factor perspective. The main guide used to develop the plan is NUREG-0711 [3].

The plan establishes the rules to be followed systematically, and it covers: sampling of fire event scenarios, testbed conditions, data collection methods, performance measures, human cognitive model, and acceptance criteria.

Sampling of Scenarios. The most representative scenarios were chosen based in three criteria: the final risk of core damage based on the plant's Probabilistic Risk Assessment (PRA); the challenges that the operation crew faces from the human factor perspective; and the type of equipment and instruments malfunctions that lead to a more difficult operation.

Simulation Environment. The so-called testbed conditions must translate the real situation with the highest fidelity to ensure the most reliable data and results.

Firstly, the validations took place in the full scope simulators of the plants, this ensured the high fidelity of the Human-System interfaces and procedures. Real operation shifts of the plants participated in the validation sessions. The local actions were based on time estimations previously made. Each scenario was performed by, at least, two different crews; however, it is planned to be repeated by different crews in the future. Finally, the most important one due to the nature of the events, the operation crews were not aware of any kind of information related to the validation and the scenario.

During the validation sessions human factors observers witnessed the operation crew and followed their human performance. In this case, a minimum of two HFE observers were needed, each following two operators. However, this number depends on the number of crew operators and the size of the control room, which might vary on different nuclear plant technologies.

Data Collection Methods. There were three ways to collect the data: observer annotations, post-scenario questionnaires and the exportation of the simulation console data.

During the scenario, each observer follows the behavior and actions of the operators and annotate any observation remarkable for analysis. Also, a questionnaire related to the scenario was fulfilled by the operators right after the validation session. Finally, the simulation console data provided objective data to the engineers about trends and operator's actions over the control panels.

Performance Measures. During the validation, the following measurements were collected by the observers and later analyzed.

- Real time to complete relevant tasks
- Timeliness of actions
- Accuracy
- Completeness of actions
- Omitted actions

Human Cognitive Model for NPP Control Room Operations. The annotations and implications from data collected through the simulation and questionnaires were analyzed according to a human cognitive model adapted from O'Hara's Model, which was discussed in NUREG-2114 [5] (Fig. 1).

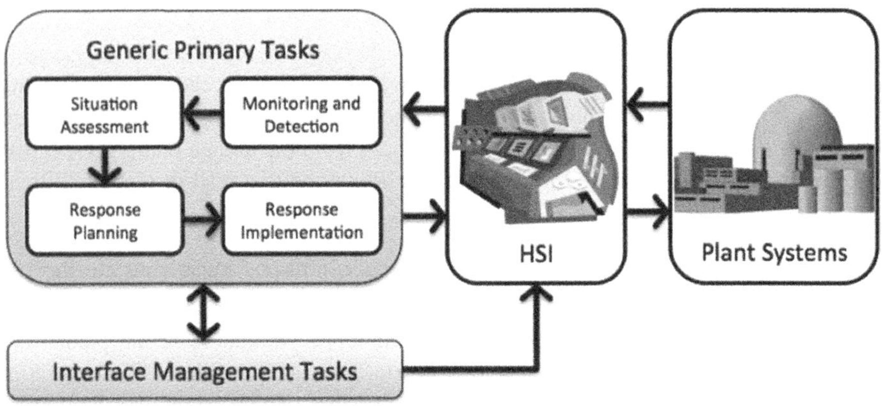

Fig. 1. Picture from NUREG-2114 [5]. O'Hara's cognitive model of NPP control room operations, adapted from O'Hara et al. (2008).

The model considers a set of elements of the human behavior that were used to evaluate and classify the data. Four elements are related to primary tasks (actions over equipment in the hardware panels and software displays) and interface management tasks (navigation, aids, diagrams):

- Monitoring and detection
- Situation assessment
- Response planning
- Response implementation

In addition, the next three elements were also considered. Firstly, the situation awareness defined as the operator's perception of the plant status versus the real plant status. Secondly, the workload defined as the number of actions to be performed at the same time. For example, following two procedures in parallel, receive a phone call while checking an annex of a procedure, etc. And thirdly, the efficiency of the teamwork within the shift and reliable communications with the local operators during the scenario.

Acceptance Criteria. Finally, three general criteria were set prior to the validation sessions to accept (or deny) the success of the tests:

- Once the fire has been initiated, the operation shift brings the plant to a safe condition with the current procedures and Human-System Interfaces (HSIs).
- During the development of the scenario, no facts that increase the risk of human error have been identified. Tasks were completed on time and main plant parameters were kept within the safety limits.
- The sequence of tasks during the scenarios does not introduce such excessive workload that prevent the operators from monitoring the plant, being able to maintain a correct situation awareness at any time.

3 Preparation of Fire-Event Scenarios

Design of Scenarios. After the sampling process, the specific design of scenarios started. Some of the scenarios selected were:

- Fire in the electrical building causing a spurious safety injection and several safeguards train B equipment failures
- Fire in the safeguards building causing several train A equipment failures and auxiliary feedwater unavailability
- Fire in the safeguards building causing inadvertent openings of safety valves and failures in safeguards system
- Fire in electrical buildings that causes the loss of all safeguard buses

Any equipment, cable and instrument affected by the fire was considered during the design of the scenario. Conservative assumptions were made to be always in the safe side. For example, failure of valves in the worst side regarding the transients that they can cause, and failure of instruments to the more confusing position for the operators.

Finally, a list of equipment affected was prepared and provided to the simulation team. This list included the type of failure for each component and the time in which it occurs from the beginning of the fire. Generally, each scenario showed several equipment failures at specific time steps.

Simulation Development of Scenarios. The first step for the simulation team was to filter and make an analysis of the list resulted from the design. Comparing to a normal training session in which two or three malfunctions are activated, the simulator should be able to handle hundreds of malfunctions at the same time.

Due to the nature of the scenario, the simulation team decided to use a tool from the simulation console that is preferably intended for planning instructors' lessons. This tool is called *TeamTutor* and it is usually used by the instructors to plan a simulation session, from the beginning to the end, without the need to actuate over the instructor console. In this case, the tool allowed the simulation team to plan the sequence of malfunctions in the time steps (Fig. 2).

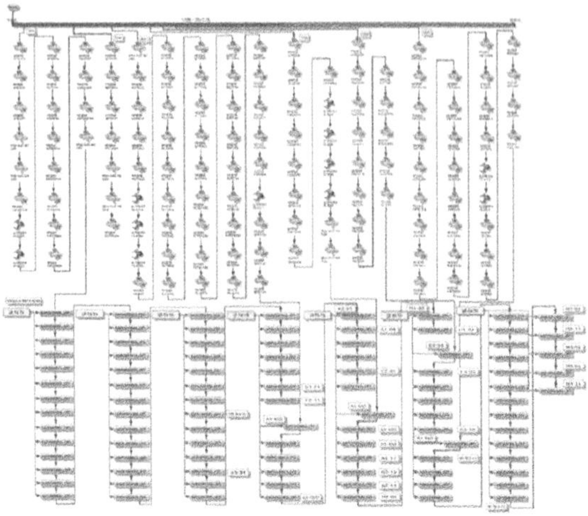

Fig. 2. Screenshot taken from *TeamTutor* tool. The figure corresponds to one of the scenarios validated, showing a timeline in the upper part and the cascade of malfunctions that were introduced in the simulation.

The simulation team also selected, according to the human factor engineer suggestion, the required plant parameter trends to be drawn during the scenarios, in order to extract them from the simulation console for analysis.

Human Factors Engineers Preparation and Materials. The human factors observers were prepared for a period of three weeks to get familiar with the scenarios. During this time the HFE team tested the scenarios in the simulator together with the simulation team, detecting and sharing the potential critical actions of the operator's actuation. In this way, the HFE observers got the key points of the scenarios in advance. Also, they prepared the sequence of actions of the operators over the different HSIs using the procedures that were expected to be followed.

Prior to the validation sessions, the HFE observers also created a notebook to collect the annotations, including the fields: description of the action, time of the action, equipment tag and location within the HSIs, parameter or result indication from the action, operator who performed the action, and remarks.

The annotations were focused on every aspect that could impact the human behavior:

- Procedures
- Staffing (size of the crew)
- HSIs: panels, alarms, displays, workstations
- Communications systems
- Operational aid systems
- General behavior and human factors techniques of the operation shifts
- Training Program

Finally, a questionnaire with ranked questions was prepared and provided to each operator after the sessions.

4 Results and Lessons from the Validation Sessions

In this section some of the results and insights from the sessions are explained. All scenarios were successfully validated, fulfilling the acceptance criteria. The operators confirmed that the scenarios increased the workload comparing to other type of scenarios. Several decisions and actions were discussed and included as lessons learned from the activity.

- The first decision made was to include the scenarios in the training programs right after the validation, as these new scenarios poses a new challenge to the operators.
- The most critical part of the scenario regarding the workload, is the beginning of it, when the fire is announced, the first wave of equipment fails, and several procedures should be followed at the same time. At this moment, the operators shall have all the tools available to understand the situation. This means that aids to locate the fire and identify potential affected equipment should be easy to find. These aids can be software or hardware based. A list of the potential equipment to be affected by the fire should be available to the operators.
- Sometimes the reactor does not trip automatically due to the fire, clear instructions shall be provided to perform the manual trip in the case of fire. In some scenarios, the manual trip of the reactor could lead to a transient that create a worse situation, the shift supervisor shall instruct the team to follow a strategy according to the plant situation.
- Internal communications and external communications become critical. As the scenarios causes many demands of attention to the operator, the shift supervisor shall prepare a strategy based on the global situation and periodically check verbally the main lines of it with the rest of operations crew. Regarding the external communications with the local operators and the fire brigade, the actuations were accomplished in a reasonable time. However, the plants shall be able measure and additionally ensure local activity timing by performing local validations.
- Another relevant fact that raised was that the operators rarely failed to understand the situation, even when indications and instruments were not available or showing false data. This was possible by making double checks of the redundant indications and trends provided by other instruments. Besides, it also shows the benefits of a tough training. Only once, a local actuation to check a post-Fukushima mobile instrument cabinet was demanded.
- Finally, even with the large amount of input demands coming, the operators were quick to identify the main plant issues and solve them. For example, failures over pressurizer and steam generator safety valves and failures related to the loss of auxiliary feedwater system and safety injection system. The response actuation on that cases was timely and precise.

5 Conclusions

So far, the most interesting conclusions of the NFPA 805 transition of the Spanish NPPs from the Human Factors perspective are:

- A different type of scenarios to challenge the operators was prepared. Similar scenarios, with hundreds of malfunctions, are highly recommended to be included in the Training Programs of any kind of nuclear power plants.
- The procedures are valid to conduct the plant to safe shutdown in any of the simulated scenarios (despite minor changes and additions).
- The operators are overloaded by the number of alarms appearing and different procedures to be used in parallel as the fire develops, so an operational strategy, beyond procedures, shall be developed in real time.
  Generally, the shift supervisor shall consistently create this strategy and periodically communicate it to the crew precisely.
- The beginning of the fire represents the most demanding part in terms of workload of the scenario. Operation aids, such as specific displays and applications might help during this part. However, the function of these applications shall be clearly bounded as an excessive information can be self-defeating. The use of these applications and the annex procedures should be defined by priority.
- During the validation sessions it was observed that it is common to reduce the human performance efficiency due to a loss of situation awareness. The operations crew shall share, communicate efficiently, and confirm their hypothesis to perfectly understand what is happening and actuate accordingly.
- A correct management of the local crew is essential to conduct the plant to safe state. This fact is influenced by the location of the fire and the availability of auxiliary operators, fire brigade, local control panels, plant routes and rooms.

References

1. Nuclear Regulatory Commission NUREG-0800, Chapter 18 Human Factors Engineering (revision 3)
2. Nuclear Regulatory Commission NUREG-0711, Human Factors Engineering Program Review Model (revision 3)
3. Nuclear Regulatory Commission NUREG-0700, Human-System Interface Design Review Guidelines (revision 3)
4. Nuclear Regulatory Commission NUREG/CR-7190 Workload, Situation Awareness and Teamwork (2015)
5. Nuclear Regulatory Commission NUREG-2114 Cognitive Basis for Human Reliability Analysis (2016)
6. International Electrotechnical Commission 60964, Nuclear power plants - Control Rooms-Design (2019)
7. International Electrotechnical Commission 61839, Nuclear power plants - Design of control rooms - Functional analysis and assignment (2000)

Data Essential for Human Factors Simulation Modeling

Optimizing Human Performance Data for Enhanced Modeling and Analytics

Seth Elkin-Frankston[1,2(✉)], Erika Hussey[1,2], and Tad Brunyé[1,2]

[1] U.S. Army Combat Capabilities Development Command (DEVCOM) Soldier Center, Natick, MA, USA
{seth.elkin-frankston.civ,erika.k.hussey.civ, thaddeus.t.brunye.civ}@mail.mil

[2] Center for Applied Brain and Cognitive Sciences, Medford, MA, USA

Abstract. A recent proliferation of wearable sensors, advancements in modeling and simulation, and a greater appreciation for individual differences have enabled the Department of Defense (DoD) to generate and collect more human performance data than ever before. Despite trends in human performance research the DoD has maintained a conservative approach to data by limiting effective data access and distribution. However, as the strategic importance of a data-centric organization has become more apparent, a cultural shift within the DoD has begun to prioritize data as an asset. This paper describes efforts to develop a Soldier-centric human performance database including a generalized approach and framework to actively support innovative human performance research and analytics. The successful completion and utilization of a dedicated human performance data management system will provide an interoperable database that facilitates research and development for researchers and operational communities across current and future applications.

Keywords: Human factors · Human-systems integration · Human performance · Database management system

1 Introduction

The Department of Defense (DoD), including the Army, Air Force, Navy, and Defense Advanced Research Projects Agency (DARPA) has historically taken a conservative approach to data. Whether due to privacy or security requirements, technical constraints, or a lack of data-focused prioritizations, effective data access and distribution has remained limited. As the world becomes increasingly reliant on rapid access to trusted data, this position will impact critical applications including implementation of data-driven decision aids, individualized training regimens, modeling and simulation, and artificial intelligence and machine learning. In response to the cultural shift, the DoD has begun to prioritize data as a strategic asset through improved data workflows and infrastructure [1].

Early efforts to transition the DoD to a data-centric organization have prioritized weapon systems, sensor platforms, logistic information, and business systems. To date,

J. L. Wright et al. (Eds.): AHFE 2021, LNNS 264, pp. 203–208, 2021.
https://doi.org/10.1007/978-3-030-79763-8_24

human performance data have not benefitted from the same level of database development and engineering. This state of engineering contrasts with the state of the science, with multiple high-priority DoD research and development efforts aimed at sensing, predicting, optimizing and enhancing warfighter cognitive and physical performance. Human performance data collected in these efforts have generally not been stored, organized, or shared across services and organizations with modern distributed databasing solutions. This may be due to several factors, for example human performance data can be difficult to define or categorize, while data may include real-life performance data and personal identifiable information (PII) it is generally not considered health data or protected under the Health Insurance Portability and Accountability Act (HIPAA). Therefore, distinguishing human performance data from health data can be challenging. Human performance data is also high density without well-specified structure or marking, making it difficult to curate and leverage for deriving actionable insights without deep expertise. Despite these challenges, vast quantities of human performance data are collected throughout the military across many research institutes and operational units. A recent informal survey estimates that a minimum of 65 current human performance related databases exist across the DoD at large. Unfortunately, many of these databases are not fully documented or interoperable with external data sources. These practices can silo or make data inaccessible – thereby reducing the overall value and impact of high-priority inter-service and inter-agency defense research and development programs.

To address these challenges, our work aims to develop an initial operating capability for a Soldier-centric cloud-based human performance database in support of current and future research efforts. We describe a general approach and framework to develop an enabling data management capability that both utilizes infrastructure and guidance provided by the DoD Data Strategy, and actively supports innovative human performance research and analytics.

2 General Approach

This work was initiated to directly support the U.S. Army's Measuring and Advancing Soldier Tactical Readiness and Effectiveness (MASTR-E) program run by the Combat Capabilities Development Command (DEVCOM) Soldier Center (Natick, MA). MASTR-E is an applied research and development program that aims to quantify and characterize cognitive, physical, social-emotional, and health-related factors for enabling real-time state assessment, future performance prediction, and cognitive and physical enhancement. Over a period of several years, this program will collect multi-modal datasets from hundreds of Soldiers engaged in laboratory, simulation, and field exercises. From these data collections, it is predicted that each Soldier will produce over 10 million data points across multiple sensor streams over the course of multi-hour and multi-day data collection events. Across participating Soldiers, this will equate to several billion data points that need to be labeled, organized, stored, and shared and accessed for collaborative analysis and visualization.

To address the many challenges presented by this level of effort, we set out to develop the capability to support the need for agile data collection, organization, analytics, and standardization, while ensuring compatibility with future DoD capabilities and security requirements. This effort included establishing a design strategy, identification of resources and capabilities, and initial implementation.

2.1 Design Strategy

Our approach aimed to operate within the constraints of DoD security requirements; for this reason, design decisions were highly influenced by guiding principles and capabilities outlined by the DoD Data Strategy [1]. We also looked to patient-centered Electronic Health Record systems (EHRs) as inspiration to ensure a maximal degree of accessibility and interoperability with future research efforts.

The DoD have identified eight foundational principles that aim to underlie all essential capabilities to support a data-centric organization [1]. These principles include: (1) *Data is a Strategic Asset*: data is both costly to obtain and is a valuable commodity that should be curated and strategically leveraged; (2) *Collective Data Stewardship*: data must be governed by established policies throughout the entire data lifecycle; (3) *Data Ethics*: data must be ethically used; (4) *Data Collection*: electronic data systems must be used to collect data as close to the point of creation as possible; (5) *Enterprise-Wide Data Access and Availability*: the DoD must make data available to authorized users; (6) *Data for Artificial Intelligence Training*: data must be managed to support current and future Artificial Intelligence/Machine Learning (AI/ML) efforts, including algorithm development, training, and validation; (7) *Data Fit for Purpose*: data that is collected should serve a purpose and have appropriate context to make sense of the data; (8) *Design for Compliance*: data must be governed in a manner that facilitates end-to-end usability throughout the data lifecycle.

While these guiding principles underlie any trusted data management system, they are operationally- and mission-focused and not specifically designed with human subjects research data in mind. Therefore, to optimize our approach for human performance, we turned to specific capabilities of EHRs. EHRs are patient-centric data systems that securely manage complex health related information and make that information instantly available to healthcare providers and external EHRs. While EHRs are often supported by complex and often arcane data standards and requirements (e.g., Health Level Seven (HL7) Fast Healthcare Interoperability Resources Specification (FHIR) [2]), it can be helpful to think of a Soldier as a patient for the purposes of designing a human performance-centric database. In the same way that a patient will have blood work ordered when seeing a healthcare provider, a Soldier will complete a 2-mile run as part of the Army Physical Fitness Test (APFT). This model provides a simple framework for managing and controlling performance data on a Soldier-by-Soldier basis.

Furthermore, the model offered by EHR systems can be leveraged to develop complex and accurate datasets to enhance modeling and analytics by providing a standardized approach for collecting and handling human performance data. For instance, when blood samples are drawn for laboratory testing, the World Health Organization (WHO) defines specimen management protocols that ensure samples are effectively managed, including their receipt, identification, documentation, transportation, storage, and access. These standard operating procedures also include elements of quality assurance and sample rejection, and once a sample is analyzed it falls into the purview of EHR data governance procedures. Several helpful insights can be drawn from the decades of research and policy development for specimen management and EHR data governance. For example, in the human performance domain, there exists a need for 1) hierarchically structured data and transparent mappings between notes and metadata, demographic information, and rich sensor data; 2) standardized reference ranges that allow for a rapid categorical assessment of sensor data (e.g., within range, low vs. high); 3) standardized nomenclature and labeling practices to facilitate interpretability and comparability across individuals, data collection settings, and institutions; 4) mechanisms for surveying and monitoring data quality; and 5) tiered scalability to support innovative data sciences approaches such as machine learning and artificial intelligence [3]. Because these and other data governance needs have been identified in the well-established EHR domain, they can be readily adopted to the nascent human performance databasing domain [4].

2.2 Resources and Capabilities

Although numerous commercial and open-source data management systems exist, few of these options are available for use within the DoD due to lack of support for stringent security regulations. Therefore, we rely on the Army's recently deployed cloud data environment, Modernization Application & Data Environment (MADE). MADE is the Army Futures Command (AFC) nascent cloud environment built on top of Microsoft Azure cloud computing services. In many ways MADE functions like the commercial Microsoft Azure service except that there are additional levels of security to comply with DoD regulations, for example completion of Risk Management Frameworks (RMFs) and Security Technical Implementation Guides (STIGs).

2.3 Initial Implementation

Our first task was to build a tangible data model. We initiated the goal of identifying every piece of data that was intended to be collected throughout the duration of the multi-year research effort. This involved developing a comprehensive data dictionary designed to store and describe the format, structure, relationship, and associated metadata for every human performance metric collected from lab and field studies conducted as part of the MASTR-E program. The task of creating an initial data dictionary was divided into three phases. First, a list of all anticipated metrics and associated data fields to be collected were requested from individual researchers. Second, information about datatype, format, min-max range, etc. were requested. Finally, definitions, associated metadata, and relationships were added to the data dictionary. Breaking this into phases was designed to simplify the task and reduce the risk of human-error. Each phase provided an opportunity to error check and request additional information as required.

While the data dictionary ensures that any data created is tagged, stored, and cataloged, it is also used to dynamically generate new data models. These capabilities ensure that the data dictionary and data management products will remain extensible to support not only future data types, but future human performance efforts. In total our data dictionary includes information from 267 assessment measures and 2,095 individual metrics (see Fig. 1). This level of detail not only supports good data governance and integrity, it provides an unprecedented level of detail for future analysts to facilitate modeling and simulation efforts.

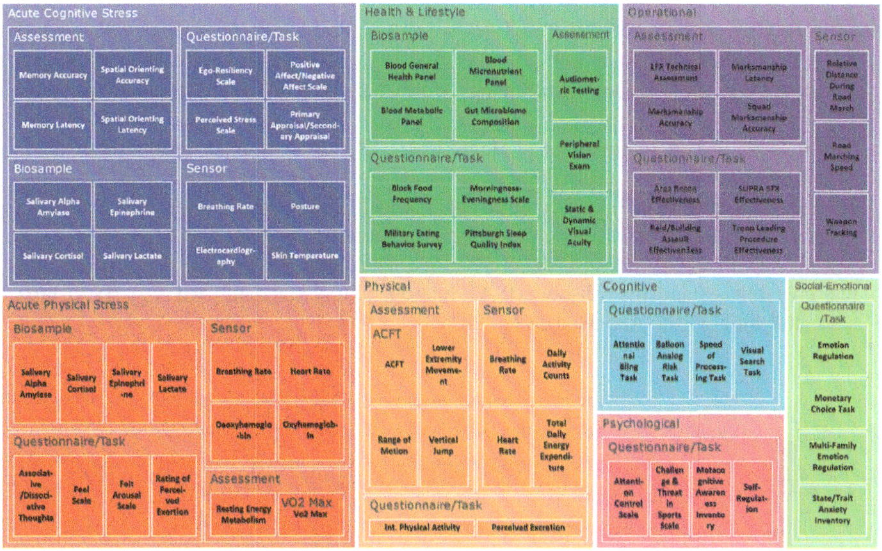

Fig. 1. Representative example of measures, data types, and domains included in the MASTR-E data dictionary

Next, we successfully demonstrated the ability to collect and store diverse human performance data. These demonstrations confirm that representative human performance data can be readily ingested, stored, and rapidly accessed by authorized users. With native integration of analytical software (e.g., Python, R), we also verified functionality for roll based authentication, and the ability to analyze data across Army Science and Technology efforts, a capability aimed to maximize the utility of complex datasets.

3 Conclusion

The successful completion and utilization a cloud-based Soldier-centric human performance research capability aims to deliver an interoperable, extensible, and accessible database that facilitates enhanced modeling and analytics for use by researchers and the operational community in the current and future operational landscape. In addition to the MASTR-E team at the DEVCOM Soldier Center, the database is being designed, developed and populated in collaboration with defense scientists and engineers across all branches of the armed services, ensuring transparency, distributed access, and information sharing. The benefit of a unified database and standard that is always accessible, comprehensive, and curated is the ability to derive unique insights and support AI/ML across increasingly complex operations. For example, access to a standardized human performance dataset can provide a rich data source to create human avatars that are more realistic and respond to environmental changes in real-time. The DoD has made a commitment to transition towards a data-centric organization. Although development can be slow at times, it is important to demonstrate even incremental progress.

References

1. Norquist, D.L.: The U.S. DoD Data Strategy, Unleashing Data to Advance the National Defense Strategy (2020)
2. Health Level Seven International - Homepage | HL7 International. https://www.hl7.org/. Accessed 12 Jan 2021
3. Ahmad, F.S., et al.: Challenges to electronic clinical quality measurement using third-party platforms in primary care practices: the healthy hearts in the heartland experience. JAMIA Open **2**, 423–428 (2019). https://doi.org/10.1093/jamiaopen/ooz038
4. Interoperability Standards Advisory (ISA) | Interoperability Standards Advisory (ISA). https://www.healthit.gov/isa/. Accessed 26 Jan 2021

A Hybrid Simulation Model for an Efficient and Flexible Shop Floor System

Kashif Zia[1(✉)], Umar Farooq[2], and Alois Ferscha[3]

[1] Faculty of Computing and Information Technology, Sohar University, Sohar, Oman
kzia@su.edu.om

[2] Department of Computer Science, University of Science and Technology, Bannu, Pakistan
umar@ustb.edu.pk

[3] Institute of Pervasive Computing, Johannes Kepler University, Linz, Austria
ferscha@pervasive.jku.at

Abstract. One of the goals of the future manufacturing systems is autonomic and self-organized shop floors. This requires flexible manufacturing units in terms of their availability and ability so that extremely personalized products, even of lot size equal to one, can be produced. In this paper, we propose a framework, which focuses on grouping cohesion of manufacturing units and efficient routing of products to resources. Towards this, we integrate an established cultural evolution model to achieve desirable flexibility of resources considering the requirements of the products that are being manufactured. A hybrid discrete simulation environment is implemented, in which cultural evolution among the resources is modeled using discrete-time, whereas, for routing, a simple queuing system is modeled using the discrete-event method.

Keywords: Industry 4.0 · Resource flexibility · Routing flexibility · Personalized production · Cultural dissemination · Group coherence

1 Introduction

The manufacturing industry is changing rapidly due to the recent developments in Cyber-Physical Systems (CPS) [1]. Industry 4.0 [2] represents the technological endeavors destined for a future that is highly mechanized. One of the major focuses of Industry 4.0 is monitoring factory floors using a synchronized cyber computational space [3]. Due to this change, a paradigm shift towards configuration-free, autonomic, and self-organized shop floors is evident, enabling factories to evolve from mass production to mass customization. Mass customization, however, encompasses many different scales and even prolongs to personalized production [4] or even production on demand [5]. Thus, it asks for an extremely flexible Manufacturing System (FMS).

According to [6], the components of an FMS should be flexible in terms of their availability and ability. The two essential features, resource flexibility, and routing flexibility are considered essential for achieving the lot size flexibility. The former believes in the flexibility of servers to act on multiple tasks while the latter is about the flexibility

J. L. Wright et al. (Eds.): AHFE 2021, LNNS 264, pp. 209–216, 2021.
https://doi.org/10.1007/978-3-030-79763-8_25

of products to be served by many servers. Towards this, the related mechanisms such as scheduling [7], resource and time optimization [8], and constraints satisfaction [9] proposed in the distributed intelligence domain are tailored to cater the self-organization of processing units (servers) and the units being processed (products). However, we have adopted a more generalized approach, learned from social sciences; namely, the mechanism of emergence of cohesive groups and culture.

The main aims of FMS – routing and resource flexibility – without doubt, need self-organization of system components and processes to achieve lot size flexibility. The body of work considering the aspect of personalization in FMS, however, is quite lean and requires further attention. Realizing this, several projects are already in progress [10, 11]. One significant effort endeavors to develop cognitive products [12] and production systems incorporating human-like capabilities such as perception, understanding, interpretation, learning, reasoning, planning and, hence, cognition-based acting. The project is about complex cognitive modalities of humans, products, and machines and their interrelationships. In this paper, we argue that one does not need to have high-level cognitive capabilities to be effective. At a scale of a population or a group, a very basic level of cognition of interacting agents may result in a desirable global situation. We just need to find the conditions in which this may happen.

Agent-based modeling (ABM) [13] has emerged as a valuable method for modeling distributed self-organizing systems. In this paper, we propose an agent-based model for the formation of productive cohesion in groups (so that servers and products can work together effectively) and the possible emergence of helpful global culture. This model integrates an established cultural evolution model into a hypothetically simulated world, to establish conditions that lead to effective resource and routing flexibility of servers and products. The motivation is to establish the effectiveness of an FMS through the reconfiguration of servers, without letting them leave their positions. The proposed mechanism of constrained cultural coherence is expected to achieve it due to an increase in the diversity of the culture. The related servers are also on average expected to end up in close vicinity. This will not only ensure an increase in servers' availability but also decrease the mobility of products in search of suitable servers.

This paper implements the proposed mechanism as a hybrid discrete-simulation environment, in which cultural evolution among the servers is modeled using a discrete-time method while routing is modeled as a simple queue using a discrete event method. However, the real contribution of the paper is the integration of manufacturing processes with cultural considerations, which naturally fits into the problem. In our view, this is a novel approach of real significance. However, the work reported in this paper is just a proof-of-concept. We need to have more thorough experiments to measure the efficiency of the proposed model in challenging settings such as environments with inflow and outflow points, more in-depth capabilities, and richer relationships between tasks.

2 Culture, Diversification and FMS

A culture is a multi-feature system, which evolves with time. Coherence is one of the major characteristics of a culture and it is obtained as a result of people trying to acquire a similar behavior (often termed as a trait) in a certain context (often termed as a feature). Relating these concepts to FMS, we need to conceptualize the features and traits of servers and products. Although a product can also be considered as a cultural entity, it is not the case for the purpose of this paper. Only a server is a cultural entity.

Servers are flexible, initially having some randomly chosen features and a randomly chosen trait against a feature. For example, a server may have the ability to perform one, two, or more tasks T1, T2,... with certain levels of precision P1, P2,.... Here, a tuple consisting of n values is a set describing capabilities of a server. Thus, the set {P2, P1, P3} can be interpreted as: this server can perform task 1 with precision 2, task 2 with precision 1, and task 3 with precision 3. The server is unable to perform any additional task. Further, all products have a sequential list of requirements. For example, a product with set {P1, P1, P2} requires task 1 with precision 1, followed by task 2 with precision 1, finally followed by task 2 with precision 2. The question is: would cultural diversification be able to generate a physical layout of the shop floor that would ensure the availability of capable servers with minimal mobility for all the products in the system? Technically, what are conditions, which lead us to an acceptable (and approximate) solution to the problem?

Axelrod provides evidence in [14] for such a simple configuration of cultural descriptions, which can result in a locally coherent but globally polarized culture in response of localized interactions of participating entities. Our intuition is that unbounded coherence between cultural groups would not help in this scenario. The reason is that limitless coherence has no control over where the boundaries of the global polarization would occur, which is not compatible with a system, which seeks the economy of resources and optimizations in several dimensions. Thus, we trying to find conditions, which end up in approximately acceptable structuring in terms of coherence (limited in nature) vs. polarization. To achieve this, we took motivation from Centola et al.'s work [15], which uses a random drift to deviate a highly coherent environment.

3 Model

3.1 Cultural Diversification Model

The model of cultural diversification (or coherence) was originally presented by Axelrod [14], termed as Axelrod model. Consider the cells in the simulation environment (shown in Fig. 1) are people, which are randomly assigned feature-trait values. The model operates in discrete time, in which, at each iteration, a random cell is selected. Let the culture of this cell is called Culture-A. The selected cell randomly selects another orthogonal neighbor (at immediate up, down, left, or right) cell. Let the culture of that cell be Culture-B. A pair-wise similarity between Culture-A and Culture-B are then calculated. Then, with probability p, a decision is made to alter Culture-A. Culture-A is altered by replacing an item of Culture-A with a different item of Culture-B, where an item is a member of a culture set such as [2, 0, 2]. The purpose of applying the cultural coherence model in re-configurable production lines of numerous production units (servers) is to reduce the diversity of servers' capabilities (as shown by different colors in Fig. 1). However, the model should not run for an unlimited time, as this would deplete the cultures to a very few or even one, thus, severally damaging the flexibility of the system. Therefore, we propose to apply a threshold below which the diversity of the system should never go. Further, we integrated the cultural coherence model with the process queuing model, detailed next.

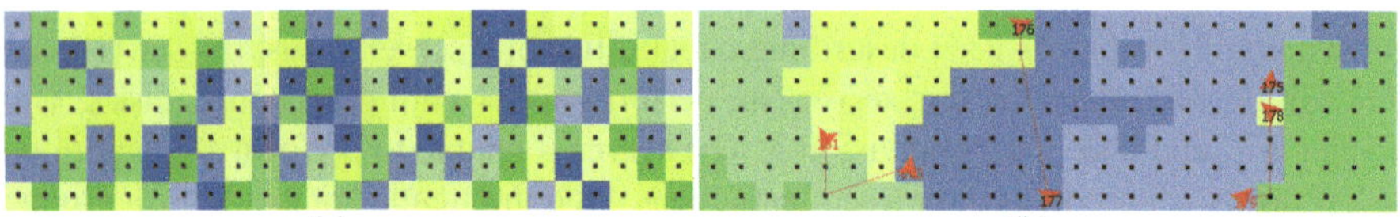

(a) (b)

Fig. 1. (a) original servers represented as cells, (b) application of cultural diversification model producing groups of smaller number of cultures (shown with different colors) and products are shown in red color.

3.2 Process Queuing Model

The purpose of the production line is to manufacture products. A product is a mobile agent that starts its journey at one of the designated entry points. It has to complete a list of tasks. The tasks are performed by the servers. The cardinality of the set of tasks is equal to the servers' set of capabilities. A culture is assigned to each of the products randomly, but compatible with servers' culture. Starting at the entry point, a product traverses to different servers. The process of manufacturing is complete when all the tasks that were needed to the performed have successfully finished. The process queuing model is event-driven. At the start, a list of timings is generated using an exponential distribution that tells when a new product would be spawned into the system. When a product is generated, it is spawned onto a randomly selected entry point, followed by a task execution process at appropriate servers. Different states (of products and servers) and a simple M/M/n queuing mechanism expedite this process. A graphical representation of the model is given in Fig. 2. The process is subdivided into four sequential sub-processes, namely notify, arrive, traverse, and serve.

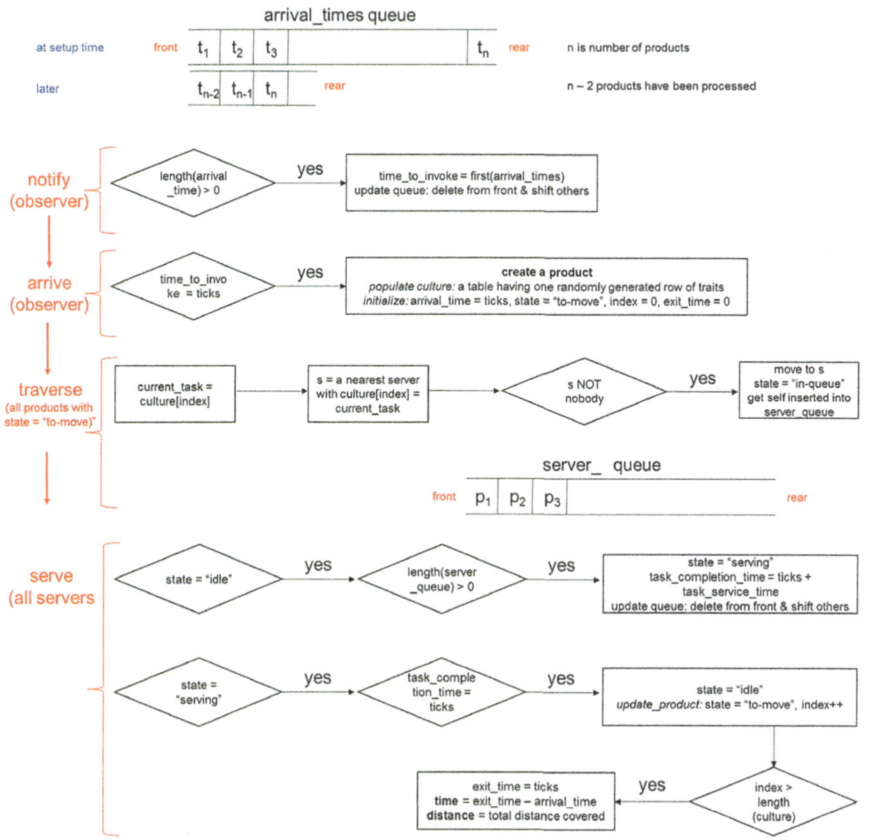

Fig. 2. Graphical representation of the model.

4 Simulation

The cultural diversification model (Axelrod's as well as proposed) is run for a certain number of iterations. We have termed it a settling time, which has three possible values; 5000 iterations, 20000 iterations and 50000 iterations. The lesser the number of iterations used, the more diverse the overall culture would be. The cumulative diversity averaged across all the servers is termed the diversity index. Also, a random configuration is also considered for which, the settling time is not applicable. Hence, a random configuration would have the highest diversity index. The model of cultural diversification is run before the introduction and execution of the products. The process model (of products' tasks execution) is also run at each simulation iteration, but the spawning of new products is governed by pre-specified timing (hence it is event-based). The total time a product remains in the system is averaged for all the products and reported as the simulation result time. Similarly, the total distance a product has traversed in the system is averaged for all the products and reported as simulation result distance. There is the possibility that even after a very long time, some products are not able to finish all their tasks due to the non-availability of servers, simulation results termed as products not served.

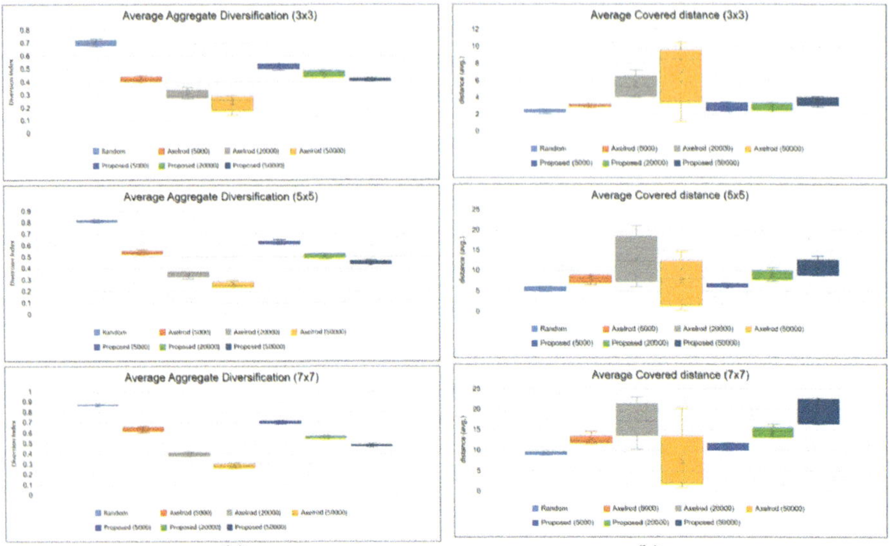

(a) (b)

Fig. 3. (a) Diversity Index and (b) Distance of all possible (model × settling time) situations; with cultural specification 3 × 3 (top), 5 × 5 (middle), and 7 × 7 (bottom).

The cases examined can be represented with a power set of cardinality (diversification model × cultural specification × settling time), where:

- diversification model = random, Axelrod or proposed.
- cultural specification = 3 × 3, 5 × 5, 7 × 7 feature-trait pairs.
- settling time = 5000, 20000, or 50000 iterations.

The simulation results are shown in Fig. 3. It can be seen that in the case of 3 × 3 specification (3 features and 3 traits), and random model, the diversification index is the highest, followed by the proposed model of restricted diversification. This increases the diversity when compared to Axelrod's model. For Axelrod's and the proposed model, the diversification decreases as the settling time increases. A similar pattern can be observed in the case of 5 × 5 specification (5 features and 5 traits) and 7 × 7 specification (7 features and 7 traits). However, there are two more trends that can be observed:

- As the features-traits number increases, the diversification index climbs up. This is obvious because we have more culture with increase in this number.
- The variation in different simulation runs decreases with increase in features-traits number. Maybe, this does not have any other but statistical significance.

Now the question is how these trends of diversification of cultures would impact the process of product execution. It can be observed that the least distance is for the random model. Overall, if the diversification is more, the distance would be less. This can be observed by looking at settling time; less settling time leads to less distance. Also, when we compare Axelrod's model with the proposed model, there are two observations:

- In most cases, the distance covered using the proposed model is less than that of Axelrod's model.
- A large variation in some cases (particularly in Axelrod's model) is due to products not able to complete the tasks, or in other words, not able to move at all due to the non-availability of suitable servers. This is particularly true if settling time is high.

The above is also the reason that Axelrod's model with settling time equal to 50000 iterations seems to perform better than the proposed model, which is not the case. This happens due to products not able to move at all.

Lastly, we also examined the average time of products spent in the system. It was observed that the proposed model reduces the variation between different simulation runs when compared to Axelrod's model. Overall, time was on the lower side in the case of the proposed model. Similar to distance, a dramatic drop in the case of Axelrod's model with higher settling time is was observed due to products not able to move, which is prohibited by the proposed model.

5 Conclusion

Manufacturing systems of the future need to have flexible resources and routing to produce extremely personalized products, even of lot size equal to one. What we have seen is that flexible manufacturing systems can be realized without moving the servers (processing units) by enabling reconfiguration of capabilities of servers based on the dissemination of culture concept. However, Axelrod model has a focus on coherence of cultural groups, which most of the times ends up in one or very few cultures. If we equate such an instance of a culture with a single capability of a server, we are left with extremely limited cultures and products cannot complete their production life cycle.

Hence, we proposed to have a constrained cultural coherence mechanism by introducing a threshold. This tiny development has a significant impact on the increase in diversity of the culture along with related servers being in close vicinity to each other on average. This not only ensured an increase in servers availability as a whole, but also managed to decrease the mobility of products in search of suitable servers.

In summary, the simulation results revealed that the products need to move less in more diversified cultural groups when looking for suitable resources. It was also observed that when more time is provided for cultural cohesion, the cultural groups become increasingly coherent due to homophily. For scenarios, which require diversification of resources, we need to find a balance between coherence and diversification. In future, we intend to include an autonomous learning system, which would help servers learn and change their configurations on the fly based on product types, requirements and trajectories.

Acknowledgments. The authors would like to acknowledge support by FFG funded Pro2Future under contract No. 6112792.

References

1. Rajkumar, R., Lee, I., Sha, L., Stankovic, J.: Cyber-physical systems: the next computing revolution. In: Design Automation Conference, pp. 731–736. IEEE (2010)
2. Lasi, H., Fettke, P., Kemper, H.G., Feld, T., Hoffmann, M.: Industry 4.0. Bus. Inf. Syst. Eng. **6**(4), 239–242 (2014)
3. Lee, J., Bagheri, B., Kao, H.A.: A cyber-physical systems architecture for industry 4.0-based manufacturing systems. Manuf. Lett. **3**, 18–23 (2015)
4. Hu, S.J.: Evolving paradigms of manufacturing: from mass production to mass customization and personalization. Procedia CIRP **7**, 3–8 (2013)
5. Matt, D., Rauch, E., Dallasega, P.: Trends towards distributed manufacturing systems and modern forms for their design. Procedia CIRP **33**, 185–190 (2015)
6. Ogunsakin, R., Mehandjiev, N., Marín, C.A.: Bee-inspired self-organizing flexible manufacturing system for mass personalization. In: Manoonpong, P., Larsen, J.C., Xiong, X., Hallam, J., Triesch, J. (eds.) SAB 2018. LNCS (LNAI), vol. 10994, pp. 250–264. Springer, Cham (2018). https://doi.org/10.1007/978-3-319-97628-0_21
7. Zhang, J., Ding, G., Zou, Y., Qin, S., Fu, J.: Review of job shop scheduling research and its new perspectives under industry 4.0. J. Intell. Manufact. **30**(4), 1809–1830 (2019)
8. Beruvides, G.: Artificial cognitive architecture with self-learning and self-optimization capabilities. Case studies in micromachining processes (2017)
9. Ezpeleta, J., Colom, J.M., Martinez, J.: A petri net based deadlock prevention policy for flexible manufacturing systems. IEEE Trans. Robot. Autom. **11**(2), 173–184 (1995)
10. Zia, K., Ferscha, A., Trendafilov, D.: Conceptualization of cultural diversity for efficient and flexible manufacturing systems of the future. In: CogSci, pp. 1290–1295 (2019)
11. Zia, K., Ferscha, A., Trendafilov, D.: Importance of coordination and cultural diversity for an efficient and flexible manufacturing system. arXiv preprint arXiv:1905.08355 (2019)
12. Trendafilov, D., et al.: Cognitive products: system architecture and operational principles. https://pro2future.at/start-en/
13. Bonabeau, E.: Agent-based modeling: methods and techniques for simulating human systems. Proc. Natl. Acad. Sci. **99**(Suppl. 3), 7280–7287 (2002)
14. Axelrod, R.: The dissemination of culture: a model with local convergence and global polarization. J. Conflict Resolut. **41**(2), 203–226 (1997)
15. Centola, D., Gonzalez-Avella, J.C., Eguiluz, V.M., San Miguel, M.: Homophily, cultural drift, and the co-evolution of cultural groups. J. Conflict Resolut. **51**(6), 905–929 (2007)

Applying Numerical Simulation to Predict Effect of Brace Wear for Scoliosis

Queenie Fok[1,2](✉) and Joanne Yip[1,2]

[1] Institute of Textiles and Clothing, The Hong Kong Polytechnic University, Hung Hom, Hong Kong, Special Administrative Region of China
lh-queenie.fok@connect.polyu.hk, tcjyip@polyu.edu.hk

[2] Laboratory for Artificial Intelligence in Design, Hong Kong Science Park, New Territories, Hong Kong, Special Administrative Region of China

Abstract. Finite element analysis (FEA) is a common numerical method used to obtain approximate solutions for complex problems in engineering and mathematical physics. This study focuses on the application of an FEA to predict the biomechanical performance of a textile brace called the anisotropic textile brace that is used to treat scoliosis, and simulate the effects of the brace on a scoliotic spine. The simulation process consists of 6 steps: 1) creating the geometrical models, 2) defining the material properties, 3) meshing the models, 4) defining the initial and boundary conditions, 5) numerically processing the setting displacements and loading of the garment, and 6) post processing and validating the results. It is found that after wearing the brace, the movement of the simulated spine and changes in the curvature of the spine obtained from the FE model show a similar trend as those from the in-brace x-rays. The correlation coefficient of the vertebral body displacement is 0.74. This demonstrates that the model can be used in future studies to demonstrate the efficacy of other medical orthoses for treating scoliosis.

Keywords: Scoliosis · Soft brace · Flexible brace · Biomechanics · Human factors · Simulation

1 Introduction

The word 'scoliosis' is derived from the ancient Greek word *skoliosis*, which literally means bending or twisting. Clinicians have defined scoliosis as a medical condition in which the spine is deformed and shows a lateral curvature that is greater than 10° on a standing radiograph [1]. Bracing is a common and conservative type of treatment for scoliosis. However, there are many different approaches to bracing in terms of design and there are many types of braces, including symmetrical, asymmetrical, hard, soft, and night braces [2]. It is well acknowledged that the in-brace correction of spinal curvature is the most important criteria to determine the efficacy of bracing [3]. However, clinicians and researchers usually face substantial challenges in the lack of volunteer subjects to test their bracing devices and concerns around repeated radiographic examinations due to unnecessary radiation exposure.

J. L. Wright et al. (Eds.): AHFE 2021, LNNS 264, pp. 217–223, 2021.
https://doi.org/10.1007/978-3-030-79763-8_26

One of the solutions to remedy these issues is the use of a finite element analysis (FEA), which is a common numerical method for solving engineering and mathematical problems. The idea behind the use of an FE method is to subdivide a structure into simpler elements which are connected at the nodal points. and results in a set of simultaneous algebraic equations to solve the unknowns [4]. An FEA overcomes the limitations of using human subjects and the cost of constructing physical prototypes, while performing virtual experiments to optimize the design. Several studies have used an FEA to simulate the effectiveness of braces and even to optimize the design of braces [5, 6].

This study demonstrates the application of an FEA to predict the biomechanical performance of a textile brace called the anisotropic textile brace and simulate its effect on a scoliotic spine.

2 Methods

This study was approved by the Human Subject Ethics Review Committee of The Hong Kong Polytechnic University (HSEARS20141220002). The model of the brace was originally developed for an adolescent female patient with scoliosis who has an S-curve; i.e., she has a right thoracic left-lumbar curve. The subject was 13 years old at the time of the study, 150 cm in height, 45.4 kg in weight, a curvature of 25° from T5 to T11, and 22° from T11 to L3. She participated in the study with the consent of her parents. She then underwent an initial clinical study with the anisotropic textile brace (Fig. 1). The experimental protocols included two radiographic examinations and two three-dimensional (3D) body scans (both of which involved one with the brace donned and one without the brace) and pressure measurement at the brace-torso interface. The two sets of radiographic images comprised a lateral view and a standard posteroanterior view. The x-rays and 3D scanned images were collected to obtain the geometric information of the skeletal frame and torso of the subject. Moreover, an in-brace radiographic examination was conducted to validate the model.

The numerical simulation process in this study involves 6 steps: 1) creating the geometrical models, 2) defining the material properties, 3) meshing the models, 4) defining the initial and boundary conditions, 5) numerical processing of setting displacements and loading of the garment, and 6) post processing and validating the results.

The geometrical model was composed of three 3D digital parts which are the spine, torso and brace models. The spine model includes the vertebral bodies, intervertebral discs, rib cage and pelvis which were constructed from the coronal and sagittal x-ray images by using Solidworks 2012 x64 Edition (Dassault Systèmes Solidworks Corp., USA). The torso model was constructed from the 3D scanned body images by using Geomagic Studio 2012 (Geomagic Inc., USA). The brace model consisted of two parts which are the elastic textile panel, and the hinged back bone which was built based on the geometric model of the torso by using Solidworks 2012 x64 Edition (Dassault Systèmes Solidworks Corp., USA).

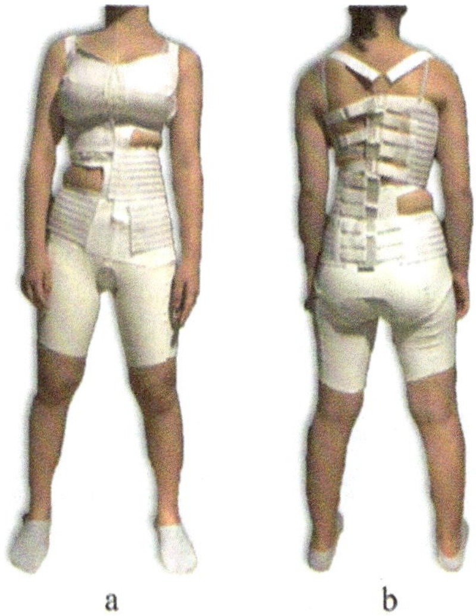

Fig. 1. Asymmetric textile brace on subject: (a) front and (b) back views

The material properties, which are the Young's modulus and Poisson's ratio of the vertebral body, intervertebral discs, ribcage, pelvis, body tissues and hinge material of the brace, were defined based on the definitions in previous studies [7–10]; see Table 1. As for the brace material, the Young's modulus of the textile material was determined from the results of stretch and recovery tests by using an Instron 4411 tensile strength tester while the Poisson's ratio was based on the values in Zhou et al. [11].

Table 1. Material properties for FE model

Material property	Vertebral body	Rib cage	Pelvis	Intervertebral disc	Body tissue	Hinge material of brace	Textile material of brace
Young's Modulus (MPa)	10,000	10,000	10,000	1	0.05	0.59	2,070
Poisson's ratio	0.3	0.3	0.3	0.3	0.4	0.4	0.3

The mesh element used in the FE model is a 10-node linear quadratic tetrahedron with three degrees of freedom at each node. The mesh size is 40–46 mm. The FE model consists of 39,290 elements in total including 8,827 elements for the vertebral bodies, rib cage and pelvis, 4,345 elements for the vertebral discs, 23,804 elements for the body tissues, 273 elements for the hinge material of the brace, and 2,041 elements for the textile material of the brace (Fig. 2).

This study hypothesizes that the corrective forces applied onto the torso of the body do not cause displacement of the head and legs, and thus the top surface of the T1 vertebra and the bottom surface of the torso are firmly fixed with the restriction of all degrees of freedom. In addition to the initial boundary conditions, boundary conditions for the translational displacement of the seams of the brace were applied to simulate the process of brace closure and more importantly, two loads were applied on the surface of the torso under the brace at the thoracic and lumbar regions to simulate the exertion of the corrective pads. The load from the thoracic and lumbar pads was determined from direct measurements of the subject, which is 24.4 kPa and 9.6 kPa respectively.

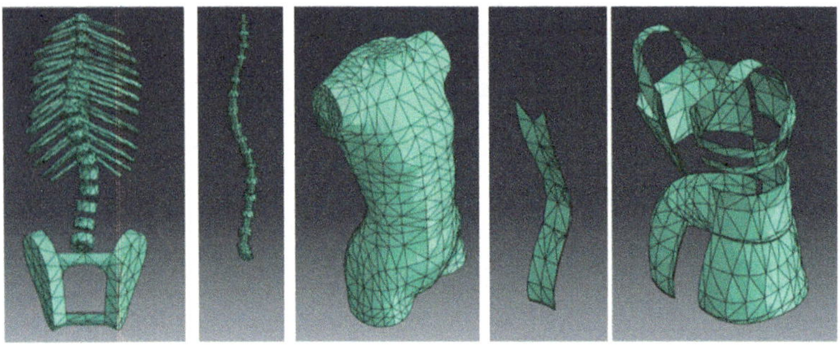

a b c d e

Fig. 2. Finite element model of (a) vertebral bodies, rib cage and pelvis, (b) vertebral discs, (c) the body tissues, (d) hinge material of brace, and (e) textile material of brace

To validate the model, a comparison between the simulated and experimental results of the scoliotic spine was conducted. The curvature of the spine in the coronal plane was considered as the parameter for comparison purposes. The curvature of the spine was plotted according to the position of each vertebral body which was determined by the horizontal distance between the vertebral body and central sacral vertical line. The correlation coefficient of the vertebral body from the FE estimations versus the measured in-brace radiographic images is calculated by using Eq. 1 [12]:

$$Correlation\ coefficient = \frac{\sum_{i=1}^{17}\left[\left(T_i - \bar{T}\right) \times \left(E_i - \bar{E}\right)\right]}{\sqrt{\sum_{i=1}^{17}\left(T_i - \bar{T}\right)^2} \times \sqrt{\sum_{i=1}^{17}\left(E_i - \bar{E}\right)^2}} \quad (1)$$

where:

T = vertebral displacement measured from the radiographic image
E = vertebral displacement estimated with the FE model
$\bar{T}$ = average vertebral displacement measured from the radiographic image
$\bar{E}$ = average vertebral displacement estimated with the FE model
i = vertebral body in the scoliotic spine.

3 Results and Discussion

The FE model showed that after donning the brace, there are relatively high levels of stress induced on the 7th and 8th ribs from the thoracic pad. Moreover, the model showed the distribution of pressure on the skin surface along the body. Apart from the pressure distribution, it was found that the thoracic vertebrae shift to the left and the lumbar vertebrae shift to the right based on the vertebral displacement results. Therefore, the movement of the vertebrae is correlated with the stress distribution of the torso.

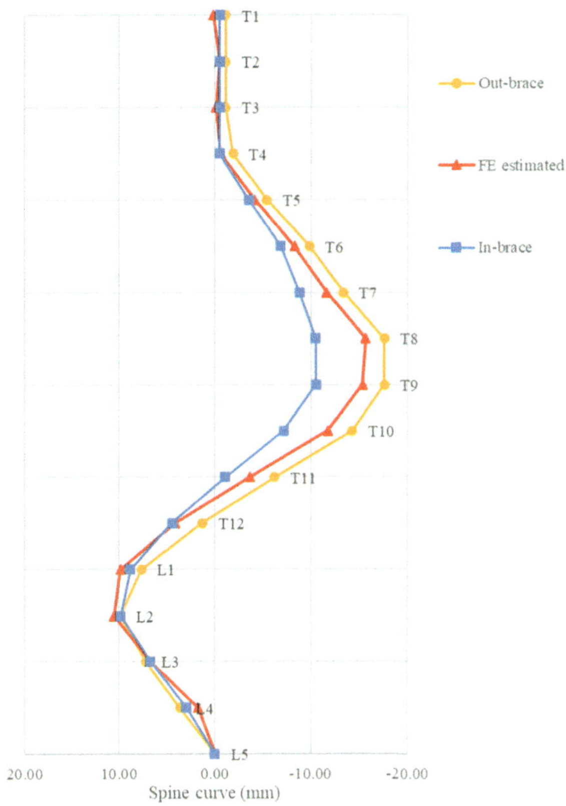

Fig. 3. Curvature of the spine of subject obtained from radiographic examination: without brace donned, with brace donned and calculated with FE model.

The movement of the simulated curvature of the spine obtained from the FE model showed a similar trend as that found on the in-brace x-ray images (Fig. 3). The correlation coefficient of the vertebral body displacement is 0.74. Nevertheless, it was also found that there is an underestimation of the magnitude of displacement of the 7th to 11th thoracic vertebral bodies. This result is in agreement with a previous finding in Périé et al. [9] on the discrepancies between FE estimation models and experimental results. Périé et al. [13] also suggested that muscle control which causes the trunk to shift away from the area of pressure is a possible explanation for the underestimation. A second possible explanation is the inaccurate mechanical properties of the model. As the FE model in this study is patient specific but the mechanical properties are defined based on previous studies, they might not be appropriate for the subject in this study. Therefore, some discrepancy in the estimated model resulted. Nevertheless, this FE model enables predictions of the trend of vertebral movement with an acceptable degree of accuracy.

4 Conclusions

This study has used a numerical simulation method to predict the effect of a brace for scoliosis. An FE model is developed to predict the biomechanical performance of the brace and simulate its effect on a scoliotic spine after donning the brace. The results of the FE model estimated curvature of the spine versus real measurements of the spine that are experimentally obtained show a similar trend which confirms that the FE model is valid and accurate, and can be used in future studies to demonstrate the effectiveness of other medical orthoses.

Acknowledgments. This research is funded by the Laboratory for Artificial Intelligence in Design (Project Code: RP1-4), Hong Kong Special Administrative Region.

References

1. Hresko, M.T.: Idiopathic scoliosis in adolescents. N. Engl. J. Med. **368**(9), 834–841 (2013)
2. Weiss, H.-R., Turnbull, D.: Brace treatment for children and adolescents with scoliosis. In: Bettany-Saltikov, J., Kandasam, G. (eds.) Spinal Deformities in Adolescents, Adults and Older Adult, pp. 1–2. InTech (2020)
3. Negrini, S., et al.: 2011 SOSORT guidelines: orthopaedic and Rehabilitation treatment of idiopathic scoliosis during growth. Scoliosis **7**, 3 (2012)
4. Dhatt, G.: Finite element method. In: Touzot, G., Lefrançois, E. (eds.) London 2012. ISTE, pp. 6–16
5. Hui, C.-L., Piao, J., Wong, M.S., Chen, Z.: Study of textile fabric materials used in spinal braces for scoliosis. J. Med. Biol. Eng. **40**(3), 356–371 (2020). https://doi.org/10.1007/s40846-020-00516-9
6. Clin, J., Aubin, C.E., Parent, S., Sangole, A., Labelle, H.: Comparison of the biomechanical 3D efficiency of different brace designs for the treatment of scoliosis using a finite element model. Eur. Spine J. **19**(7), 1169–1178 (2010)
7. Cahill, J.P., et al.: The use of a transition rod may prevent proximal junctional kyphosis in the thoracic spine after scoliosis surgery: a finite element analysis. Spine **37**(12), 687–695 (2012)
8. Cheng, F.H., Shih, S.L., Chou, W.K., Liu, C.L., Sung, W.H., Chen, C.S.: Finite element analysis of the scoliotic spine under different loading conditions. Bio-Med. Mater. Eng. **20**(5), 251 (2010)
9. Périé, D., Aubin, C.E., Lacroix, M., Lafon, Y., Labelle, H.: Biomechanical modelling of orthotic treatment of the scoliotic spine including a detailed representation of the brace-torso interface. Med. Biol. Eng. Comput. **42**(3), 339–344 (2004)
10. Cambridge University Engineering Department: Materials Data Book (2003)

11. Zhou, J., Li, Y., Lam, J., Cao, X.: The poisson ratio and modulus of elastic knitted fabrics. Text. Res. J. **80**(18), 1965–1969 (2010)
12. Chou, W., et al.: Using finite element method to determine pad positions in a boston brace for enhancing corrective effect on scoliotic spine: a preliminary analysis. J. Med. Biol. Eng. **32**(1), 29–35 (2012)
13. Périé, D., Aubin, C.E., Petit, Y., Beausejour, M., Dansereau, J., Labelle, H.: Boston brace correction in idiopathic scoliosis: a biomechanical study. Spine **28**(15), 1672–1677 (2003)

The Future of Data Standards and Storage: Harnessing Data Generation into a Standardized Repository

Jennifer Neugebauer-Sperlein[1(✉)], Samantha Chambers[1], Jerod Bernicky[2], and Scott Pridgeon[2]

[1] Weapons Branch, U.S. Army Combat Capabilities and Development Command (DEVCOM), Data and Analysis Center (DAC), Aberdeen Proving Ground, Aberdeen, MD, USA
[2] Joint Data Branch, U.S. Army Combat Capabilities and Development Command (DEVCOM), Data and Analysis Center (DAC), Aberdeen Proving Ground, Aberdeen, MD, USA

Abstract. Data storage and standards are often the last consideration, or given no consideration at all, when assessing human factors. Generally, objective and subjective data are collected and used for a specific purpose and then stored as needed by individual investigators. Data storage may include hard drives, hard copies, removable storage devices (e.g. USB sticks, DVDs), and a plethora of file types (e.g. delimited, pdfs, proprietary formats). Historically, these data have not been standardized or stored in a manner that allows for easy and widespread future access and use. This results in potential duplicate efforts and dollars, increased analysis time, and delays in decision making while waiting for new data to be generated. The cloud-based Army Experimentation Resource Data Repository (AERDR) was developed to address these issues. Here we present the repository framework and capabilities, representative data included in the repository, and challenges and considerations on standardizing data.

Keywords: Human performance · Data repository · Repository · Standards · Standardization · Data storage

1 Introduction

Data storage and standards are often the last consideration, or given no consideration at all, when assessing human factors. Particularly in the case of human performance data, generally, objective and subjective data are collected to address a specific question or goal, used for that purpose, and then stored as needed by individual investigators. Data storage mechanisms may include both physical and electronic copies. Hard drives (external and internal), removable storage devices (e.g., USB sticks, DVDs), and cloud storage are just a few examples of how electronic data may be stored. These data are not only stored in a variety of formats, but also include a variety of file types such as delimited, narrative, surveys, and time series data. An additional layer of consideration is added when storing and accessing data generated by human subjects. These data are

J. L. Wright et al. (Eds.): AHFE 2021, LNNS 264, pp. 224–228, 2021.
https://doi.org/10.1007/978-3-030-79763-8_27

subject to additional permission layers when establishing community access. Historically, these data have not been standardized or stored in a manner that allows for easy and widespread future access and use. Ultimately, this can result in duplicate efforts and dollars, increased analysis time, and delays in decision making while waiting for new data to be generated when legacy could be used. The Army Experimentation Resource Data Repository (AERDR) addresses these issues.

Here we present the AERDR repository framework and capabilities, a representative human performance study included in the repository, challenges on standardizing data, and future considerations.

2 AERDR Overview and Structure

The AERDR is a centralized, globally available repository. The AERDR is a digital solution, based in Microsoft Azure Cloud, designed to support the centralized collection, management, and storage of all curated experimentation data with the option of including raw data (Note: curated data are cleaned, analysis ready data; experimentation leads have corrected any missing data, incorrect data, or duplicate data.). Experimentation data may include different types of data, such as human performance data or materiel research data, and all file formats, such as delimited time series, pdfs, or proprietary sensor formats.

The AERDR contains three primary layers: (1) Data Staging, (2) Data Warehouse, and (3) Interaction Layer (Fig. 1 shows an illustration). Data are posted in AERDR by experimentation activity leads to a Staging Layer, which is divided into two sections. The Staging sections are for (1) raw data and (2) curated data. Any given dataset may include either both raw and curated data or curated data only.

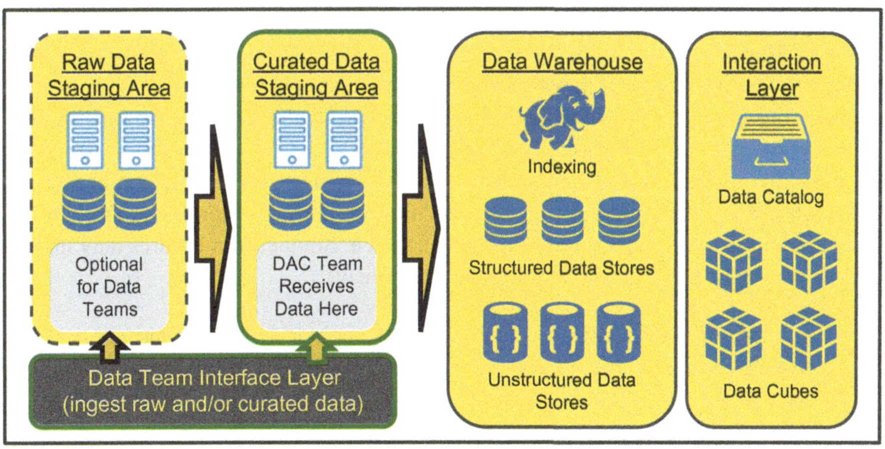

Fig. 1. Graphical illustration of the AERDR and its structure. Data collection teams upload data to either or both the raw or curated data staging areas. Those data are then moved to the Data Warehouse. The Data Interaction Layer provides a means for access, analysis, and export of data.

Associated experiment and data metadata are collected and corroborated with related metadata to paint the full experimentation picture and to enable collaboration. The Staging Layer enables Extract, Transport and Load (ETL) capabilities to support cleaning of raw data.

Once data are cleaned in the Staging Layer, data are moved to the Data Warehouse. The Data Warehouse is the permanent, final stop for data being placed and preserved in the repository portion of AERDR. The Warehouse will contain the final, read-only, analysis ready version of the curated data that will be available to the analytical community.

Lastly, the Interaction Layer provides the ability for analysts to access, analyze, and utilize the results of previously collected and preserved experimentation data. This layer provides data analytic capabilities like RStudio, Python, and Cognitive Search that enable full realization and analysis of available and preserved data. The analytical tool-set will enable the full spectrum of data discovery, query, analysis and visualization.

In summary, the AERDR provides long-term, globally available, access to preserved experimentation data along with an interactive layer to enable analysis.

3 Human Performance Data

Human performance data to be included in the AERDR include several different types of data, both raw and curated. These data may include both individual and squad level metrics to quantify the ability to shoot, move, navigate, communicate, protect, and sustain.

Methods and Data Outcomes. Individual and squad level metrics collected at different military installations can be collected from both individuals and squads. Data collection can occur on a single day, or over multiple days and included both day and night activities. Marksmanship, movement, navigation, and communication can be quantified during tasks such as physical fitness, a 6–12 mile road march, individual and team/squad obstacle course, squad build movement, casualty care and evacuation, marksmanship, and land navigation for example. Soldiers and squads can be assessed using objective and subjective measures. Objective measures may include demographics, training history, qualification scores, target engagement (hit, miss), audio recordings of rate of fire, squad communication, GPS routes via Garmin watch. Subjective measures may include cadre score cards, grade-book, and after action reviews.

Data Preparation and Curation. Data generally include both raw and curated data (Fig. 2 provides an illustration of raw to curated data). Raw data may include electronic copies, such as copies of the cadre scorecards, audio files, proprietary files, such as those provided by the GPS watches, and other original forms of the data, such as time series data.

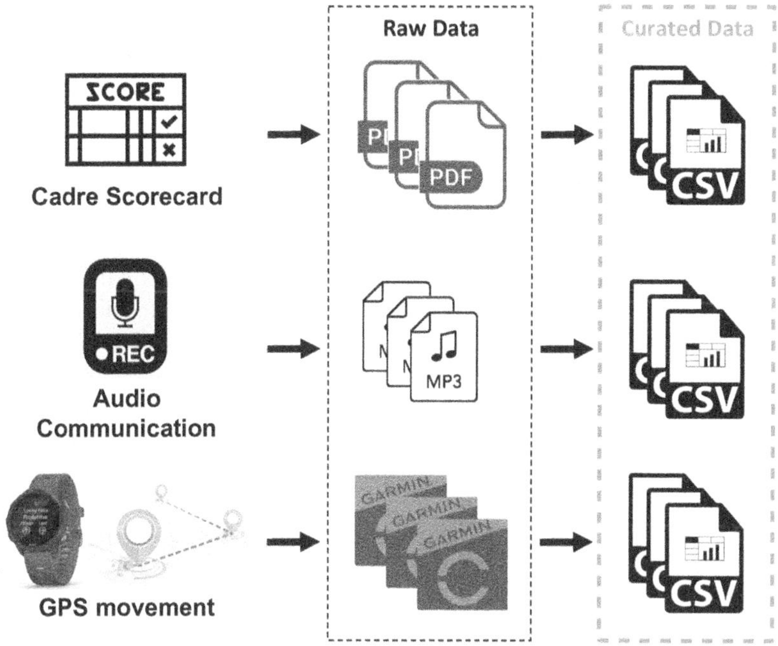

Fig. 2. Graphical illustration of three examples of metrics collected with their raw and curated data forms. The black dotted square shows the raw data and the gray dashed square shows the curated data.

Curated data is born from the raw data being processed and cleaned. For example, for the GPS movement data, as mentioned, the proprietary files from the watch are included as raw data. The export from that file creates a.csv file and is the start of the curated data file. Once created, the.csv file is examined, cleaned, and labelled as needed to create the curated file. For human subjects data, the curation process also de-identifies data as needed, including the removal of PII.

The final step to data preparation and curation is creating a detailed readme file(s). The readme file is arguably as important as the actual data files. The file provides information to the analyst accessing the data and details all pertinent information needed to use the data. The readme typically includes details such as definitions of acronyms, file type information, units of measure, the number of files included. Additional information may also be included that details specifics of data curation including the programs used to curate the raw data.

4 Adding Data to AERDR

As previously mentioned, human performance data presents unique requirements when included in an accessible data repository. Institutional Review Boards (IRB) govern human subjects research and data collection to ensure the welfare and rights of the person included in the data collection. As such, data collected from humans is subject to layers of protection when sharing and using that data. Additionally, human subjects data may include personally identifiable information (PII) that is protected from further distribution such as through a data repository. The AERDR must address both the additional access considerations and handling of PII.

To date, the final protections for accessing human subjects data through the AERDR have not been implemented. The planned approach is to incorporate two levels of data protection and access control. The first will occur during data ingestion to the AERDR and the second when data is accessed by an analyst or researcher.

First, when data are ingested to the AERDR, data team leads will address if the original protocol or research determination specified inclusion of the data in the repository. If it was not addressed in the original protocol or determination of research, the data team lead will not include these data until an amendment is submitted specifically detailing inclusion of the data in the AERDR. For future studies, the delay ingesting to the AERDR will be avoided by incorporating in the consent form or the determination of if the data collection is classified as research or non-research that the raw and curated data may be included in the AERDR.

Second, analysts or researchers accessing data that resides in the AERDR will sign a data sharing agreement. This agreement will cover specifically what data are being shared and the acceptable use of those data. The combination of these layers of protection should address the unique aspects of sharing and utilizing data collected from human subjects.

5 Future Directions and Considerations

The AERDR provides a novel and necessary first step towards standardizing and storing data for future use. While the data discussed here have been included as an initial use-case, for future inclusion of human performance, a data dictionary and documented standards for raw and curated data files will be developed. The data dictionary will include the definitions of common terms used as well as common units of measure used. Data standards will include data formatting guidelines and the documentation needed for the readme files. The combination of the data dictionary and data standards will allow for interoperability of data when accessed by a variety of analysts and scientists.

Ultimately, the AERDR protects the data generation investment and enables data access to the Army analytical community for future use. This investment will save money and time by limiting the need to recreate data for future analyses that could use legacy data made available through the AERDR. Additionally, the availability and use of these data will allow quick evidence based acquisition decisions.

Acknowledgments. The authors would like to acknowledge and thank the Soldiers that participated in the data collection efforts. Additionally, the authors would like to thank the team from Applied Information Solutions (AIS) and others with the Joint Data Branch that helped develop and execute the AERDR in a very short timeframe during exceptionally challenging work conditions.

Human Sensing in Simulation

Comfortable SCBA Weights from Biomechanical Models for Firefighting Tasks

Susan Xu(✉), Michael Hu, Jeffrey Powell, and Ziqing Zhuang

National Institute for Occupational Safety and Health, National Personal Protective Technology Laboratory, 626 Cochrans Mill Road, Pittsburgh, PA 15236, USA
mxu9@cdc.gov

Abstract. Firefighters are required to use a self-contained breathing apparatus (SCBA) for respiratory protection when engaged in a variety of firefighting duties. While the SCBA provides crucial respiratory support and protection, it is also cumbersome and heavy. Questions have arisen to address the current weight limitation listed in 42 CFR Part 84, which mandates the maximum weight of 16 kg (35.2 lbs) for an SCBA system. The objective of the overarching study is to inform a maximum comfortable SCBA weight by developing static and dynamic biomechanical models and collecting physiological burden and postural stability data to evaluate stressors for firefighting tasks. This paper presents data from the static models. Eight representative firefighting tasks (stair climb, hose carry, weighted carry, rope pull, dummy drag, hammer slam, overhaul, and hose advance) were used in the biomechanical models. Maximum comfortable SCBA weights were calculated based on three biomechanical models (abdominal extension/tension, lateral extension/tension, and abdominal rotation). This study found that the maximum comfortable weight for an SCBA was 39 lbs for the static postures of interest. Further studies with human subjects to investigate the SCBA weight limit by using physiological burden and dynamic models are underway and will provide crucial data to inform updates to the current weight limits for SCBAs allowing more features and extended durations of these devices without compromising user safety, performance, or comfort.

Keywords: Biomechanical modelling · Shoulder force · Lumbar lift force · SCBA load carriage · Firefighting hazard evaluation

1 Introduction

Firefighting is one of the most dangerous occupations that require intensive physical work in hazardous environment [1]. Firefighters work in varied and complex environments that potentially expose them to a number of hazards. To combat this, firefighters wear a turnout ensemble (i.e., protective coat and pants) and other personal protective equipment (PPE) such as a helmet, gloves, boots, and a self-contained breathing apparatus (SCBA) approved by NIOSH's Respirator Approval Program as meeting the requirements of 42 CFR Part 84 and certified by NFPA (National Fire Protection Association). Firefighters' PPE have been designed to provide protection against multiple hazards such as thermal

J. L. Wright et al. (Eds.): AHFE 2021, LNNS 264, pp. 231–238, 2021.
https://doi.org/10.1007/978-3-030-79763-8_28

threats (e.g. exposure to flame and excessive heat), toxic gas inhalation, and physical injuries (e.g. cuts, collisions, punctures, slips, falls, etc.). While the SCBA is instrumental in reducing the risk of exposure to toxic gases, it increases the load that a firefighter must carry and shifts the center of mass (COM) away from the firefighter's core, limiting the range of motion and decreasing overall gait performance [2]. The SCBA has also been attributed to increased physiological burden [3]. Firefighters were more likely to be injured during fireground operations than during any other duties. In 2019, 23,825 injuries, or 39% of all reported firefighter injuries, occurred at the fireground. The leading cause of fireground injuries was overexertion or strain (29%) [4].

The firefighting industry has seen an increase in the usage of larger (and consequently heavier) extended duration SCBA cylinders (45-min or 60-min). The addition of new safety technologies including the Personal Alert Safety System, Heads Up Display systems, and buddy lights into the SCBA has also been a factor in the overall increase in weight. While the SCBA provides crucial respiratory support and protection, it is also cumbersome and heavy. Questions have arisen to address the current weight limitation listed in 42 CFR Part 84 [5], which mandates a maximum weight of 16 kg (35.25 lb) for an SCBA. The question at hand, however, is by what means this additional weight will contribute to biomechanical variables. Research is needed to inform decisions regarding any updates to the current SCBA weight limit. The objective of the overarching study is to inform a maximum comfortable SCBA weight by (1) developing static and dynamic biomechanical models and (2) collecting physiological burden and postural stability data to evaluate stressors for firefighting tasks. The goal of the study is to utilize static biomechanical modelling of common firefighting tasks to estimate the comfortable SCBA weight limit. The static biomechanical modeling is a commonly used, low cost and quick approach for obtaining preliminary results for the research questions.

2 Methods

The methodology used in this study includes two parts: selection of firefighting tasks and development of three static biomechanical models based on the firefighting tasks of interest.

2.1 Firefighting Tasks

Eight common firefighting tasks were considered. For each task, one static posture was used in each of the models. The selected tasks included stair climb, hose carry, weighted carry, rope pull, hose advance, overhaul, hammer slam, and dummy drag and are shown in Fig. 1. Stair climb was defined as the subject climbing up the stairs. The static posture is where the left foot is at ground level and the right foot is on the first step. Both feet support the body. Hose carry was defined as the subject carrying a hose. The static posture is where the left foot is on the ground supporting the body and the right foot is just coming off the ground. A 50 ft. empty hose was modelled as weighing 30 lbs, with only a 70 in long section of the hose being lifted by the firefighter. This results in a 3.5 lbs load on the right hand at 45° from the horizontal. Weighted carry was defined as the subject holding weights with both arms. The static position is where the legs are mid stride and both arms are fully extended while carrying 45 lbs loaded on each hand, with only the right foot supporting the body. Rope pull was defined as the subject pulling a rope. The static posture is where the left leg is flexed and planted on the ground, while the right leg is planted on the ground and extended. The right and left arm are both holding the rope at chest level with 25 lbs loaded on the other end of the rope. The chest is perpendicular to the frontal plane and both feet are supporting the body. Dummy drag was defined as the subject dragging a dummy. The static posture is where both feet are planted on the ground and the arms are holding the dummy with the elbows bent at a 90°. The Dummy exerts 150 lbs of load on the subject's wrists and both feet are supporting the body. Hammer slam was defined as the subject slamming a hammer on a rubber tire. The static posture is where the feet are both planted on the ground, with the body facing perpendicular to the tire while both arms are holding the hammer. The hammer weights 20 lbs and both feet are supporting the body. Overhaul was defined as the subject pulling a load downwards. The static posture is where both feet are planted on the ground supporting the body and the right arm is holding the load at a higher position than the left. A 20 lbs load is applied to the hands at 45° from both horizontal and vertical. Hose advance was defined as the subject pulling a hose forward. The static posture is where both feet are planted on the ground supporting the body while the subject is crouched down. The arms are both holding the hose with 20 lbs resistance applied horizontally on the hands from performing the task. The torso is rotated 45°.

Fig. 1. Selected firefighting tasks (Photo credit: Illinois Fire Service Institute (IFSI) and Florida State University (FSU)).

2.2 Biomechanical Models

Maximum comfortable SCBA weights were calculated based on three biomechanical models. Two of the models in this study were developed to determine lateral extension/tension and abdominal rotation. The third model was abdominal extension/tension

[6]. From each model, final equilibrium equations were developed. According to a previous study [7], a comfortable SCBA weight results in a shoulder force and lumbar lift force less than 280 N and 135 N, respectively. Therefore, the maximum comfortable SCBA weights were determined using this threshold for each of eight representative static postures, which were shown in Fig. 1. Of the eight postures, stair climb, hose carry, weighted carry, rope pull, and dummy drag, had abdominal extension/tension. Hammer slam and overhaul had abdominal rotation. Hose advance had a complex posture that consisted of abdominal rotation, abdominal and lateral extension/tension. The weight of the SCBA was varied for each posture within the models, and the forces exerted on the shoulder and the lumbar lift forces were determined and compared to the comfortable limit.

Abdominal Extension/Tension Model

A biomechanical model previously developed by Kajaks [6] can calculate SCBA exerted forces that accounted for abdominal extension/tension. For a full description of this model please see Kajaks [6] and Pelot [7]. This model was used in this study to determine the abdominal extension/tension forces.

Lateral Extension/Tension Models

When a person laterally bends, the reaction force from the SCBA weight exerted on the shoulders are no longer equal and require additional calculations to determine the exact reaction force on each shoulder. Figure 2 shows a simplified view of the SCBA unit in the frontal plane. The straps are modelled as string tensions T_1 and T_2, and the SCBA belt is simplified to a point support at the bottom of the SCBA unit, we assumed that the location of this point support is unaffected with lateral bending. The support from the belt and back are represented by forces F_{Ly} and F_{Lz}. The weight of the SCBA unit is modeled as Fg, and angles θ_1 and θ_2 model the angles at which the strap extends from the SCBA unit, and θ_3 represents the angle at which the whole system is tilted at, compared to the horizontal. The length of the SCBA unit is d_1, and the vertical distance between the center of mass of the SCBA and the lumbar attachment point is d_2.

From Fig. 2, the following three static equilibrium equations were created:

$$\sum F_y = -T_1 cos\theta_1 + T_2 cos\theta_2 + F_g cos\theta_3 + F_{Ly} sin\theta_3 - F_{Lz} cos\theta_3 = 0 \quad (1)$$

$$\sum F_z = T_1 sin\theta_1 + T_2 sin\theta_2 - F_g sin\theta_3 + F_{Ly} cos\theta_3 + F_{Lz} sin\theta_3 = 0 \quad (2)$$

$$\sum M = F_g cos\theta_3 + T_2 cos\theta_2 (d_1) - T_1 cos\theta_1 (d_1) = 0 \quad (3)$$

After studying pictures of the connecting area between the straps and the SCBA unit, θ_1 and θ_2 were observed to be similar enough to be assumed to be the same and unified into θ, then T_1 and T_2 are solvable given F_{Ly}, F_{Lz}, F_g, θ_3, and θ. θ_3 was determined by the posture from the static image, and θ was determined from image analysis of a connecting point between the SCBA unit and the straps and was assumed constant throughout. F_{Ly} was assumed to be zero, and F_{Lz} was assumed to equal L_z from the abdominal extension/tension model where $\beta = 0°$, making this model an extension of the abdominal extension/tension model that assumes there is no postural bending in the coronal direction. As such, the first step of the lateral extension/tension model was to calculate shoulder force (S_z) and lumbar lift force (L_z) from the abdominal extension/tension model where $\beta = 0°$. Thus, the true inputs to this model were θ_3 and θ for the lateral model portion, and W and θ_1 from the abdominal model portion.

Since the only important effect of lateral extension/tension is the redistribution of weights on the shoulders, the output of this model should just be a force ratio between the forces applied to the shoulders by the SCBA straps (T_1 and T_2) that can be multiplied to or divided from S_z values from the abdominal extension/tension model, allowing the S_z's to account for lateral bending.

As a way to validate the model, T_1 and T_2 can be converted to S_{z1} and S_{z2} values via the following equation:

$$S_{z1,z2} = T_{1,2} sin\theta \quad (4)$$

One important note about this model is that it is very dependent on the unified strap angle (θ) value.

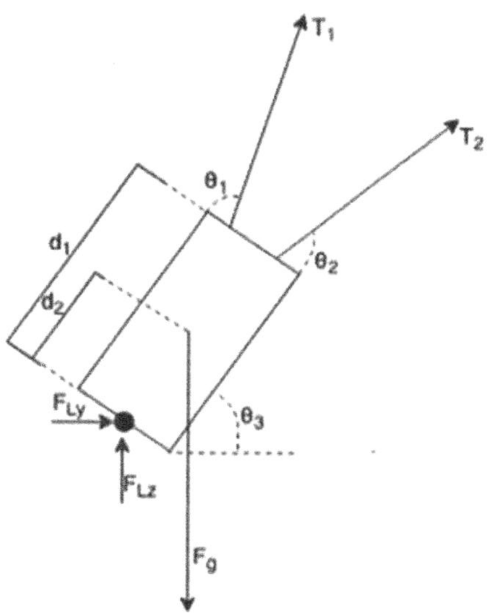

Fig. 2. Full-body diagram of the SCBA unit during lateral bending at angle θ3.

Abdominal Rotation Model

Once the static equilibrium models for both abdominal and lateral extensions/tensions were established, the static equilibrium model for abdominal rotation was created.

The abdominal and lateral extension/tension models have a sine and cosine relationship, with each model completely dominating the calculations at different degrees of rotation. Looking at the two abdominal and lateral extension models, SCBA weight is an input to both models and can be varied indpendent of the SCBA design, making it the ideal variable for which to apply the sine and cosine relationship. Thus, it was assumed the reaction forces exerted on the body by an SCBA unit in a posture with abdominal extension/ tension and rotation can be modeled by splitting the weight of the SCBA (Fg) into weight contributing to abdominal extensions/tensions (Fg_A) and weight contributing to lateral extensions/tensions (Fg_L), which are governed by the following equations:

$$Fg_A = Fg cos\theta_r \tag{5}$$

$$FgL = Fg sin\theta_r \tag{6}$$

Where θ_r is the rotation angle. Fg_A was then input into the abdominal extension/tension model with appropriate variables, and Fg_L is input into the lateral extension/tension model with appropriate variables. The forces calculated at each of the three contact points from both models are then summed together via the Pythagorean Theorem and then rotated based on the rotation angle, resulting in x-, y-, and z- components of the reaction forces applied to said posture with abdominal extension/tension and rotation.

In the case of the hose advance posture, the posture contains abdominal extension/tension, lateral extension/tension, and abdominal rotation. As such, it was assumed that once the reaction forces were adjusted for the combination of abdominal extension/tension and rotation, they only needed to be multiplied by a force ratio corresponding to the lateral bend angle to account for the lateral extension/tension.

3 Biomechanical Analysis Results

The maximum comfortable SCBA weights were calculated for each biomechanical model across postures. For the abdominal extension/tension model, the maximum comfortable SCBA weight was 39 lbs, for the abdominal rotation model, the maximum comfortable SCBA weight was 43 lbs, and for the three combination biomechanical models, the maximum comfortable SCBA weight was 40 lbs. See Fig. 3.

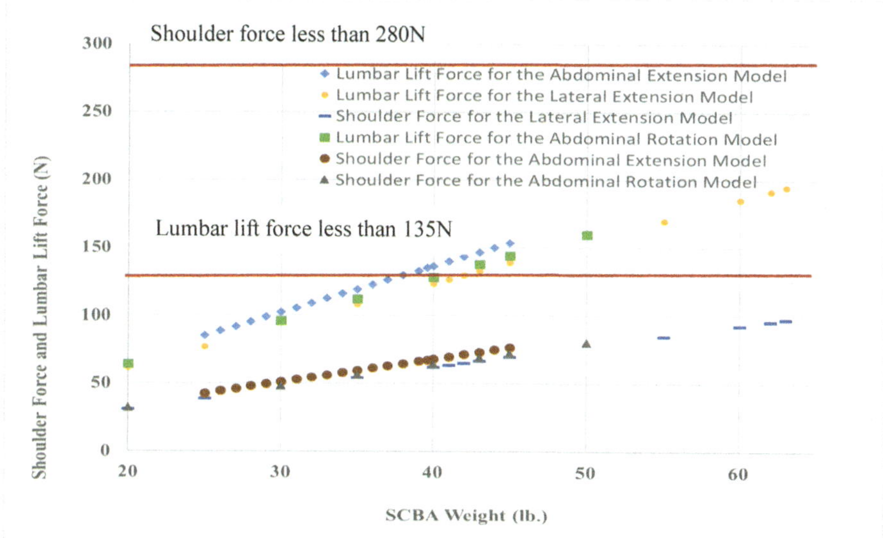

Fig. 3. Shoulder force and lumbar lift force vs SCBA weights at three models.

4 Discussion

This study was the second step of an overall research effort to understand and establish the maximum allowable weight that would prevent firefighter injury while maintaining comfort. The first step was to determine the maximum allowable weight by using a 3-Dimensional Static Strength Prediction Program (3DSSPP). Based on a biomechanical model, 3DSSPP calculated lower back (L4/L5) compression forces and the results were compared to the NIOSH guidelines. The maximum safe weight of an SCBA ranged from 35 to 75 lbs [8].

In this study, maximum comfortable SCBA weights were calculated based on three biomechanical models (abdominal extension/tension, lateral extension/tension, and abdominal rotation). Significant differences in maximum SCBA weights were found among tasks. The maximum comfortable weight for an SCBA was 39 lbs for the static postures of interest. This is heavier than the 35 lbs obtained from the 3DSSPP method.

Individual firefighter strength would impact postural stability and comfort as well as other physiological variables and was not accounted for in this study. This is a limitation of this study. Future studies are planned to investigate physical and psychophysical variables to support the evaluation of SCBA design and weight on firefighter stamina, comfort, and postural stability with human subjects. Subjects will perform firefighting activities while wearing a standard firefighting ensemble and SCBA (three different weights) in hot conditions. The human subject data will be used to compare with and validate the static and dynamic biomechanical models. These models will then be used to assess more scenarios to provide results to improve SCBA designs and develop guidance and recommendations on practical use during firefighting activities.

5 Conclusions

This study found that the maximum comfortable weight for an SCBA was 39 lbs for the static postures of interest. Further studies with human subjects to investigate the SCBA weight will provide crucial data to inform updates to the current weight limits for SCBAs allowing more features and extended durations of these devices without compromising user safety, performance, or comfort.

Acknowledgments. The authors thanks for Illinois Fire Service Institute (IFSI) and Florida State University (FSU) to support our study.

Disclaimer. The findings and conclusions in this report are those of the author(s) and do not necessarily represent the official position of the National Institute for Occupational Safety and Health, Centers for Disease Control and Prevention.

References

1. Coca, A., Williams, W.J., Roberge, R.J., Powell, J.B.: Effects of fire fighter protective ensembles on mobility and performance. Appl. Ergon. **41**(4), 636–641 (2010)
2. Park, K., Rosengren, K.S., Horn, G.P., Smith, D.L., Hsiao-Wecksler, E.T.: Assessing Gait Changes in firefighters due to Fatigue and Protective Clothing. Saf. Sci. **49**(5), 719–726 (2011)
3. Dorman, L.E., Havenith, G.: The effects of protective clothing on energy consumption during different activities. Eur. J. Appl. Physiol. **105**(3), 463–470 (2009). https://doi.org/10.1007/s00421-008-0924-2
4. NFPA's "United States Firefighter Injuries in 2019", Richard Campbell and Ben Evarts, Issued: November 2020
5. Self-Contained Breathing Apparatus U.S. Code of Federal Regulations (CFR) Title 42 Part 84 Chapter 1 Subpart H –84.89 Weight requirement
6. Kajaks, T.: Virtual ergonomics and gaming technology for posture assessment: from automotive manufacturing to firefighting. (Ph.D.), McMaster University (2017). http://hdl.handle.net/11375/21970
7. Pelot, R.P., Rigby, A., Stevenson, J.M., Bryant, J.T.: A static biomechanical load carriage model. In: Soldier Mobility: Innovations in Load Carriage System Design and Evaluation, pp. 25(1)–25(12). Kingston, ON: RTO MP-056 (2000)
8. Xu, S., Hu, M., Powell, J., Zhuang, Z.: Biomechanical modeling and 3D simulation of firefighting tasks. In: Cassenti, D.N., Scataglini, S., Rajulu, S.L., Wright, J.L. (eds.) AHFE 2020. AISC, vol. 1206, pp. 174–179. Springer, Cham (2021). https://doi.org/10.1007/978-3-030-51064-0_23

Perceptive Patient: Important Factors for Practical Emotion Sensing in Conversational Human-Computer Interactions and Simulations

Thomas B. Talbot[1(✉)], Matthew Hackett[2], and William Pike[2]

[1] Medical Virtual Reality Laboratory (MedVR) at the University of Southern California Institute for Creative Technologies, 12015 Waterfront Drive, Playa Vista, CA 90094-2536, USA
talbot@ict.usc.edu

[2] DEVCOM Soldier Center, 12423 Research Parkway, Orlando, FL 32826, USA
{matthew.g.hackett.civ,william.y.pike.civ}@mail.mil

Abstract. Much was learned in the development of an emotion sensing virtual standardized patient (VSP) which employs various inputs to obtain an emotional read and VSP-oriented judgment of the human participant during a simulated medical interview. Such a virtual patient changes behavior based on its sensing and alters its behavior and degree of honest disclosure to the human 'doctor' accordingly. We have discovered several important human factor and technology strategies to produce practical and impactful affective simulations. It is important to recognize overpromises of AI vendors and fallacies associated with affective computing. Essential technologies for affective applications include natural language understanding (NLU), computer vision, auditory processing of speech tonal characteristics and contextual assessments. Important factors for effective sensing include methods to identify context of the conversation in progress, a model of the VSP's emotional state and use of evocative stimuli to provoke an emotional reaction when conducting measurements.

Keywords: Affective computing · Computer vision · Emotion detection · Human factors · Human-systems integration · Human-computer dialogue · Medical simulation

1 Introduction

The MedVR laboratory has constructed several virtual human applications that employ emotion sensing AI for both clinical evaluations and medical simulations. One development, Perceptive Patient, is an emotion sensing virtual standardized patient (VSP) which employs various inputs to obtain an emotional read and VSP-oriented judgment of the human participant during a simulated medical interview. Depending on sensory inputs, the virtual patient alters its behavior and degree of honest disclosure to the human 'doctor' accordingly (aka user). We have discovered several important human factor and technology strategies to produce practical and impactful affective simulations [1].

J. L. Wright et al. (Eds.): AHFE 2021, LNNS 264, pp. 239–246, 2021.
https://doi.org/10.1007/978-3-030-79763-8_29

2 Fallacies of Emotion Sensing

It is important to understand the overpromises of AI service vendors and fallacies associated with affective computing. One fallacy is that detection of facial expressions (Facial Action Units/FAUs) is equivalent to detection of emotion. It is not, for a variety of reasons. One FAU or a combination of FAUs being detected only represents a signal that a certain facial expression is present on camera. Often, these FAUs do correspond to certain expressions of emotion, but they are also sometimes present just transiently or in other situations in which they may not be easy to interpret. A person could be smiling for a variety of reasons; they could be sharing a good moment or teasing or threatening. Interpreting that the user is 'happy' is an oversimplification. Perhaps they are nervous? An oral smile lacking eye musculature changes may be a 'fake' or forced smile as compared to being genuine. Lacking context of the conversation in progress and its recent history makes even detection of a smile a difficult proposition to interpret emotionally. Overall, the concept of 'easy' emotion sensing is a fallacy. There are additional sources of data to be considered, technical challenges to overcome and sophisticated scientific approaches to be employed before one may have confidence in user data revealing emotional information.

3 Essential Technologies for Affective Computing

Essential technologies for affective applications include natural language understanding (NLU), computer vision, auditory processing of speech characteristics and contextual assessments.

Natural Language Understanding. NLU provides the basis for conversational interaction, the ability to sense the context of the current conversation and the ability to perform linguistic analysis to detect emotional content. The first component of NLU, is typically speech recognition, which as of this writing has fairly good accuracy and is a widely available, if imperfect, technology.

A more critical component is the NLU classifier system. The classifier takes user speech acts and interprets them. Such speech acts could be greetings, insults, responses to computer questions or questions for the computer. The NLU classifier is the most critical component of a system like Perceptive Patient, providing the basis for computer responses and subsequent interactions. The classifier also provides conversational context. Once the system has a classified statement from the user, one can now calculate 1) the VSP response and the 2) conversational context.

It is also possible to conduct linguistic analysis of the user statement for additional context such as evaluating the emotional tone of the statement, word choices and use of exceptional words such as technical jargon or profanity. LIWC (Linguistic Inquiry & Word Count) [2–4] is a basic, yet popular tool for the collection of linguistic statistics from text. It provides statistical outputs regarding a set of text, which can be built up as a conversation progresses. LIWC analysis can output useful context cues such as positive emotion (love, delight, affection), negative emotion (anger, anxiety, sadness), certainty vs. tentativeness, and conversation topicality.

Conversational AI capability which governs VSP content and responses is an essential element to providing a simulated conversation. The design of the virtual agent and conversational repertoire is pre-authored and must provide a naturalistic conversation that flows well. Such conversations typically focus on either the user asking questions of the VSP or the converse, where the VSP/virtual agent drives the conversation. Systems that place the virtual agent as the driver have an advantage in affective computing as the agent drives the conversation, the conversation's context is automatically a known factor and there are numerous opportunities to expose the user to emotionally provocative questions and statements. These critical moments represent the optimal opportunity to glean data from the user.

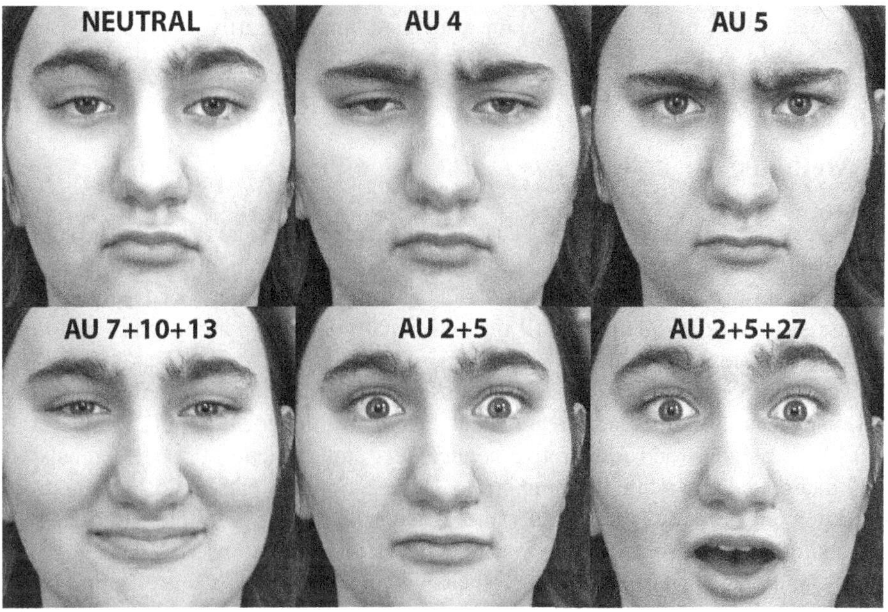

Fig. 1. Pictured above are the following Facial Action Units (FAUs/AUs) and combinations. AU 4 (Brow Lowerer) is associated with worry, confusion & concentrated effort. Adding AU 5 (Upper Lid Raiser) appears confrontational, serious or angry. AU 7 (Lid Tightener), AU 10 (Upper Lip Raiser) & AU 13 (Cheek Puffer) produce a genuine smile. AU 2 (Outer Brow Raiser) with AU 5 produces a surprised expression which is further accentuated if AU 27 (Mouth Stretch or AU 26 Jaw Drop) is added.

Computer Vision. Computer vision is employed to examine the user's face, for example; eye-contact between the user (Fig. 1) and the VSP character. This technique affords the opportunity to read facial action units (FAUs or AUs) which can reveal emotional expressions and reactions. Such reactions are interpretable when context is known. The Facial Action Coding System (FACS) was developed as a system to taxonomize human facial movements [4–6].

Auditory Processing. Auditory processing can analyze tone of speech and detect cheerfulness, exasperation, anger, and anxiety [7]. Affective qualities of the user's speech include the rate of speech, variability in speech frequency, speech energy (relative volume & forcefulness), fraction of speech vs silence while speaking, the fundamental frequency of speech and speech prosody [8]. It is also useful to evaluate user response time to provocative stimuli from the VSP. Delays typically indicate that a response requires thought or that the stimuli was emotionally provocative.

Contextual Assessments. Reading context is vital to tracking progression of the interview, as is the commission of appropriate or inappropriate acts during the simulation which provide additional context and basis for assessing judgment. Contextual assessments are based on an understanding of the current topic, performance in the current scenario and other measurements taking which are not necessarily based upon sensory data.

4 Methods

The Perceptive Patient is based on a custom technical architecture (Fig. 2). The starting point included a fully operational VSP system (USC Standard Patient) and an operational visual system to read FAU data from a webcam (using MultiSense or OpenFace 2.0) [9] and some custom additions. All the additional interpretive architecture had to be constructed from scratch.

The MultiSensing (Fig. 2) system obtains visual and auditory information including eye contact and FAUs. L1 signals are facial data in raw numerical form (Example: AU3 is 0.12), L2 signals add interpretive context (ex: smile fraction 0.3), L3 signals are fully interpreted (ex: while listening, a shocked reaction was present 25% of the time). The Interpreter processes L2 signals into L3 signals, adding in a time-based analysis and additional interpretation. The conversation is divided into user response to VSP, user formulating a question and user speaking, with the Cycle Controller informing the interpreter as to context of the conversation's current state. The Standard Patient system contains a full conversational VSP and handles natural language processing, VSP responses and assessment. The Judgement Engine updates a model of the VSP's affective impression of the human user on a turn by turn basis as the user-VSP conversation progresses.

A Working Approach. The Perceptive Patient interview runs in real-time and is interpreted by the system as a turn-based strategy game. Each question/response dyad is a considered a turn, at which point an assessment is recorded. Within each dyad, real-time data is collected regarding user appearance. These short-term impressions are recorded and minimized to "L3 signals" which summarize the emotional expression, level of attention and length of time such a signal was present.

Melding a Variety of Data. The signals employed in the Perceptive affective assessment includes more than just computer vision data. The primary input is based around

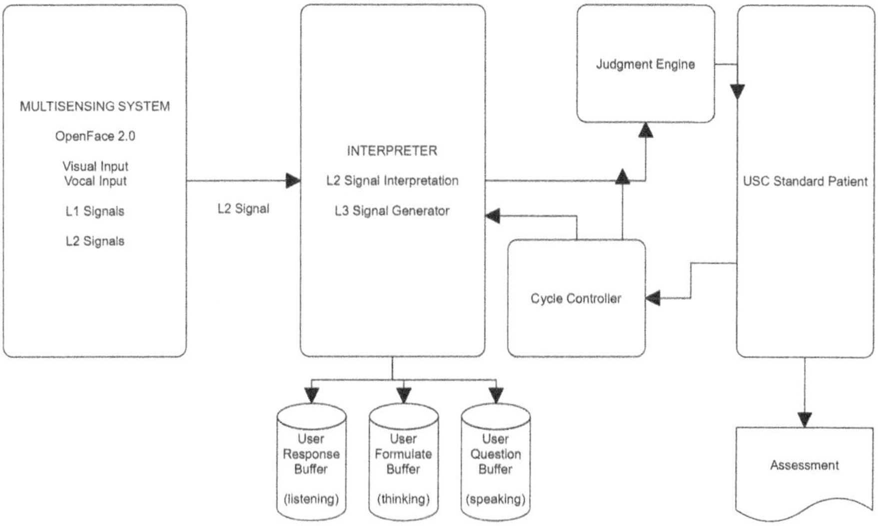

Fig. 2. Pictured above is the architecture employed by the Perceptive Patient affective virtual patient prototype.

the speech act. The actions and performance reflected by what the user says drive the scenario. The choice of questions represents on-target behaviors. The fact that the user greets the virtual patient, asks them questions about them as a person, asks open-ended questions and finally asks questions that correctly address the patient's concern are all on targe behavior which will positively impact the Judgment Engine's model. On target topical statements give the impression of authority. Practicing good social skills will improve likability.

Perceptive conducts a limited linguistic analysis. Training language statements are tagged as effective, ineffective, judgmental, profane or cheating (1–2 word questions such as "headache" or "depression"). This language tagging data can influence the likelihood of an honest response for sensitive questions and it can also affect user authority & likability.

Empathy is an important measurement. Serious conversations often evoke histories with concerning and upsetting information. When such a statement is made, the Perceptive system is on the alert for an empathetic response from the user. The user's performance in this regard will affect the judgment engine's compassion & likability variables. This is an example of using VSP statements to reflect provocative material off the user. Employing provocative statements allows the system to focus on the user's reaction and creates the opportunity to update the judgment model accordingly. Such probative approaches are more important in scenarios where the user is asking questions and the agent is responding, as user directed conversations are more challenging to assess than computer directed dialogue.

Finally, sensory signals are employed by the prototype. Eye contact is considered an important measurement of interest or distraction. Eye contact is most critical when the user should be listening to the patient, is moderately important when the user is speaking and should be disregarded during prolonged pauses when the user is formulating their next question to the VSP.

Interruption detection is an excellent tool to provide negative behavioral inputs to the judgment model. The prototype we've developed does not make use of voice analysis due to effort constrains and because there is very little literature on vocal parameters and conducting patient interviews.

Conversation Assessment. Assessments are conducted every turn and update the Judgment Engine. A transcript (Fig. 3) is appended which may then be presented upon conclusion of the scenario. When a new dyad is encountered, input from the Judgment Engine may inform the VSP how to respond to a sensitive question. If one such question is posed, the VSP will query a Trust variable which is formulated from a combination of impressions, which include a longitudinal assessment of the user's authoritativeness, likability & compassion. If the Trust variable remains under a certain threshold, the VSP will provide evasive or factitious answers. When the threshold is exceeded, the VSP will respond honestly.

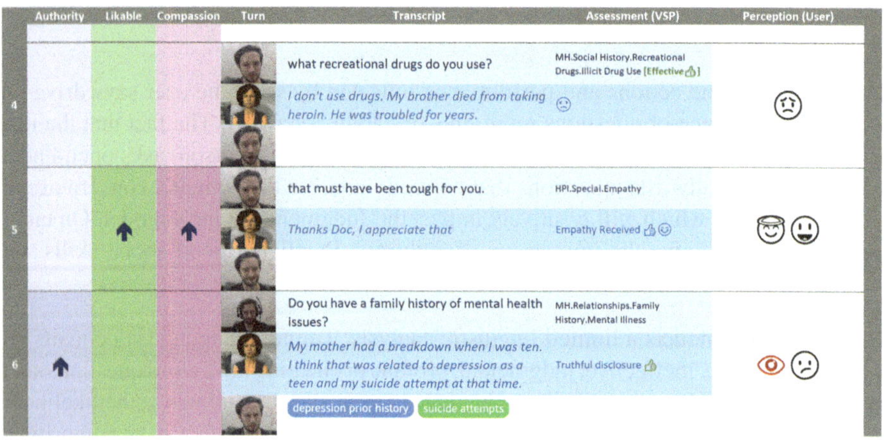

Fig. 3. Affective indicators are included in the after-action feedback conversation transcript. User facial expressions and eye contact levels are indicated by emojis. In this example, linguistic analysis (Question 4, leftmost column) indicates that the user asked a question in a very effective manner (direct, yet open an nonjudgmental). The response was met by an empathy probe by the VSP (Question 5), which the user passed by providing empathy. This improved the likability and compassion values in the VSP judgment model. In Question 5, the question about family mental health reveals desired information (with interview assessment objectives discovered listed in colored boxes). A positive VSP impression of the user at the time this question was asked yielded a truthful response as opposed to an evasion. The user had a concerned expression upon hearing this information and eye contact was reduced at that time.

5 Factors for Effective Emotion Sensing and Interaction

There are several important factors which contribute to effective emotion sensing in conversational interactions between humans and computers. The most important is to know the context of the conversation, which allows your sensing application to know the intention behind speech acts and context of user reactions to VSP. The next step is to build an emotion model which stores and modifies the VSPs impression of the human user it is interacting with as the interview progresses. Such models can be state machines and scoring systems (such as our judgment engine) or they could be sophisticated AI engines which employ user training data or other types of behavioral analysis and projections.

Useful techniques include virtual human characters that express emotion, both verbally and nonverbally through facial expressions and body movements. This, along with well written dialogue, can create a greater sense of engagement between the user and the VSP.

Another technique to enhance engagement and measure the emotional connection between the user and virtual agent is to address entrainment. Entrainment represents small reciprocal nonverbal behaviors that form a silent but emotionally important part of a conversation. With good entrainment, the human user will reciprocate changes in body posture, facial expressions and smiles; each will subtlety mirror the other. This sort of nonverbal interplay occurs largely unnoticed between people all the time. A proactive affective system can increase this entrainment with avatars that reflect the user's posture, and some select expressions. We can state that the first time a computer avatar smiles back at you in response to your own smile, you will experience a sense of delight very much like eliciting a smile of delight from an infant. The affective system's assessment of user entrainment is as telling and informative about the user as is observing any person's behavior around a baby.

Useful emotion detection design patterns include detection of user engagement and user-VSP entrainment, human reaction detection following provocative stimuli and supporting use of nonvisual indicators of emotional perception such as simulation performance or linguistic analysis of user statements.

It is essential to understand the purpose behind attempting to sense emotional context of a conversational simulation so one is able to guide interactions towards detectable pathways in the conversation which provide emotional stimulus, necessary provocation, technology to detect reactions and an assessment logic to decode the reaction and modify the VSPs impression of the human and how they interact with him going forward [10].

6 Conclusion

The communication between a clinician and the patient is critical to assessing a patient's well-being and to determining a differential diagnosis. However, without training appropriate bedside manner and interpersonal skills, a healthcare provider is ill-equipped for these tasks. Affective computing enables providers to practice these skills with emotive patients, enabling risk-free repetition to improve these soft skills. Further, a perceptive and emotive patient also increases realism of the scenario, improving user immersion and potentially engagement as well.

Virtual patients represent the most recent area in which we have employed affective computing. We have also made use of affective computing for conducting psychological evaluations for depression [10], training for public speaking performances and to advance the science of affective computing and virtual human technology. The number of potential applications is probably quite large. In order to provide practical affective applications which function reliably and make affective technology inclusion worthwhile it is important for the technology to continue to advance with more robust cloud-based AI tools and to employ state of the art methods to make use of sensory data along with the all-important measures to determine conversation context. Future affective computing applications will likely be easier to create, work online and in mobile settings and include a built-in repertoire of emotional reactions and techniques as part of a virtual human agent behavioral repertoire.

Acknowledgments. The project or effort depicted was or is sponsored by the U.S. Army Research Laboratory (ARL) under contract W911NF-16-C-0034, and that the content of the information does not necessarily reflect the position or the policy of the Government, and no official endorsement should be inferred.

References

1. Talbot, T.B., Hackett, M.: Perceptive patient: a multimodal sensing virtual standardized patient experiment to create an emotionally judging ai-based patient responsive to non-verbal behaviors and other affective indicators. International Meeting on Simulation in Healthcare (2020)
2. Hamm, J., Kohler, C.G., Gur, R.C., Verma, R.: Automated Facial Action Coding System for dynamic analysis of facial expressions in neuropsychiatric disorders. J. Neurosci. Methods **200**(2), 237–56. doi:https://doi.org/10.1016/j.jneumeth.2011.06.023. PMC 3402717. PMID 21741407
3. Tausczik, Y.R., Pennebaker, J.W.: The psychological meaning of words: LIWC and computerized text analysis methods. J. Lang. Soc. Psychol. **29**(1), 24–54 (2010). https://doi.org/10.1177/0261927X09351676
4. Ekman, P., Friesen, W.: Facial Action Coding System: A Technique for the Measurement of Facial Movement. Consulting Psychologists Press, Palo Alto (1978)
5. Tian, Y., Kanade, T., Cohn, J.F.: Recognizing action units for facial expression analysis. IEEE Trans. Pattern Anal. Mach. Intell. **23**(2), 97–115 (2001)
6. Pantic, M.: Machine analysis of facial behaviour: naturalistic and dynamic behavior. Philosoph. Trans. Roy. Soc. B: Biol. Sci. **364**, 3505–3513 (2009)
7. Giannakopoulos, T.: pyAudioAnalysis: an open-source python library for audio signal analysis. PLoS ONE **10**(12), e0144610 (2015). https://doi.org/10.1371/journal.pone.0144610
8. Tawari, A., Trivedi, M.M.: Speech emotion analysis: exploring the role of context. IEEE Trans. Multimedia **12**(6), 502–509 (2010). https://doi.org/10.1109/TMM.2010.2058095
9. Baltrusaitis, T., Zadeh, A., Lim, Y.C., Morency, L.: OpenFace 2.0: facial behavior analysis toolkit. In: 13th IEEE International Conference on Automatic Face & Gesture Recognition Xi'an, 2018, pp. 59–66, (2018). https://doi.org/10.1109/FG.2018.00019
10. Stratou, G., Morency, L.P.: MultiSense—context-aware nonverbal behavior analysis framework: a psychological distress use case. IEEE Trans. Affect. Comput. **8**(2), 190–203 (2017). https://doi.org/10.1109/TAFFC.2016.2614300

Effects of Non-driving Task Related Workload and Situational Awareness in Semi-autonomous Vehicles

Shruti Amre[1] and Ye Sun[2](✉)

[1] Department of Cognitive and Learning Sciences, Michigan Technological University, 1400 Townsend Drive, Houghton, MI 49931, USA
smamre@mtu.edu
[2] Department of Mechanical Engineering – Engineering Mechanics, Michigan Technological University, 1400 Townsend Drive, Houghton, MI 49931, USA
yes@mtu.edu

Abstract. Recent advances in autonomous and semi-autonomous vehicles have resulted in more efficient vehicle maneuvers, which causes drivers to be less involved in operational driving, thus leading to increased drivers' engagement in other non-driving related activities. High levels of automation arise concerns such as a decrease in situational awareness (SA) and driving performance as drivers are likely to undertake a non-driving related task during vehicle maneuvers. For cars with conditional or higher automation (SAE Level 3 or up), there is a need to study and monitor drivers' level of engagement/involvement with the vehicles. Considering the key factors for safety when drivers perform a non-driving related task (NDRT) during conditional automation, issues concerning drivers' SA and takeover performance in absence or failure of automation need to be addressed. In this study, we designed and conducted driving simulator experiments to investigate drivers' SA while they were performing a NDRT during the conditional automated driving. We also investigated drivers' takeover time and takeover quality in different NDRTs. We found that drivers' vigilance levels are different while performing different NDRTs during semi-automation. Our results show that the response time was typically between 5–8 s and the deviation of distance to the center of the lane ranged from −0.66 m to −0.3 m which might suggest the variability in takeover quality during different NDRTs.

Keywords: Autonomous vehicles · Situational awareness · Conditional automation · Takeover time · Takeover quality

1 Introduction

1.1 Background

Human errors caused 94% of all traffic crashes in the United States according to the data from the National Highway Traffic Safety Administration (NHTSA) [1]. With numerous new scientific advancements in the automotive industry in recently years, understanding

J. L. Wright et al. (Eds.): AHFE 2021, LNNS 264, pp. 247–254, 2021.
https://doi.org/10.1007/978-3-030-79763-8_30

human factors in autonomous or semi-autonomous vehicles is significant. The Society of Automotive Engineers (SAE) International has defined six levels of automated driving systems with Level 0 referring to no automation and Level 5 referring to full automation - commonly referred to as self-driving vehicles [2, 3]. Drivers are actively engaged in vehicle maneuvers in Level 1 and Level 2 vehicles whereas for Level 3 or higher ones, the vehicle can handle certain or all parts of trips on its own and is normally consider as autonomy or semi-autonomy. Level 2 automation combines two or more functions to relieve driving, such as adaptive cruise control and automatic emergency braking. However, drivers are still fully engaged with vehicle maneuvers. In Level 3 automation or up according to SAE definition, although drivers are still seated in the driver's seat, they are not actually driving when the automation features are being used. According to the NHTSA, Level 3 and higher are qualified as automated driving systems (ADS) [4]. They can essentially drive themselves, however, drivers can intervene and take control as needed. While automated vehicles have high potentials to save lives and reduce injuries for both drivers and passengers as well as bicyclists and pedestrians due to the advanced technologies, it is important to understand driver's role and interaction with vehicle systems [5, 6]. High levels of automation arise concerns such as a decrease in situational awareness (SA), driving performance with drivers most likely to undertake a non-driving related task [7]. In recent years, there have been several accidents involving Level 2 or 3 semi-autonomous vehicles, which were mainly caused by driver distraction on non-driving related tasks (NDRTs) during conditional automation when the advanced driver assistive features were being used [8].

Therefore, the goal of this study is to understand drivers' SA in performing NDRTs during driving semi-autonomous vehicles and study takeover time and takeover quality in conditional automation condition.

1.2 Literature Review

In order to design solid experiments for measuring SA, takeover time and quality for NDRTs, we first summarized the state-of-the-art measuring strategies and experimental designs. A review of existing driver control transitions to assess takeover time was conducted in 2017 [9] in which the authors also conducted their own study to determine automation to manual takeover times while the participants were reading newspapers or monitoring the system. They found that the total time to take over ranged between 1.97–25.75 s [9]. A study conducted in [10] highlighted age as a factor influencing takeover performance in automated vehicles. The study included younger drivers $<$ 28 years of age and older drivers $>$ 60 years of age that participated in the driving simulator study involving two conditions - with and without a secondary task (a cellphone conversation) in different traffic densities (low, medium and high). They measured the quality of takeover as a measure of maximum longitudinal or lateral accelerations and time to collisions. Gold et al. [11] conducted a study measuring takeover time in which they measured two different takeover times with a highly engaging secondary task (i.e., placement of tablet was different for two conditions, center console and handheld). They used two degrees steering wheel angle and 10% breaking pedal positions to measure reaction times after the automation alarmed the user to take over. They found that the reaction time was between 5–7 s. They used maximum acceleration after takeover as a measure of the quality of takeover between the experimental and baseline conditions. The results did not show any significant differences between the two experimental conditions, however, they twice as high as the baseline condition for participants with reaction time of 5 s and thrice as high for participants with 7 s of reaction time.

Another study by Louw et al. [6] employed a monitoring condition versus a secondary task using an iPad during automation to determine driver reaction in the form of takeover performance. They used maximum lateral and longitudinal acceleration, time to first steer and time to lane change as measures to study performance. Driver reactions to avoid collisions were measured using steering, and braking. They found that regaining vehicle control was challenging for participants who were distracted during automation deeming them "further out of the loop". de Winter et al. [12] demonstrated that it is crucial for drivers to be mindful while using automation until automation is perfected. They used different levels of automation alongside a secondary task to determine driver workload and response time to an intervention. They used NASA-TLX scores on the divided attention task along with mean speed, maximum speed and standard deviation of lateral position to measure driving performance. Results varied depending on the level of automation. Zeeb et al. [13] used writing emails, reading a new text and watching a video clip as levels of engagement on a secondary task. Participants had to take over driving while engaging in secondary tasks. Besides having people perform tasks while taking over, the study also uses an absolute steering wheel angle velocity > 0.075 degrees per second, absolute steering wheel acceleration > 5 degrees per second as a measure of takeover performance. The brake pedal position was similar to other studies. They used deviation from the center of lane and lateral acceleration over the periods of 5 and 10 s.

2 Methods

2.1 Overview

Most drivers perform non-driving related tasks such as using their cellphones, using the in-vehicle display system, talking to passengers, etc. while driving. These secondary tasks induce workload, thereby, decreasing human performance. In the context of highly automated vehicles, drivers are more likely to be involved in non-driving related secondary tasks. Understanding driver involvement in non-driving related tasks and workload becomes critical in the situations involving drivers resuming control of vehicles - referred to as "takeover". Thus, in this study, we are using a driving simulator to investigate driver SA, takeover time and quality while they are performing non-driving related tasks. Three scenarios were using in the experiments, baseline, checking Google Map, and playing cellphone games during semi-autonomous driving and transition, which are corresponding to low, moderate, and high workload. Additionally, we designed a survey to measure SA and proposed our methods for takeover measurement. The workload was measured by using a perceived workload self-rating scale.

2.2 Participants

Twenty-three subjects (eleven females and twelve males) from within the Michigan Tech community volunteered to participate in this study. They were in the age group of 18–30 years old (mean age = 23.43, SD = 2.35). All participants were experienced drivers with a valid driver's license and at least a year of driving experience. The study was approved by the Michigan Tech Institutional Review Board. All the subjects were screened for simulator sickness prior to the study.

2.3 Experiment

The objectives of this study were to determine SA while the drivers performed a NDRT in a semi-autonomous vehicle simulated environment and investigate their takeover time and takeover quality during transition. For measuring the SA, we designed a SA survey the participants need to answer questions pertaining to objects presented in the simulation to determine vigilance during the virtually simulated drive. Additionally, we used NASA-TLX [15], a task load self-rating scale to determine perceived workload. Our research questions are as follows:

- Are drivers vigilant enough to be aware of their environment while performing a non-driving related task in a vehicle involving conditional automation?
- How do different NDRT secondary tasks impact takeover quality and takeover performance during driver request takeover intervention from autopilot mode of conditional automation?

Table 1 Description of variables

Variables	Description
NASA-TLX	Measures workload in six subjective subscales: mental demand, physical demand, temporal demand, performance, and frustration recorded for game and navigation experimental conditions
SA	A newly designed survey with questions about road sign and stationary object recognition
Takeover time	Reaction time measured in seconds from the end of conditional automation to steering wheel angle > 2°
Takeover quality	Offset deviation from the center of the lane after takeover from conditional automation

The experiments consisted of a baseline condition and two non-driving related tasks - using Google Map and playing cellphone games - during semi-autonomous driving and transition, which were corresponding to low, moderate, and high workload conditions. Each participant performed the NDRTs during conditional automation. Both tasks were administered on an iPhone 6 s. Participants were made aware that they were expected to be vigilant of their surroundings while performing the NDRTs. The order of these tasks was counterbalanced. The dependent variables measured in this study were workload, SA, takeover reaction time and takeover quality. The detailed description is listed in Table 1.

2.4 Materials

Driving Simulator and Scenarios. We used the NADS miniSim in our study. NADS miniSim is a PC-based Quarter Cab driving simulator which provides a horizontal FOV 138 to 180° and a vertical FOV 27 to 50° [14]. The ISAT and miniSim software were used to create the scenario design and run simulations respectively. We used the ISAT to create a basic rural setting involving curvatures and passing traffic as seen in Fig. 1. In the scenario, the semi-autonomous driving and transition triggers were embedded. For the three experiments of baseline, navigation, and gaming, we modified the exception of locations of road signs and other objects like runners, deer, and dogs along with the specifics of conditional automation to minimize learning effects.

Table 2 SA survey with multiple choices

Scenarios	Survey questions
Baseline	How many white cars did you see while driving? a) 1 b) 3 c) 5 d) 0 How many deer did you see while driving? a) 1 b) 3 c) 5 d) 0 Did you see a person running on the side of the road? a) yes b) no Have you seen this sign while driving? a) yes b) no
NDRTs (Navigation and gaming)	Have you seen this sign while driving? a) yes b) no How many deer did you see while driving? a) 1 b) 3 c) 5 d) 0 How many white cars did you see while driving? a) 1 b) 3 c) 5 d) 0 Did you see a person running on the side of the road? a) yes b) no

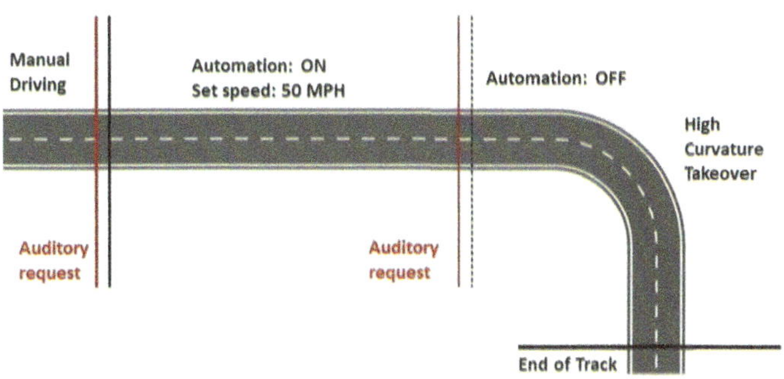

Fig. 1. Driving scenarios with conditional automation takeover model.

SA Surveys. We designed the SA surveys which included questions about road signs and stationary objects presented in the scenario while the participants were engaged in the non-driving related tasks. There were three surveys for the three conditions respectively. The questions on the game and navigation experimental conditions were the same but in a different order. The survey is shown in Table 2.

2.5 Procedure

After signing the informed consent form, participants were asked to fill out basic demographic information. Participants had to clear the simulator sickness test which involved manual driving in the simulator for five minutes in a basic city scenario. Following this, they were assigned to the baseline condition in which the conditional automation starts and stops. Participants began with manual driving until conditional automation took over. Participants were explicitly asked to sit back and be aware of the surroundings while the car was in autonomous mode. They were then asked to fill out the baseline condition SA survey. Then, depending on the order of assignment, participants either performed the game or the navigation tasks as discussed in the above sections. At this point they were engaged in NDRT in which they either played a game of 2048 or look up a list of places on Google Maps application until they heard an auditory stimulus intervention demanding them to takeover driving. Following each of these tasks, participants were asked to fill out the respective SA surveys. The entire experimental procedure lasted for about 30 min.

3 Results

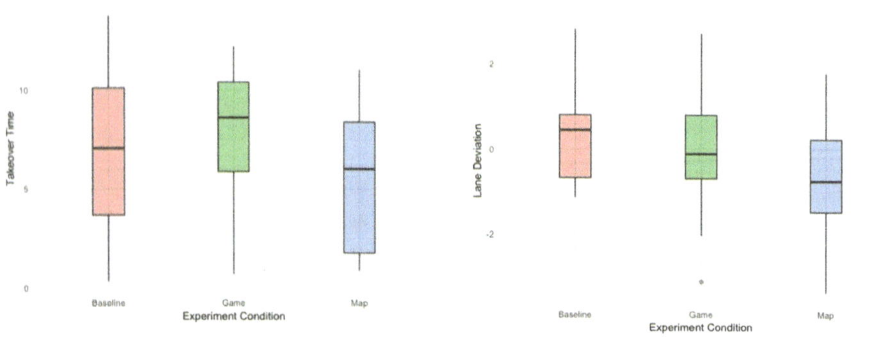

Fig. 2. Takeover time calculated from the moment automation stops to an increase in steering wheel angle > 2° and takeover quality measured as lane deviation analyzed for 20 s after automation shuts off

Data were analyzed using the R studio software. While visually analyzing the lane deviations obtained from the driving simulator, we found that two participants were driving on the left side of the road and others had missing data points. Only 16 participants were considered for this analysis due to missing data. Data was submitted to a repeated-measures analysis of variance (ANOVA) (Fig. 2).

We found that people took over faster in the map navigation condition compared to the gaming and baseline conditions as seen in Table 3. The gaming condition was more engaging in comparison to the other two conditions. A repeated-measures ANOVA revealed no significant effect of the type of task on workload. This may suggest that either game or navigation conditions failed to capture the change in workload.

Table 3. Results of takeover time

Experimental condition	Mean (s)	Standard Deviation (s)
Baseline	6.80	4.38
Game	7.41	4.08
Navigation	5.47	3.89

We also measured overall workload using the NASA-TLX at the end of the drive. We did not find any significant differences for workload across either the gaming or the navigation condition with $F = 0.956$ and $p = 0.336$. Our assumption was that the gaming condition would be more engaging, thus, making the task of takeover challenging. We did find slower response times for takeovers in gaming condition in comparison to the navigation condition which suggests that the participants were immersed in the task. However, the workload results do not reflect this assumption. We measured the SA following the drive and we did not find any significant differences across the baseline, gaming or the navigation condition $F = 3.106$, $p = 0.0547$. The participants were likely not very aware of their surroundings while they were engaged in the NDRTs despite being asked to be aware of the driving environment before the start of their respective session.

4 Discussion and Conclusion

Considering takeover time, our results were similar to previous findings reported in the literature review section of this paper. Our study reported the mean takeover time of an average 5 s for the navigation condition and 7 s for the gaming condition. The gaming condition was designed to be more engaging and hence, reflects in slower takeover times in comparison to the navigation condition. The faster response times in the navigation condition may be due to the fact that most current drivers regularly use one or the other form of navigation during day-to-day driving tasks and hence, the participants may have had experience with managing take-over from current technology equipped in most vehicles such as the cruise control.

Considering workload and the SA assessment, we do not see major differences between the game and maps tasks. This may have happened due to the fact that there was no element of driving involved while performing the game and map tasks. Participants were entirely focused on the secondary task, which was either gaming and navigation. One of the reasons we did not see major differences between the groups maybe due to the fact that our study was under powered.

In conclusion, we found the SA changed as the engagement level of the non- driving related task changed. Future work will involve analyzing correlations, and driving simulator data using multidimensional scaling techniques and hierarchical clustering to analyze driver takeover performance at different time intervals after the takeover from the autonomous mode. It will also be worthwhile to compare driving simulator data with the surveys to establish relevance.

References

1. NHTSA: Critical Reasons for Crashes Investigated in the National Motor Vehicle Crash Causation Survey. DOT HS 812 115. Washington, DC: National Highway Traffic Safety Administration (NHTSA), U.S. Department of Transportation (2015)
2. National Highway Traffic Safety Association: Federal Automated Vehicles Policy: Accelerating the Next Revolution in Roadway Safety (2016)
3. S.A.E.: Definitions for Terms Related to Driving Automation Systems for On-road Motor Vehicles (j3016) (2016). https://saemobilus.sae.org/content/J3016_201806/
4. Merat, N., Jamson, A.H., Lai, F.C., Daly, M.C., Carsten, O.M.: Transition to manual: driver behaviour when resuming control from a highly automated vehicle. Transp. Res. Part F: Traffic Psychol. Behav. **27**, 274–282 (2014)
5. Liu, P., Yang, R., Xu, Z.: How safe is safe enough for self-driving vehicles? Risk Anal. **39**, 315–325 (2019)
6. Louw, T., Merat, N., Jamson, A.: Engaging with Highly Automated Driving: To Be or Not To Be in the Loop? (2015)
7. National Transportation Safety Board: Collision between a Sport Utility Vehicle Operating with Partial Driving Automation and a Crash Attenuator (2020). https://www.ntsb.gov/news/press-releases/Pages/NR20200225.aspx
8. Naujoks, F., Purucker, C., Wiedemann, K., Neukum, A., Wolter, S., Steiger, R.: Driving performance at lateral system limits during partially automated driving. Accid. Anal. Prev. **108**, 147–162 (2017)
9. Eriksson, A., Stanton, N.A.: Takeover time in highly automated vehicles: noncritical transitions to and from manual control. Hum. Factors **59**, 689–705 (2017)
10. Körber, M., Gold, C., Lechner, D., Bengler, K.: The influence of age on the take-over of vehicle control in highly automated driving. Transp. Res. Part F: Traffic Psychol. Behav. **39**, 19–32 (2016)
11. Gold, C., Damböck, D., Lorenz, L., Bengler, K.: Take Over! How long does it Take to Get the Driver Back into the Loop? In: Proceedings of the Human Factors and Ergonomics Society Annual Meeting, vol. 57, pp. 1938–1942 (2013)
12. de Winter, J., Stanton, N., Price, J.S., Mistry, H.: The effects of driving with different levels of unreliable automation on self-reported workload and secondary task performance. Int. J. Veh. Des. **70**, 297–324 (2016)
13. Zeeb, K., Buchner, A., Schrauf, M.: Is take-over time all that matters? the impact of visual-cognitive load on driver take-over quality after conditionally automated driving. Accid. Anal. Prev. **92**, 230–239 (2016)
14. https://www.nads-sc.uiowa.edu/minisim/
15. Hart, S.G., Staveland, L.E.: Development of NASA-TLX (Task Load Index): results of empirical and theoretical research. Adv. Psychol. **52**, 139–183 (1988)

Sensing to Learn: Deep Learning Based Wireless Sensing via Connected Digital and Physical Experiments

Lan Zhang[1(✉)], Xianhao Chen[2], Yawei Pang[3], and Xiaoyong Yuan[1]

[1] Michigan Technological University, Houghton, MI 49931, USA
{lanzhang,xyyuan}@mtu.edu
[2] University of Florida, Gainesville, FL 32611, USA
xianhaochen@ufl.edu
[3] Nanjing University of Information Science and Technology, Nanjing 210094, China
003209@nuist.edu.cn

Abstract. With the advancement of wireless technologies and sensing methodologies, wireless sensing empowers wireless hardware with the additional ability to learn the target location, activity, gesture, and vital signs. By analyzing the target's influence on surrounding wireless signals, deep learning-based wireless sensing has attracted great attention due to its excellent performance in extracting discriminative sensing patterns. Nevertheless, it is labor-intensive to generate massive amounts of data for deep sensing study. Most existing efforts are focused on better exploiting the acquired sensing information, which, however, can hardly address the knowledge limitation fundamentally. In view of this, we propose an innovative learning framework by strategically integrating digital and physical experiments, alleviating data collection's intensive efforts. Specifically, one adaption module is introduced to connect the digital simulator and field tests for producing high-quality digital sensing data. This framework is expected to enable automatic, reliable, and efficient sensing data generation for future wireless sensing studies.

Keywords: Deep learning · Digital simulator · Experiment · Field test · Machine learning · Wireless sensing

1 Introduction

Wireless technologies have achieved a great success in data communications over the past few decades. The ubiquitous availability of wireless signals nowadays has simulated a surge in wireless sensing analysis [1–5], which braces a wide range of compelling applications, such as the human-machine interactions on smart home appliances, health monitoring, elderly care, and safety surveillance. These applications are enabled by active researchers from different perspectives, including pinpointing the positions of the target, tracking or tracing the target's trajectory, recognizing the regular or specific pattern of the targets (e.g., indoor occupancy monitoring or gait detection) [1]. Comparing

J. L. Wright et al. (Eds.): AHFE 2021, LNNS 264, pp. 255–262, 2021.
https://doi.org/10.1007/978-3-030-79763-8_31

with conventional sensing technologies, such as those based on wearable sensors or surveillance cameras, wireless sensing via radio frequency (RF) hardware provides a convenient, cost-effective, and non-intrusive approach with inherent privacy benefits.

In essence, wireless sensing can be modeled a pattern recognition problem by monitoring the target's influence on surrounding wireless signals [2–4]. Due to the "multipath" nature of wireless transmissions, signals sent from the transmitter will bounce off the surrounding environments, like wall, roof, furniture, and human body, leading to distinct patterns of the received signals regarding the sensing environments [3–6]. Conventional machine learning techniques, such as support vector machine (SVM) and k-nearest neighbor (KNN), have been used to address such sensing pattern recognition, which, however, can hardly handle complicated tasks due to the limited capability of feature extraction from noisy wireless signals. Fortunately, recent years have witnessed the success of deep learning, allowing machines to learn automatically from complicated and massive amounts of data [2, 7]. Nevertheless, it is by no means trivial to train such deep sensing models. First, due to the large parameter size of deep neural networks (DNN), massive sensing data is required for both training and testing. Second, due to the complexity of wireless sensing demands, a well-established deep sensing model may need to be updated or retrained for reliable performance in different scenarios or applications. Such labor-intensive efforts, especially in labeling-based supervised learning, significantly restrict deep sensing from practical implementation. Moreover, most existing efforts are focused on better exploiting the acquired sensing knowledge, such as from similar sensing tasks or virtual sensing data [1–3], which can hardly address the knowledge limitation for deep sensing study thoroughly.

In view of this, we propose an innovative learning framework by strategically integrating digital experiments with the labor-intensive field tests for automatic, efficient, and reliable deep sensing applications. Specifically, digital experiments will be conducted in the surrogate field test scenario via digital simulators, such as Ray-tracing software [8–11]. By leveraging Ray-tracing techniques, digital experiments can capture the multipath propagation effects in customizable environments. One challenge here is how to construct a desired digital environment to mimic the field test scenarios. To address this issue, we devise an adaptation module by observing the field test environment to automatically generate the corresponding digital map, facilitating digital sensing data generation with high quality. In the following, we first present a general deep learning-based wireless sensing workflow, elaborating key challenges of the deep sensing processing in Sect. 2. Section 3 introduces the proposed learning framework, including two experiment modules and one adaption module, to facilitate sensing data collection efforts. Section 4 finally concludes this paper.

2 Deep Learning Based Wireless Sensing

A general workflow of deep learning-based wireless sensing is illustrated in Fig. 1. As a typical deep learning process, this workflow is mainly composed of three critical phases: sensing data collection, deep model construction, and implementation. The three phases interact with each other, facilitating a broad range of wireless sensing applications.

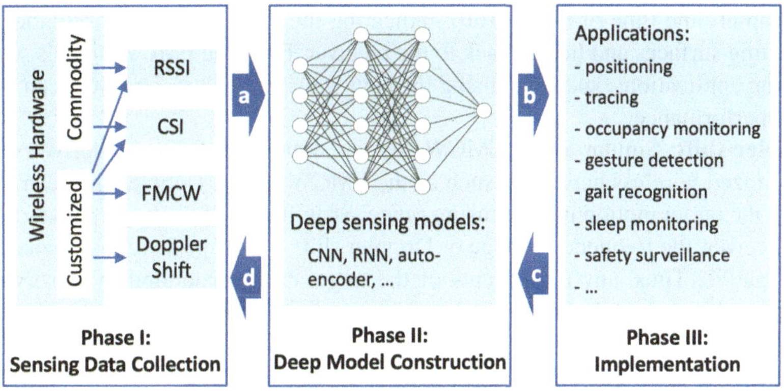

Fig. 1. Workflow of deep learning-based wireless sensing.

2.1 Phase I: Sensing Data Collection

Until now, wireless sensing hardware can be roughly divided into commodity devices and customized hardware. Like the commercial off-the-shelf (COTS) WiFi products, commodity devices have a great potential to be pervasively adopted in many of our daily applications. While when handling more challenging tasks, such as differentiating multiple targets even behind the wall [1–3], customized hardware (e.g., universal software radio peripheral (USRP)) can provide better sensing granularity. For different types of wireless sensing hardware, the extracted physical layer properties from wireless signals are quite different; different wireless sensing applications may require different physical layer properties as well. In the following, we briefly introduce four commonly implemented wireless properties.

- **Received signal strength indicator (RSSI)**: Almost all wireless hardware can obtain the radio signal strength (RSS) that indicates the path loss of a wireless signal regarding the propagation distance. One representative RSS-based sensing application is positioning. Research has shown that human body can introduce signal attenuation within a sensing area, leading to the variation of RSSI. However, the RSSI can only provide coarse-grained sensing information, whose stability also leads to unreliable sensing performance in many sensing applications.
- **Channel state information (CSI)**: CSI can capture the combined propagation effect, like scattering, fading, and attenuation with distance. Due to the multipath nature of wireless transmissions, CSI, i.e., the amplitude and phase information, can provide fine-grained wireless sensing, such as the object tracing or tracking. Unlike the widely applicable RSSI, CSI can be obtained only from a few commodity devices with modified driver, such as Intel 5300 network interface controller (NIC) [13]. With more compelling wireless sensing applications, this situation may be changed in the near future.
- **Frequency modulated carrier wave (FMCW)**: To gain more precise sensing performance, researchers have utilized customized wireless hardware, such as USRP defined radio platform, to capture wireless properties, e.g., FMCW signals. FMCW

can capture the time of flight (ToF), indicating the time used from the sender to the reflecting surfaces and finally back to the receiver. By using FMCW signals, wireless sensing applications, such as sensing through walls and gait recognition, can achieve great performance.

- **Doppler shift**: Similar as the FMCW signals, doppler shift can be derived only by customized wireless hardware, such as the FMCW Radar. Doppler shift can characterize the target motion in the sensing range. Specifically, if the target moves towards the receiver, the frequency change or Doppler shift will be positive; otherwise, it will be negative. Thus, any movements of the target can be captured by observing the Doppler shift, which has been used to detect walking or running activities.

2.2 Phase II: Deep Model Construction

As one of the most successful artificial intelligence (AI) technologies, deep learning has attracted great attention due to its excellent feature extraction and analyzing capability [7]. Various deep learning models, such as convolutional neural network (CNN), recurrent neural network (RNN), and auto-encoder, have been used in various wireless sensing applications [2]. In general, physical layer properties measured in Phase I are collected over a continuous period of time, which are used to construct deep sensing models for Phase III. For example, CNN is most commonly applied to analyze visual imagery, i.e., two-dimensional data. By acquiring the RSSI over a continuous time, a two-dimensional RSSI dataset can be established, which can be used to train a CNN model. While when sensing applications require to learn relationship involved in sequential time, such as target tracking or tracing, RNN is a more suitable choice. Besides, auto-encoders provide an unsupervised solution to deal with classification-based sensing tasks without labeling requirements. Hence, suitable deep learning models need to be selected, which is jointly determined by the wireless properties and the sensing targets in regards to Phase I and Phase III, respectively.

2.3 Phase III: Implementation

According to the above discussions on wireless hardware, physical layer properties, and deep learning models, a wireless sensing system can be established by selecting appropriate components for specific implementation. Given a broad range of applications, wireless sensing can be roughly divided into device-based and device-free applications. Specifically, device-based applications require users to carry devices with wireless capability, such as smartphones. By observing the received wireless properties, users can estimate their activities with high accuracy, which can be used in applications, such as robotic self-tracking and vibration detection in smart industry. Differently, device-free applications rely on existing infrastructures, such as WiFi routers, by analyzing the fluctuation pattern on wireless properties affected by the targets. Considering the ubiquitous wireless signals as well as the convenience of implementation, device-free sensing has a great potential for applications, like safety surveillance, elderly care, and emergency rescue [4].

3 Learning-Based Wireless Sensing Experiments

Although the above workflow can be generalized to various deep learning-based wireless sensing applications, it is a labor-intensive process to learn and maintain the desired sensing performance. Specifically, to take advantage of deep learning techniques, adequate samples need to be collected to feed a deep model's training and testing. Meanwhile, to maintain the sensing performance, more data is needed to timely fine-tune a deep model. Even a deep sensing model is well-trained, it can hardly be implemented into another scenario, even for the same sensing task. To alleviate such deep learning efforts, Wang et al. conducted a comprehensive study by introducing two typical types of strategies: learning from the similar sensing tasks and generating virtual samples [2]. Nevertheless, those studies only focused on exploiting the acquired sensing information, which can hardly address the knowledge limitation problem in a fundamental method. In view of this, we propose an innovative learning framework by jointly implementing digital and physical experiments to alleviate physical experiments' efforts thoroughly.

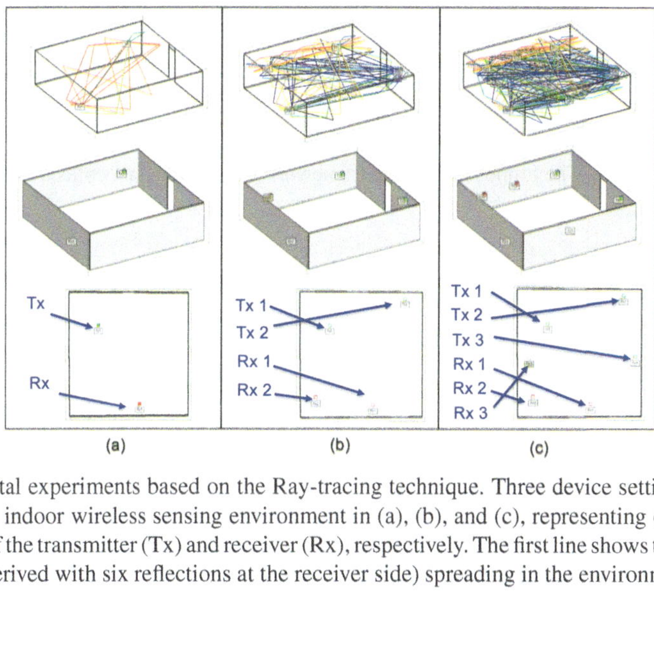

Fig. 2. Digital experiments based on the Ray-tracing technique. Three device settings are illustrated for an indoor wireless sensing environment in (a), (b), and (c), representing one, two, and three pairs of the transmitter (Tx) and receiver (Rx), respectively. The first line shows the detectable rays (e.g., derived with six reflections at the receiver side) spreading in the environment.

3.1 Digital Experiments

Given the labor efforts in field tests, digital experiments, like Ray-tracing simulations, have been considered to facilitate deep learning-based wireless sensing experiments. Ray-tracing technologies, initially applied to optical propagation problems, have been popular in the last two decades for field predication of RF signals in man-made environments, like indoor and urban scenarios [10]. Due to the intrinsic capability of simulating multipath propagation, Ray-tracing technologies become a promising solution to provide site-specific and reliable field prediction, characterizing RF channels multidimensionally, i.e., in time, space, and polarization domains. There are many Ray-tracing simulators nowadays, either free or commercial ones, that can record the physical layer properties of RF signals interacting with the virtual objects in digital worlds, which has been already used for wireless sensing studies [8–11]. Here, we consider the ALTAIR FEKO WINPROP as an example [12]. As illustrated in Fig. 2, the simulator can automatically generate the detectable signal rays spreading in the environment. Particularly, the environment is customizable, such as the building materials, layout, and hardware properties, which can be probability used to conduct the surrogate field tests in a digital way.

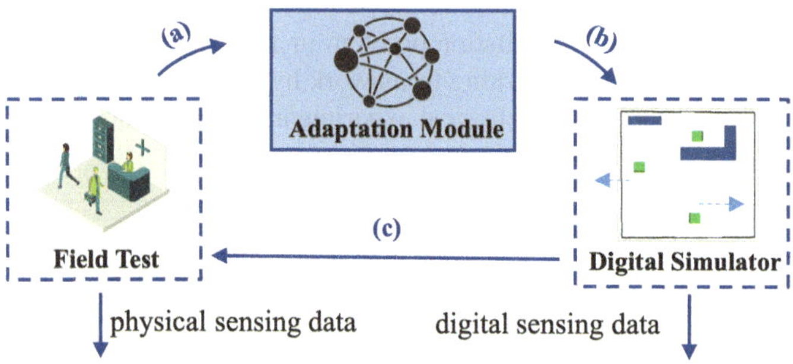

Fig. 3. The proposed learning framework for sensing data collection.

3.2 Integrated Learning Framework

To effectively leverage the feasibility and convenience of digital simulations, we propose an innovative learning framework by integrating the digital simulator and field tests for wireless sensing data collection. As illustrated in Fig. 3, the digital simulator and field tests are connected by the adaptation module. Specifically, the adaptation module will first acquire the information of the field test in (a), which can be an image or a video about the current field test. This information will be learned in the adaptation module to transform into the desired information, i.e., a digital map, for digital experiments. The digital map is expected to mimic the field test scenario, such as the building materials (e.g., brick, glass concrete, and wood), layout, number and locations of the sensing hardware. Most Ray-tracing software supports manually customizing the digital map, which, however, is still based on labor efforts. Hence, we proposed to construct digital maps via advanced machine learning techniques, such as the visual simultaneous localization and mapping (SLAM) [14], which can automatically construct and update a map of an unknown environment while tracking the locations of objects within it. By leveraging the received digital map from (b), Ray-tracing experiments can be conducted to generate the desired digital sensing data. Both the generated physical and digital sensing data will be used to construct the deep sensing models. Meanwhile, to monitor and maintain the quality of digital sensing data, the digital simulator will update its output to the field test module in (c), which can further facilitate the operations in the adaptation module. We expect the proposed learning framework will inspire more thorough research along this line.

4 Conclusion

This paper proposed an innovative learning framework to alleviate data collection's intensive efforts in deep learning-based wireless sensing applications. By introducing the workflow of wireless sensing via deep learning, we first elaborated on the challenges of constructing deep sensing models. A viable solution was proposed to strategically integrate the labor-intensive field tests with the computationally-feasible and reliable digital experiments. By leveraging advanced machine learning techniques, such as visual SLAM, this framework is expected to achieve automatic, reliable, and efficient sensing data generation for future wireless sensing.

Acknowledgments. The work of Lan Zhang and Xiaoyong Yuan was supported by Michigan Technological University COE project: COVID-RESEARCH.

References

1. Liu, J., Liu, H., Chen, Y., Wang, Y., Wang, C.: Wireless sensing for human activity: a survey. IEEE Commun. Surv. Tutor. **22**(3), 1629–1645 (2019)
2. Wang, J., Gao, Q., Ma, X., Zhao, Y., Fang, Y.: Learning to sense: deep learning for wireless sensing with less training efforts. IEEE Wirel. Commun. **27**(3), 156–162 (2020)
3. Liu, K.R., Wang, B.: Wireless AI: Wireless Sensing, Positioning, IoT, and Communications. Cambridge University Press, Cambridge (2019)
4. Wang, J., Gao, Q., Pan, M., Fang, Y.: Device-free wireless sensing: challenges, opportunities, and applications. IEEE Network **32**(2), 132–137 (2018)
5. Di. Domenico, S., De. Sanctis, M., Cianca, E., Giuliano, F., Bianchi, G.: Exploring training options for RF sensing using CSI. IEEE Commun. Mag. **56**(5), 116–123 (2018)
6. Zhang, L., Yan, L., Lin, B., Ding, H., Fang, Y., Fang, X.: Augmenting transmission environments for better communications: tunable reflector assisted mmwave wlans. IEEE Trans. Veh. Technol. **69**(7), 7416–7428 (2020)
7. Yuan, X., Feng, Z., Norton, M., Li, X.: Generalized batch normalization: towards accelerating deep neural networks. In: Proceedings of the AAAI Conference on Artificial Intelligence, vol. 33, no. 01, pp. 1682–1689 (2019)
8. Vaca-Rubio, C.J., Ramirez-Espinosa, P., Kansanen, K., Tan, Z.H., de Carvalho, E., Popovski, P.: Assessing Wireless Sensing Potential with Large Intelligent Surfaces. arXiv preprint arXiv: 2011.08465 (2020)
9. Yang, C.F., Wu, B.C., Ko, C.J.: A ray-tracing method for modeling indoor wave propagation and penetration. IEEE Trans. Antennas Propagation. **46**(6), 907–919 (1998)
10. Fuschini, F., Vitucci, E.M., Barbiroli, M., Falciasecca, G., Degli-Esposti, V.: Ray racing propagation modeling for future small-cell and indoor applications: a review of current techniques. Radio Sci. **50**(6), 469–485 (2015)
11. Yun, Z., Iskander, M.F.: Ray tracing for radio propagation modeling: Principles and applications. IEEE Access. **3**, 1089–1100 (2015)
12. Winprop, altair engineering, inc. https//www.altairhyperworks.com/winprop

13. Halperin, D., Hu, W., Sheth, A., Wetherall, D.: Tool release: gathering 802.11 n traces with channel state information. ACM SIGCOMM Comput. Commun. **41**(1), 53 (2011)
14. Fuentes-Pacheco, J., Ruiz-Ascencio, J., Rendón-Mancha, J.M.: Visual simultaneous localization and mapping: a survey. Artif. Intell. Rev. **43**(1), 55–81 (2012). https://doi.org/10.1007/s10462-012-9365-8

Virtual Simulation Experiment of Curtain Wall Design Based on Behavior Logic of Construction

Manhai Li(✉) and Chenguang Ma

Chongqing University of Posts and Telecommunications, Chongqing 400065, China
limh@cqupt.edu.cn

Abstract. The work aims to propose the virtual simulation experiment of curtain wall design, by analyzing the difficulties of curtain wall design and the feasibility of virtual simulation technology to solve the problems. The most important thing is to organize these contents reasonably, such as professional knowledge, virtual scene, logical structure, interface interaction, quality evaluation and etc. Through the exploration and practice of virtual simulation experiment of curtain wall design, a design principle was illustrated, which is the virtual simulation experiments should be based on behavior logic of construction. By combining the theoretical knowledge with practice organically, students can more effectively understand the points, as well as the basic steps and principles of operation.

Keywords: Curtain wall design · Virtual simulation experiment · Behavior logic · Virtual reality technology · Experimental teaching

1 Introduction

1.1 The Dilemma of Curtain Wall Design

The curtain wall is the outer layer of the building wall, with its supporting structure. Different materials and installation methods of the curtain wall will make the building present different visual effects. For example, the glass pyramid at the entrance of the Louvre Museum in France is a classic work of curtain wall, reflecting the perfect combination of curtain wall and building. curtain wall also undertakes the role of reducing building energy consumption and improving building comfort. With the continuous development of construction engineering technology, the techniques and materials of curtain wall are becoming more advanced. On the basis of ensuring the construction safety and aesthetics, new technologies and new materials should be reasonably used, including new materials of insulating glass and coated glass, new technology of annular double-layer structure glass curtain wall, and new technology of photoelectric curtain wall using solar energy.

Since 2001, curtain wall has been growing rapidly at a rate of about 7.5 million square meters every year. Up to 2017, the total output value of curtain wall projects in China has reached 370 billion yuan [1]. With the rapid development of curtain wall industry, the demand for professional talents is increasing. The curtain wall design needs to keep pace

J. L. Wright et al. (Eds.): AHFE 2021, LNNS 264, pp. 263–269, 2021.
https://doi.org/10.1007/978-3-030-79763-8_32

with the times and integrate architectural appearance, function, and efficiency. However, nearly half of the staff engaged in curtain wall construction are migrant workers. Due to their lack of professional theoretical knowledge and practical skills, it is difficult for them to adapt to the current needs of the development of the curtain wall industry. Embarrassingly, there is almost no specialized curtain wall design major in domestic colleges and universities.

Therefore, speeding up the cultivation of high-tech application-oriented talents to meet the needs of the front-line curtain wall industry has become an urgent task. The solution is to strengthen the specialized training, continuously improve the comprehensive ability of the designers, and let the designers have the quality of matching function, so as to better learn new knowledge, use advanced design means, and maintain the innovation of curtain wall design. As we know, the curtain wall design needs to adopt the experimental teaching method, but the experimental teaching has many practical problems, such as high investment, high risk, difficult action and difficult access. On the one hand, the cognitive learning and practical experience of the material and technology of curtain wall design need to be completed in the curtain wall construction site. However, the construction site itself is a high-risk place with the strictest safety precautions. The curtain wall construction unit will not allow teachers and students to enter the construction site. On the other hand, the site of physical laboratory not only needs large space, but also needs to be equipped with a strict safety management system and corresponding safety management personnel. As a result, the construction cost of physical teaching laboratory is very high, which is prohibitive.

1.2 The Advantages of Virtual Simulation

With the development of information technology, virtual reality technology has sprung up and rapidly penetrated into various fields. For example, simulation exercises training in the military field, virtual simulation design of goods in the manufacturing industry, somatosensory games in the entertainment industry, clinical surgery training in the medical field and so on. These systems can not only truly and vividly present the products, but also are helpful for users to understand the structure and function of products by multi-directional and real-time observation. For example, Peking University applied virtual simulation technology to geology and developed a comprehensive and international immersion virtual simulation laboratory for geology, which improved students' interest and enthusiasm in experimental teaching [2].

Virtual simulation experiment includes operating object, operating environment, and scene roaming, which mainly creates a realistic experimental condition or atmosphere, by the deep integration of professional knowledge and information technology [3]. It focuses on solving problems such as lack of real experimental project conditions or actual operation difficulties, involving the high-risk or extreme environment, high cost, high consumption, irreversible operation, large-scale comprehensive training, etc. at the same time, it is also a teaching mode conducive to autonomous learning, active experiment, sharing and opening. Using virtual simulation technology to carry out the experimental teaching of curtain wall design, it can not only avoid the problems brought by the social practice of course teaching, but also fill the gap of the lack of offline physical laboratory construction due to the high cost, so as to meet the rigid needs of curriculum teaching. Relying on the high-level, innovative, and challenging attributes of new material engineering and advanced engineering technology, the virtual simulation experiment teaching can improve the level of curtain wall design education.

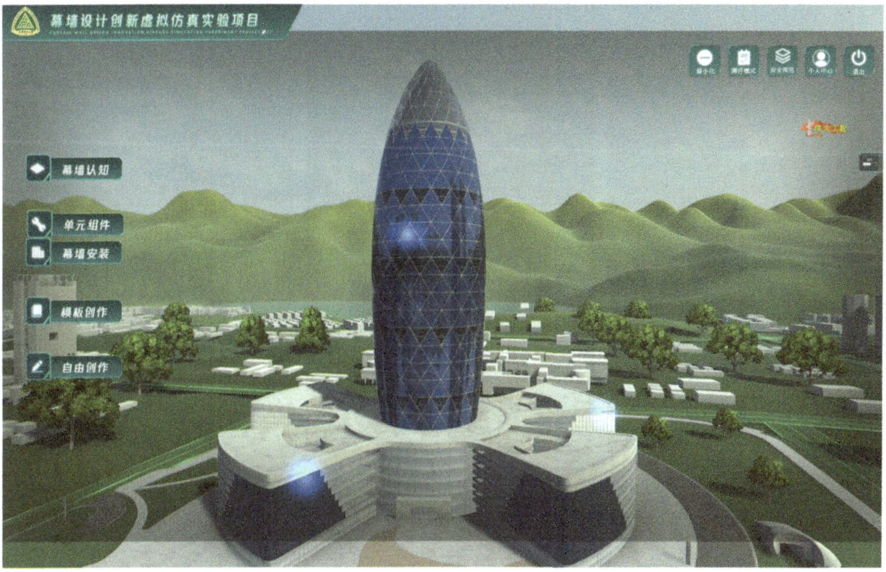

Fig. 1. Virtual simulation experiment of curtain wall design

The main interface of the virtual simulation experiment of curtain wall design is shown in Fig. 1. As a learning platform, it not only can present the complex structure of the curtain wall in detail, but also can show the installation progress of curtain wall in a realistic experimental condition. This experimental teaching can realize the design of different experimental schemes, emphasizing the details of the experiment and strengthening knowledge integration. Here students can carry out technical scheme design, parameter selection, condition configuration, and virtual simulation operation. The kind of teaching model can ensure that the learning can be carried out subsequently and correctly only when the experimental parameters are correctly configured.

2 Virtual Simulation Experiment of Curtain Wall Design

2.1 The Contents of Virtual Simulation

The goal of the virtual simulation experiment of curtain wall design is to train students to have the organic combination of material, engineering, technology and art, and to cultivate high-level applied talents with comprehensive abilities to solve complex problems in the curtain wall industry. In the real experimental teaching process of curtain wall design, the characteristics of different panels will be introduced. For example, glass curtain wall can organically combine different elements, so that the building has a variety of colors, especially when the external environment changes. While in the virtual simulation experiment, the whole process of assembling will be demonstrated. Taking the glass curtain wall as an example, not only the construction of three-dimensional geometric models of columns, embedded parts, and beams should be considered, but also the mathematical and physical models of their coordinates and relative motion should be regarded. The virtual simulation of three different materials, as shown in Fig. 2.

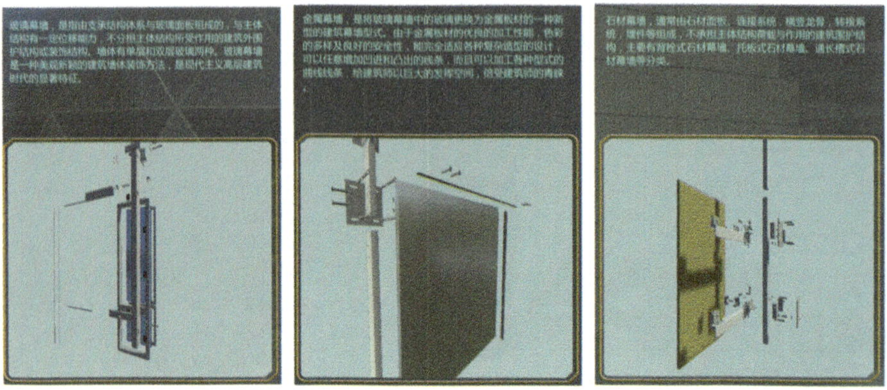

Fig. 2. Virtual simulation effect of three common materials curtain wall

The curtain wall is usually composed of a panel and supporting structure at the back of the panel. The common panel is glass, metal, stone, etc., and the supporting structure is usually steel structure, aluminum beam, etc. The contents of the curtain wall which need virtual simulation should be sorted out firstly. Take the glass panel as an example, the materials of curtain wall are shown in Table 1.

Table 1. The materials of the glass curtain wall

Major categories	Specific materials
Panel material	Tempered glass, low radiation coated glass, heat reflective coated glass, laminated glass, high strength cesium potassium fireproof glass, insulating glass, heat-absorbing glass, thermal insulation rock wool
Skeleton material	Aluminum alloy framework, seamless steel pipe, section steel, U-groove
Metal connector	Embedded parts, connecting corner code, connecting claw, pressing plate, core sleeve, chemical anchor bolt, adapter, expansion bolt, bolt, screw, gasket, adjusting bolt, aluminum alloy decorative strip
Sealing adhesive	Weather-resistant silicone sealants, structural silicone sealants, double-sided stickers, foam rods, insulating gaskets, rubber strips, rubber seals, sponge strips
Process operation	Surveying and setting out, assembly, drilling, anti-corrosion treatment, welding, glue injection and cleaning

As a learning platform for students to acquire knowledge and ability, the virtual simulation experiment of curtain wall design needs to evaluate students' learning effect. Therefore, this platform has established a perfect evaluation system to monitor and track students' online learning and experimental operation: 1. The platform will examine some key technical points and core processes in the experiment by pop-up dialog box, instant message and other ways, to check whether the student has mastered. 2. The platform provides several schemes of the assembly sequence of curtain wall units, and the correctness of the scheme selection reflect the students' mastery of relevant knowledge by corresponding scoring rules. 3. The platform provides online submission, online correction and intelligent evaluation to keep track of students' learning progress in time. 4. The platform provides a complete student's individual learning effect report by collecting the data, including online learning time, completion progress, success or failure of single experiment and completion efficiency.

Communication is another important event in virtual simulation experiment of curtain wall design. It is an effective way for teachers to highlight the details of experiments, transfer skills, and solve puzzles without limitation of time and place. The virtual simulation experiment of curtain wall design takes single-choice questions or multiple topics, mainly in the form of intuitive pictures and 3D animation, virtual reality glasses will be added in the future.

2.2 Based on Behavior Logic of Construction

The virtual simulation experiment of curtain wall design based on construction behavior logic means that these contents are reasonably organized according to the sequence of curtain wall construction behaviors, so that the abstract theoretical knowledge can be deeply recognized and applied creatively through the simulation and test of virtual simulation. The purpose is to enable students to master the theoretical knowledge and combine them with the construction practice. For example, during the curtain wall construction, the insulating gasket should be placed in a suitable position before connecting the aluminum alloy column, which is mainly to prevent corrosion. If the metal is directly contacted, bimetallic corrosion may occur.

Taking glass curtain wall as an example. The traditional teaching focuses on the glass curtain wall material itself, and focuses on the function, structure layout, color style and other attributes of the glass curtain wall material. However, the focus of this platform is to string the knowledge points of the glass curtain wall's own attributes and installation points according to the clues of the construction process of the glass curtain wall, namely, to correctly install the glass curtain wall, and to regard the glass curtain wall material as one of the material media of construction behavior [4], rather than the only one. In the design of glass curtain wall, first of all, we should have a basic understanding of the project progress, design situation, construction situation and so on. Once the design problems are found, we should timely adjust the design scheme and select scientific and reasonable remedial measures. The front embedded parts are made of hot-rolled steel bars. In the design of embedded parts, it is necessary to comprehensively consider various requirements to determine an optimal anchorage length. The fixed combination mode can be selected for the rear embedded parts and the front embedded parts. The inverted cone adhesive anchor bolt and galvanized steel plate can be reasonably combined to control the anchor bolt level, and the anchoring depth is controlled well. After the design scheme of embedded parts is determined, the bearing capacity of embedded parts can be tested on site, and the formal construction can only be carried out after confirming that there is no error (Fig. 3).

Fig. 3. Virtual simulation effect of three common materials curtain wall

Assembly sequence: Installation of column, glue injection and cleaning, measurement and setting out, installation of pressing plate and gusset plate, installation of glass, installation of beam, installation of embedded parts, inspection and acceptance. In this way, students can be trained to take a global view to view the design and construction of curtain wall.

3 Conclusion

Virtual simulation experiment of curtain wall design will become the development direction of education reform in the information age, and will become an important means for students to acquire knowledge and skills of curtain wall. This paper starts from the necessity and feasibility of construction, analyzes the significance and importance of virtual simulation experiment of curtain wall design, and then elaborates the design principle, construction content and related aspects of virtual simulation teaching plan from professional knowledge, virtual scene, logical structure, communication and interaction, teaching feedback, effect and quality evaluation design These suggestions and conclusions will be conducive to the construction of high-quality virtual simulation teaching resources, the application and promotion of virtual simulation teaching mode, and promote the diversified development of teaching mode and the cultivation of high-level talents.

Acknowledgments. This work was supported by 重庆邮电大学博士启动基金、人才引进基金项目-大数据产品化成本与效益分布研究 (K2020-201) and 重庆市教育科学规划项目课题-高校"社会主题型"的人才培养体系研究 (2020-GX-284) and 重庆邮电大学网络社会发展问题研究中心-网络大数据产品化的成本问题研究 (2020SKJD06)

References

1. Cui, X.: Analysis of the optimization design method of building curtain wall. Const. Des. Proj. **18**, 40–41 (2020)
2. Qiang, W., Guo, Y., Zhou, Z., Zhou, Y.: Review of virtual simulation technique in geology. Geol. J. China Univ. **26**(04), 464–471 (2020)
3. Liu, Y., et al.: Design and practice of the teaching scheme for virtual simulation experiment. Res. Expl. Lab. **036**(003), 185–188 (2017)
4. Xin, X.: Interaction design: from logic of things to logic of behaviors. Zhuangshi **1**(261), 58–62 (2015)

Digital Human Modelling and Applied Optimization

Design, Execution, and Results of an Anthropometric Surveillance Study

Joseph L. Parham[1(✉)], Hyegjoo E. Choi-Rokas[2], Todd N. Garlie[1], David Brantley[1], and Linda L. Desimone[3]

[1] US Army Combat Capabilities Development Command Soldier Center (DEVCOMSC), Natick, MA, USA
[2] US Army Combat Capabalities Development Command (CCDC), Natick, MA, USA
[3] NSRDEC, Natick, USA

Abstract. Large anthropometric surveys are expensive, time-consuming, and accordingly rare. The databases drawn from these surveys necessarily represent a snapshot of a given population's anthropometric variability. To ensure their continued validity, the maintenance of anthropometric databases should include the periodic review of current body size information and changes in demographics of the described population. This paper reports a body size review through the execution of an anthropometric surveillance study in support of the US Army's 2012 Anthropometric Survey (ANSUR II) databases. This surveillance study was a targeted data collection effort specifically designed to investigate the continued use of the ANSUR II databases. Analyses of the data from the surveillance study show that the ANSUR II databases still accurately reflect the current US Army.

Keywords: Anthropometry · Surveillance study · Military anthropometry · ANSUR II

1 Introduction

There are relatively few anthropometric survey databases with large sample sizes of both sexes and which report a large number of directly measured dimensions. Accordingly, such databases are highly valued and widely used. Anthropometric surveys are a snapshot of anthropometric variability within a population at a specific time, but it is difficult to determine when those data no longer represent the population they are drawn from. At some point, underlying demographic changes within the population as well as the accumulation of secular changes in body size and proportions will necessitate the collection of new data via a new anthropometric survey, and the old data can no longer be used to accurately reflect the population. An anthropometric surveillance study is a less extensive, targeted data collection effort intended to identify whether data from a prior, larger anthropometric survey is still valid.

The ANSUR II databases [1] were developed from body dimension data collected from over 12,000 US Army Soldiers from 2010 to 2012. In the subsequent years, the Army has undergone demographic change as well as potentially experiencing underlying

J. L. Wright et al. (Eds.): AHFE 2021, LNNS 264, pp. 273–281, 2021.
https://doi.org/10.1007/978-3-030-79763-8_33

secular change both within the Army and from the general civilian population from which Soldiers are recruited. Given the importance of the ANSUR II data within the US military scientific, research, and engineering communities, in 2018 the US Army Combat Capabilities Development Command Soldier Center executed a surveillance study to insure the validity of the ANSUR II databases for continued use. The surveillance study reported here was designed to answer the question of whether the ANSUR II data still reflect the anthropometric variability of the current US Army.

2 Study Design

The anthropometric surveillance study was designed assess the physical makeup of the US Army in 2018 compared to the US Army of 2010–2012. Detailed, careful planning is critical to such an effort's success, especially because few aspects may be altered once data collection has begun.

2.1 Dimension Selection

While ideally all dimensions would have been captured during this study, it was necessary to reduce and focus on only the most critical dimensions. Criteria for the dimension selection were: dimensions on overall body size (i.e., Weight, Stature), critical design dimensions (e.g., Chest Circumference, Waist Circumference, Crotch Height), and dimensions for capturing potential changes in body size proportions or segments.

It was necessary, of course, that the selected dimensions be a subset of those included in the ANSUR II databases, since comparisons to those databases is the purpose of this study. Practical considerations including equipment requirements and the limited time allotted for data collection further reduced the possible dimensions. Table 1 reflects the 13 measurements selected for this surveillance study. The number associated with each measurement reflects the reference number for that dimension in the ANSUR II Technical Report [1].

Table 1. Selected Anthropometric dimensions for the surveillance study.

Type	Selected dimension
Weight	91-weight
Height	75-stature, 76-Suprasternale height, 56-knee height mid-patella, 27-crotch height; 90-waist height (omphalion)
Circumference	68-shoulder circumference; 24-chest circumference; 87-waist circumference (omphalion); 16-buttock circumference
Length	40-forearm-hand length
Head	45-head breadth; 47-head length

2.2 Sample Size

The goal of the surveillance study was to detect any shifts on the central tendency of selected body dimensions measured in 2012 (ANSUR II) and measured again in 2018 (the surveillance study). Specifically, it sought to detect differences in mean values of body dimensions. To accomplish this, it would be necessary to collect a sample size sufficiently large enough to perform reliable statistical analyses and produce robust results. In order to determine the appropriate sample size, the following equations were used to reflect (1) the minimum sample size needed to estimate the $5^{th}/95^{th}$ percentiles with 95% confidence and 1% precision; and (2) the minimum sample size needed to estimate the mean of a parameter with 95% confidence and 1% precision [2, 3].

$$n \geq (1.96 * \mathrm{CV})^2 * (1.534)^2 \tag{1}$$

$$n \geq (1.96 * \mathrm{CV})^2 \tag{2}$$

In these equations n is the minimum sample size and CV is the coefficient of variation of the dimension in question. Application of these equations to data collected in ANSUR II provides the minimum sample sizes required of the surveillance study in order to estimate these parameters in the modern US Army population. Table 2 represents the minimum sample size calculated using Eqs. (1) and (2) for three dimensions important to equipment design.

Table 2. Mean, coefficient of variation, and minimum sample sizes (n)

	Male				Female			
	Mean	CV	n (1)	n (2)	Mean	CV	n (1)	n (2)
Stature (mm)	175.6	3.9	138	59	162.9	3.9	140	60
Chest circumference (mm)	105.9	8.3	616	262	94.7	8.7	690	293
Waist circumference (mm)	94.1	11.9	1,275	542	86.1	11.6	1,217	517

Unsurprisingly, the sample sizes needed to estimate the $5^{th}/95^{th}$ percentiles are larger than those needed to estimate the mean values by a factor of approximately 2.4 (1.534^2). Practically, there was no opportunity to capture the larger sample sizes, and for the purposes of this surveillance study, focus remained on estimating the means of several dimensions in the modern US Army for statistical comparison to ANSUR II means. Therefore, a target of 550 male and 550 female Soldiers was determined which would facilitate an estimate of the mean of the most variable critical design dimension, Waist Circumference, based on the results of Eq. (2) for this dimension.

2.3 Demographic Data

Collection of demographic data are critical to anthropometric data collection efforts. The demographic composition of a sample, particularly the age and race/ethnicity of the participants, often explains much or most of the variation seen in the data. Indeed, the minimum sample sizes discussed in Sect. 2.2 are only sufficient if that sample is made up of demographically diverse participants proportionally representative of the population. To that end, participants were not only measured, but also surveyed for the demographic data shown in Table 3. These demographic data are especially important considering the difficulty in gaining access to large numbers of Soldiers and the likelihood that a relatively small sample would be skewed from the larger Soldier population.

Table 3. Demographic data collected from each surveillance study participant.

Variable	Brief description
Grade/rank	Military rank (e.g., E7, sergeant first class)
Time in service	In years and months
Military occupation series (MOS)	Military job series (e.g., 68W, Combat Medic)
Component	Active duty, national guard, reserves
Age	In years
Gender	Male or Female
Birthplace	US state or foreign country
Biological race	Self-identified race to include mixed races with multiple selections allowed
DOD race	Self-identified single race selection, only one race selected

The DOD (Department of Defense) Race variable is necessary to compare sample demographics to the US Army population where often only single race selections are collected. These data were also used to compare the surveillance study anthropometric data to the ANSUR II data, as described later. Other demographic data were collected largely in order to illustrate a diverse sample from throughout the Army.

3 Study Execution (Data Collection)

Data collection for the anthropometric surveillance study commenced in March and concluded in November, 2018. During this time, teams of researchers traveled to multiple locations to collect a diverse representation of Soldiers, targeting specific sites and leveraging other data collection efforts which benefited from anthropometric data collection.

As shown in Table 4, the minimum sample size for Female Soldier participants was not reached. Indeed, access to all Soldiers is a common, difficult hurdle for military anthropometric efforts due to the varied and often unpredictable duties of this population, as well as the time needed to measure each participant.

For the 940 US Soldier participants in the anthropometric surveillance survey, each provided demographic information as described in Table 3 and each was measured to

Table 4. Surveillance study measuring sites and the sample size at each site.

Site	Female soldiers	Male soldiers	Total
Fort McCoy, WI	136	143	279
Fort Lewis, WA	–	20	20
Fort Polk, LA	16	–	16
Soldier Center, MA	–	12	12
Fort Bragg, NC	5	48	53
Springfield Armory, MA	12	31	43
Joint Base Cape Cod, MA	10	53	63
Fort Richardson, AK	72	257	329
Fort Lee, VA	41	84	125
TOTAL	292	648	940

collect the dimensions delineated in Table 1. It is important to note that the dimensions were measured by trained personnel strictly following the procedures outlined in the Measurer's Handbook [4] which accompanied the ANSUR II data collection effort.

4 Results and Discussion

The results of the anthropometric surveillance study focus primarily on the anthropometric data directly collected and compared to the ANSUR II measured data. These data and comparisons are considered in the context of allowable measuring error and the specific demographic makeup of the surveillance study sample.

4.1 Comparison of Anthropometric Measurements.

The 940 Soldiers measured as part of the anthropometric surveillance study were fewer than the study design intended due to the difficulties in gaining access to large numbers of Soldiers during their duty hours. It was especially difficult to access female Soldiers given their relative scarcity in any given US Army unit. The 648 male Soldiers measured in the surveillance study did exceed the target sample size.

While reaching sampling goals is important, the demographic composition of the sample must also be considered. A sample that skews too young or too old, or does not share the racial proportions of the population, will not accurately reflect the population it is seeking to describe. In the case of the surveillance study, the sample indeed did not demographically match the ANSUR II sample. Table 5 shows the age distribution of male participants from each anthropometric survey, while Table 6 reports race/ethnicity distributions.

Table 5. Percent distribution of participant ages of male participants from ANSUR II and the surveillance study

Age range	ANSUR II distribution	Surveillance study distribution
17–19	5.1	20.4
20–24	28.2	44.4
25–39	22.8	16.2
30–34	14.9	8.6
35–39	11.0	4.5
40+	18.0	5.9
TOTAL	100.0	100.0

Table 6. Percent distribution of participant races/ethnicities of male participants from ANSUR II and the surveillance study.

Race/Ethnicity	ANSUR II distribution	Surveillance study distribution
White	69.0	58.3
Black	15.7	17.1
Hispanic	10.8	18.0
Asian	2.9	4.6
Native American	0.7	0.8
Pacific Islander	0.8	1.2
Other	0.1	–
TOTAL	100.0	100.0

It is clear that the male sample in the surveillance study skews young. Fully 64.8% of the surveillance study participants were aged 24 years or younger, compared to just 33.3% off the participants in the ANSUR II databases. Participants 30 or older make up just 19.1% of the surveillance study sample, compared to 43.9% in ANSUR II. The surveillance study sampling deficiencies are a direct result of Soldier availability for a small-scale study where older, higher-ranking Soldiers are busier and less inclined to participate. The distribution of races/ethnicities in each survey are also noticeably different, with the surveillance study exhibiting more diversity.

It is well-known that differences in age and race significantly affect a sample's anthropometric variability [5, 6]. The surveillance study sample as a whole is therefore not directly comparable to the ANSUR II databases. The largest available subset within the surveillance study, that of males aged 20–24, was extracted for direct comparison to the corresponding subset within ANSUR II. The surveillance study sample was further statistically weighted to match the race/ethnicity demographics of the ANSUR II male database. Table 7 reports the mean values of each dimension measured in both surveys, and the difference between them. The difference values (Δ) were computed by subtracting the mean values of ANSUR II from surveillance study. Positive values indicate the mean values of surveillance study are greater than those of ANSUR II. Those difference values were compared to the measurers' allowable error in collecting those dimensions.

Table 7. Comparison of mean values for Male Soldiers Aged 20–24 in ANSUR II and the surveillance study. The surveillance study data are statistically weighted to match the ANSUR II race/ethnicity proportions. The bolded values indicate that the difference between means exceeds the allowable error for that dimension.

Dimension unit: mm	ANSUR II $n = 1{,}150$	Surveillance study $n = 285$	Δ	Allowable error
	Mean	Mean		
Weight (kg)	82	82	0	0.3
Stature	1757	1759	2	6
Suprasternale height	1436	1434	−2	5
Waist height	1061	1063	2	7
Forearm-hand length	480	478	−2	4
Shoulder circumference	1168	1198	**30**	**12**
Chest circumference	1029	1015	−14	14
Waist circumference	897	878	**−19**	**12**
Buttock circumference	1004	1007	3	12
Crotch height	849	842	−6	10
Knee height mid-patella	488	498	**10**	**6**
Head length	199	198	−1	2
Head breadth	154	154	0	2

It is important to further compare the differences in means to the allowable measuring error for each dimension. The allowable error is the maximum expected difference in a measured dimension collected by two highly-trained measurers. The allowable error for each dimension is listed in the ANSUR II technical report [1], and those values for dimensions included in the surveillance study are reported in Table 7.

In 10 of 13 cases, the differences between the means of measured dimensions are less than the allowable error, indicating that there is no real, discernible difference between them. Further statistical analysis of these differences is therefore unnecessary.

In three instances, the difference between the means is greater than the allowable measurer's error: shoulder circumference, waist circumference, and knee height mid-patella. While these are not immediately explainable, it is possible that unintended changes in methodology may have affected collection of these measurements in the surveillance study. In any case, it is highly unlikely that the differences seen in these dimensions truly reflect any change in the overall body size and proportion of the population. Indeed, if they were indicative of population change, similar observable differences would surely have been detected in other measured dimensions that are closely correlated.

5 Conclusions

Relatively few anthropometric survey databases with large sample sizes of both sexes and a large number of directly measured dimensions exist. The ANSUR II databases [1] were developed from data collected from over 12,000 US Army Soldiers from 2010 to 2012. Changes in Army demographics since this time and potential underlying secular changes from the population from which Soldiers are recruited could mean the reference database is no longer truly representative of the whole Army population. It is critical to monitor any large anthropometric survey at specific intervals in order to recognize any populations changes in body size and shape that could significantly impact clothing equipment and design impact Soldier performance and safety.

Given the importance of the ANSUR II data within the US military scientific, research, and engineering communities, this surveillance study was undertaken to insure the continued validity of the ANSUR II. The surveillance study reported here was designed to answer the question of whether the ANSUR II data still reflect the anthropometric variability of the current US Army. Comparative results from this study suggest that broad measureable differences between the ANSUR II database and the surveillance study are not seen and indicate that the data from ANSUR II continue to be valid and representative of the overall US Army population for design and engineering purposes.

The surveillance study and its results reported here represent the first effort of continued scrutiny of the ANSUR II databases. A demographic analysis to determine if the racial/ethnic makeup of the US Army has changed sufficiently to affect design parameters is a reasonable follow-up to this study, and may be conducted at any time, independent of the surveillance study.

Acknowledgements. This research was funded and executed under the US Army DEVCOM Soldier Center research program 15-060. The authors would like to thank those Soldiers who participated in ANSUR II or the surveillance study for their time and effort, as well as for the sacrifices they make every day for our country.

References

1. Gordon, C.C., Blackwell, C.L., Bradtmiller, B., Parham, J.L., Barrientos, P., Paquette, S.P., Corner, B.D., Carson, J.M., Venezia, J.C., Rockwell, B.M., Mucher, M., Kristensen, S.: 2012 anthropometric survey of U.S. army personnel: methods and summary statistics. Technical report, Natick Soldier Research, Development and Engineering Center, Natick, MA. NATICK/TR-15/007 (2015)
2. Sokal, R.R., Rohlf, F.J.: Biometry: The Principles and Practices of Statistics in Biological Research. W.H. Freeman, New York (2011)
3. ISO 15535:2012 General requirements for establishing anthropometric databases. International Organization for Standardization, Geneva (2012)

4. Hotzman, J., Gordon, C.C., Bradtmiller, B., Corner, B.D., Mucher, M., Kristensen, S., Paquette, S., Blackwell, C.: Measurer's handbook: US army and marine corps anthropometric surveys, 2010–2011. Technical report, Natick Soldier Research, Development and Engineering Center, Natick, MA. NATICK/TR-11/017 (2011)
5. Greiner, T.M., Gordon, C.C., Carson, E.A.: Anthropometric differences among occupational groups in the U.S. army. Technical report, Natick Soldier Research, Development and Engineering Center, Natick, MA. NATICK/TR-95/016
6. Walker, R.A.: The impact of racial variation on human engineering design criteria. In: Gordon C.C. (ed.) Race, Ethnicity, and Applied Bioanthropology, pp. 7–21. American Anthropological Association NAPA Bulletin 13, Washington D.C (1993)

Spatial Needs for Non-driving Related Activities

Martin Fleischer(✉) and Rongqian Li

Chair of Ergonomics, Technical University of Munich, Boltzmannstraße 15, 85747 Garching, Germany
{martin.fleischer,rongqian.li}@tum.de

Abstract. Due to the temporary abandonment of driving tasks in conditionally automated driving (SAE Level 3), the driver is able to devote the available time and attention to non-driving related activities, while the vehicle takes on more and more driving tasks. The aim of this work is to investigate how much space the driver needs for non-driving activities and to what extent the space requirements of different non-driving activities differ. A mock-up study is carried out at the Chair of Ergonomics at the Technical University of Munich, in which test subjects carry out various non-driving related activities. The longitudinal distance between the center of the steering wheel and the subject's chest is measured, which serves as a measure of the required space of non-driving activities. The data shows significant differences concerning the factor "activities".

Keywords: Automated driving · Non-driving related activities · Human factors · Car interior · Spatial needs

1 Introduction

Nowadays, individual mobility is taken for granted as part of everyday life. Due to the widespread use of smartphones and technological advances, new mobility concepts are emerging. Against this background, automated driving is being researched both at universities and in the automotive and service industries.

Vehicle drivers usually perform non-driving activities such as operating the smartphone and listening to music [1–3] in addition to driving the vehicle. At the same time, inattentive driving poses potential dangers in many traffic situations and should therefore be avoided as far as possible. In contrast, the highly and fully automated driving concept brings a new form of interaction between driver and vehicle. With an increasing degree of automation, the active vehicle driver is transformed into a passive vehicle passenger, while driving the vehicle requires less and less attention from the driver. The vehicle takes over the driving tasks of the active driver, allowing the driver to devote his time and attention to non-driving activities [4]. This brings great possibilities [5], but also uncertainties for the designers of car interiors [6]. Participant studies and digital human models such RAMSIS [7] are used to evaluate the spatial needs of the passengers in vehicles with automation less then SAE Level 3 [8].

This paper deals with the study of the required space of a driver performing non-driving activities in a highly automated vehicle (SAE Level 3). This forms a basis for

J. L. Wright et al. (Eds.): AHFE 2021, LNNS 264, pp. 282–287, 2021.
https://doi.org/10.1007/978-3-030-79763-8_34

understanding this new driving concept and will support further development of the vehicle interior of automated vehicles.

2 Methodology

The study was conducted to evaluate the needed frontal distance *d* between passenger torso and the next environmental object (Fig. 1). In this study a modular ergonomic mock-up was used, where the steering wheel is movable and thus d is adjustable. Distance d is defined as the horizontal distance between steering wheel center and torso of the participant. Participants were instructed to conduct several activities, while the steering wheel was continuously moved away from them. The steering wheel started in a position where the rim was barely touching the participant. The participants were asked to stop the steering wheel two times. First when they deemed d as "sufficient" to execute the task and a second time when d was considered to be "comfortable". This factor is called the nature of the spatial needs with the two characteristics sufficient and comfortable.

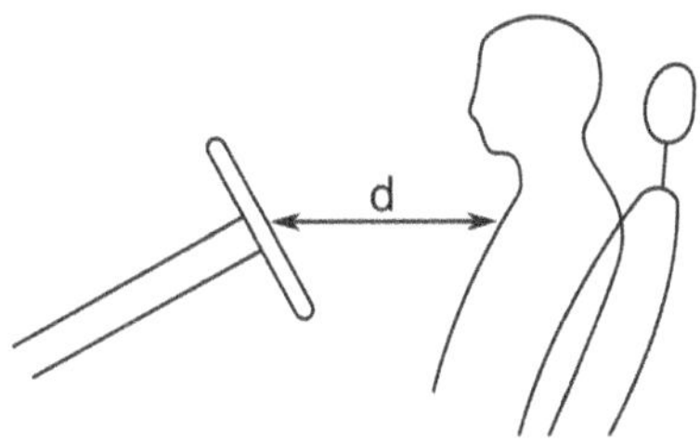

Fig. 1. Dependent variable of the experiment; Distance d between torso and steering wheel

30 participants were invited to a mock-up study and asked to carry out a total of eleven activities.

1. Make a phone call
2. Observe landscape
3. Using smartphone
4. Eating and drinking
5. Talking to the co-driver
6. Relaxing
7. Reading a book
8. Using a tablet
9. Reading a newspaper
10. Paperwork (writing into a DIN A4 ring binder)
11. Working with laptop.

The task "reading a book" was repeated three times to evaluate the reliability of participants' responses in regards to the sufficient and comfortable distances. Overall the experiment was structured in a way that every participant read part of a book, executed six randomized activities, read in the book, executed four randomized activities and then concluded the experiment with reading a book.

The participants' age ranged from 18 to 43 years (mean = 25.9 years; SD = 5.5 years) with a mean body height of 1735 mm (SD = 89 mm). The participants were rewarded with 10 € for their one-hour effort. 15 men and 15 women participated.

The data is evaluated in two different statistical procedures. In Procedure the distance d is put into relation to the eleven "activities" and the "nature of the spatial needs" ("sufficient"/"comfortable"). For the activity "reading a book" only the first repetition was used. That way every task in the procedure is executed for the first time. Procedure B looks into the relationship of the dependent variable d and the predictors "repetitions" and the "nature of the spatial needs". The three repetitions of "reading a book" were used to evaluate if participants can reliably estimate d for "sufficient" and "comfortable" needs. Both procedures were carried out through a two-way repeated measures ANOVA with a Bonferroni corrected post-hoc test.

3 Results

3.1 Procedure A

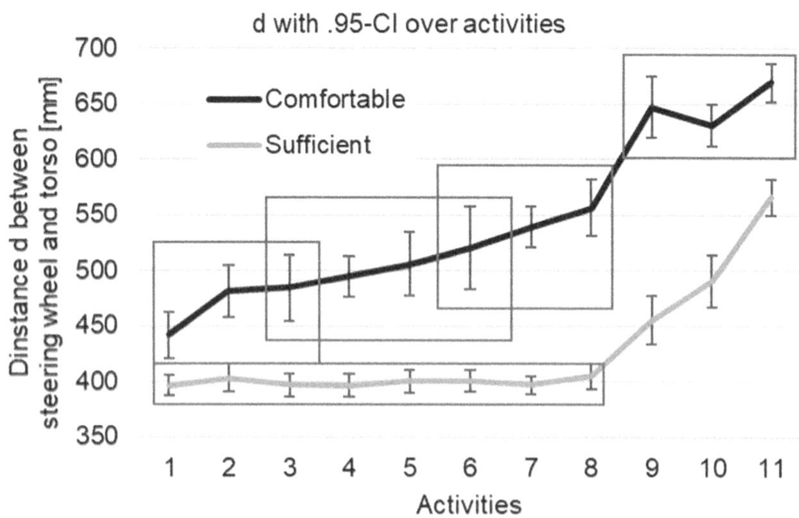

Fig. 2. Distance d with the .95-CI with the nature of spatial needs separated over the activities. The non-significant clusters are marked in blue boxes 1. Make a phone call; 2. Observe landscape; 3. Using smartphone; 4. Eating and drinking; 5. Talking to the co-driver; 6. Relaxing; 7. Reading a book; 8. Using a tablet; 9. Reading a newspaper; 10. Paperwork; 11. Working with laptop

The executed activities as well as the nature of the spatial needs have significant influences. All differences between the natures of the spatial needs were significant in every activity. The smallest difference is between a sufficient and a comfortable d while "making a phone call" with 45 mm, .95-CI [26 mm, 64 mm] and the biggest difference is during "reading a newspaper" with a difference in d of 192 mm, .95-CI [158 mm, 226 mm]. For the activities during the sufficient nature of spatial needs there is a cluster of the activities 1 through 8 that are not significantly different that can be clearly seen in Fig. 2. For the comfortable nature of spatial needs the significant differences are not that easy to cluster, as the clusters overlap.

3.2 Procedure B

The reliability of estimating the sufficient and comfortable d for "reading a book" is evaluated throughout the three repetitions. For the sufficient nature there a no significant effects of the repeated measurements ($F(2.28, 28) = 2.254, p = .124$). For the comfortable condition the influence of the repetitions are significant with $F(1.673, 48.504) = 16.315$, $p < .001$. The detailed data can be seen in Fig. 3, where those phenomena can be seen. The biggest mean difference is between the first and the third measurement with 41 mm, .95-CI [19 mm, 63 mm], $p < .001$.

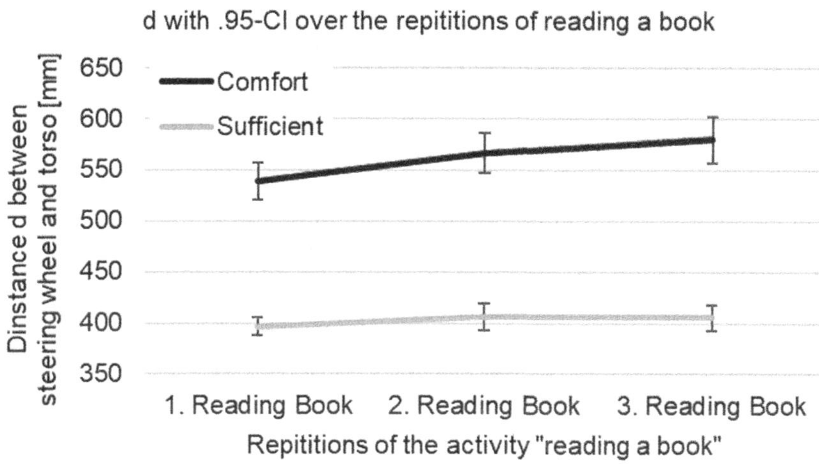

Fig. 3. Distance d with the .95-CI with the nature of spatial needs separated over the three repetitions of "reading a book"

4 Discussion

4.1 Methodology

The study did not evaluate different anthropometric factors as body height or belly circumference. Those factors might play a major role especially when looking at persons with a high Body Mass Index (BMI). On the same side it is discussable if the distance d is the best possible metric. The distance between the lower end of the steering wheel to the stomach might be more restricting in some aspects (see Sect. 4.2).

The repetitions were only evaluated for one activity as the experiment would have been much too long. That way it cannot be said safely that the reliability is consistent for every activity.

The setup is static this might lead to smaller space being perceived as comfortable.

4.2 Data

Looking at the data there is a clear difference between sufficient and comfortable space. It can be proclaimed that there is a difference between being barely able to perform an activity and being able to comfortably perform the same activity. For the sufficient nature a floor effect can be witnessed as the smallest distance d is not zero but around 400 mm. This might occur because the participant were able turn their torso sideways to perform the activity besides the steering wheel. The trend is broken as soon as the objects used become a certain size, which seems to be around the size of a newspaper, a DIN A4 ring binder and a 15.4″ Laptop. In the comfortable nature there is also a hike in the required distance for those three activities. In the comfortable condition there is also a trend that the bigger the objects become, the more space is wanted by the participants.

Judging the reliability of the sufficient nature is not quite possible as the floor effect probably prevents deviation. For the comfortable condition the differences are significant, but looking at the biggest difference between means it is 7.3% from the mean distance. This is not drastic and an asymptotic nature can be seen after the third repetition. On the other hand, it can be concluded that participants tend to want more space if they have to estimate a comfortable space more often.

5 Conclusion

There is an impact of the activity on the required space, but many activities can already be executed in conventional vehicles as the steering wheel in this setup is much closer to the passenger compared to a real car. This can be seen from the floor effect. The space needed to comfortably perform an activity is primarily effected by the size of the used object.

The data presented in this study probably displays an underestimation of the sufficient and comfortable space due to the static setup.

Acknowledgements. This study was conducted in the context of the project INSAA funded by the Federal Ministry of Education and Research of the Federal Republic of Germany.

References

1. Dingus, T., Klauer, S., Lewis, V., Petersen, A., Lee, S., Sudweeks, J., Perez, M., Hankey, J., Ramsey, D., Gupta, S., et al.: DOT HS 810 593: the 100-car naturalistic driving study phase II. Results of the 100-Car Field Experiment (2006)
2. Pfleging, B., Schmidt, A.: (Non-)driving-related activities in the car: defining driver activities for manual and automated driving. https://www.hcilab.org/wp-content/uploads/chi15-ws-non driving-activities.pdf

3. Huemer, A.K., Vollrath, M.: Ablenkung durch fahrfremde Tätigkeiten - Machbarkeitsstudie. Bericht zum Forschungsprojekt FE 82.376/2009. Berichte der Bundesanstalt für Straßenwesen (2012)
4. Yang, Y., Klinkner, J.N., Bengler, K.: How will the driver sit in an automated vehicle? – the qualitative and quantitative descriptions of non-driving postures (NDPs) when non-driving-related-tasks (NDRTs) are conducted. In: Bagnara, S., Tartaglia, R., Albolino, S., Alexander, T., Fujita, Y. (eds.) Proceedings of the 20th Congress of the International Ergonomics Association (IEA 2018), vol. 823, pp. 409–420. Springer, Cham (2019)
5. König, M., Neumayr, L.: Users' resistance towards radical innovations: the case of the self-driving car. Transp. Res. Part F: Traffic Psychol. Behav. **44**, 42–52 (2017)
6. Yang, Y., Fleischer, M., Bengler, K.: Chicken or egg problem? New challenges and proposals of digital human modeling and interior development of automated vehicles. In: Di Nicolantonio, M., Rossi, E., Alexander, T. (eds.) Advances in Additive Manufacturing, Modeling Systems and 3D Prototyping, Proceedings of the AHFE 2019 International Conference on Additive Manufacturing, Modeling Systems and 3D Prototyping, July 24–28, 2019, Washington D.C., USA, vol. 975, pp. 453–463. Springer, Cham (2020)
7. van der Meulen, P., Seidl, A.: Ramsis – the leading cad tool for ergonomic analysis of vehicles. In: Duffy, V.G. (ed.) First International Conference on Digital Human Modeling, ICDHM 2007, Held as Part of HCI International 2007, Beijing, China, July 22–27, 2007, Proceedings, vol. 4561, pp. 1008–1017. Springer, Berlin (2007)
8. On-Road Automated Driving (ORAD) Committee: Taxonomy and Definitions for Terms Related to Driving Automation Systems for On-Road Motor Vehicles. SAE International, 400 Commonwealth Drive, Warrendale, PA, United States (2018). https://saemobilus.sae.org/content/j3016_201806

A Computational Approach to Evaluate Headborne Equipment Impacts to Field of View Over a Population: A Case Study

Peng Li(✉), K. Blake Mitchell, and J. David Brantley

US Army DEVCOM Soldier Center, Natick, MA 01760, USA

Abstract. This paper presents a computational approach to evaluate the field of view (FOV) of a prototype bomb suit helmet. The presented approach computes the FOV based on three-dimensional scans of head and helmet. The variation of the FOV is also evaluated based a large anthropometric database. By identifying four variables to relate cornea location to helmet position, their influence to the FOV variation is computed, which demonstrates the potential for virtual testing of a large population.

Keywords: Field of view · Helmet · ANSUR II database · 3D scan · Head borne equipment · Vision · Human performance · Equipment assessment · Human factors · Human system integration · Digital human modeling

1 Introduction

The evaluation of field of view (FOV) degradation is an important human factors task in product design, as well as training and simulation [1–3]. Usually, FOV evaluations require multiple human subject volunteers, an approach that can be time-consuming and often suffers from limited sample size. Using a virtual human for FOV evaluation in a digital human modeling (DHM) environment is an alternative approach developed to overcome the drawback of real human testing [4–6]. Three-dimensional (3D) scanning technology has made realistic digital human modeling possible and 3D scanning of the human face has been used in FOV tasks [4]. However, population wide FOV testing needs a large sample size to reflect human variation.

Head borne systems (i.e., helmets, goggles, face shield, etc.) are likely to restrict FOV due to their proximity to the eyes. Testing FOV of military head borne systems is difficult with traditional perimetry devices as their bulkiness prevents proper head placement and modern test apparatus focus on significant degradations caused by medical issues and therefore do not capture the natural range of field of view of a healthy eye. Alternative methodologies have been developed to simulate the line of sight for FOV evaluation, such as laser shadowgraphy [1]. But such device requires a laser source and photography equipment to capture image of the field of view from the eye location, which is not easy

J. L. Wright et al. (Eds.): AHFE 2021, LNNS 264, pp. 288–295, 2021.
https://doi.org/10.1007/978-3-030-79763-8_35

to set up inside of a head borne system. Clearly computational approaches are more flexible and inexpensive [4]. In this paper, we present a computational approach to a FOV evaluation of a prototype bomb suit helmet based on 3D scans and facial information from a large anthropometric survey database.

2 Materials and Methods

2.1 Individual FOV Test of the Helmet

The helmet tested is available in four sizes: small, medium, large and x-large based on the current US Army standard issue ballistic helmet (advanced combat helmet (ACH)) size prediction table (Table 1). In order to fit each helmet size to its target population we segmented the 2012 Anthropometric Survey of U.S. Army Personnel (ANSUR II) [7] population into four corresponding groups according to their predicted ACH size. For each helmet size group, an average head model was generated from the associated 3D head scan database (due to excessive hair in female head scans only male heads were used to create the average models). These head models provided reference cornea locations for corresponding sized helmet during FOV testing.

Table 1. ACH size prediction chart.

Size	Head length	Head breadth	Head circumference
Small	≤184 mm	≤162 mm	≤538 mm
Medium	184–198 mm	≤162 mm	538–573 mm
Large	198–210 mm	≤162 mm	573–597 mm
x-Large	>210 mm	>162 mm	>597 mm

Figure 1 plots ACH size coverage on the ANSUR II subpopulation, and Table 2 shows mean values of the four head measurements within each ACH size group, separated by sex (no females fall into the x-large size). The generated average male head models were adjusted to match the dimensions of the mean male head in Table 2 for each helmet size. Figure 2 shows the corresponding four head models. These head models are aligned to Frankfort plane[1] so that the variation of cornea location can be decomposed into vertical (superior/inferior), lateral and anterior/posterior directions for the population testing (Sect. 2.2).

[1] See Fig. 6. Frankfort plane is defined by left and right tragions and orbitales 1) and coincided with the X-Z plane in our 3D coordinate system where the Z axis pointing from the left to the right and the X axis pointing to the front and the origin is located at the middle point between two tragions.

Computing the field of view of the helmet is based on ray-tracing originating from the cornea. The ACH system is comprised of a seven pad suspension system to create standoff between the helmet shell and the wearer's head. This system comprises of a front and back square pad, two oblong pads on both the right and left side of the helmet, and a circular crown pad at the center of the helmet. When a helmet is properly donned, the top of the head should touch the crown pad and the forehead should touch the front pad. Before conducting the virtual fit with the head models, the prototype helmets are scanned on three different sized headforms to determine actual reference touch points. Then the average head models were aligned to the corresponding headform scans. With the helmet scan and head model in good alignment, FOV was computed by tracing the line of sight, starting from the cornea position towards infinite. The maximum angular range of unblocked line of sight is considered as the view range. A front view of the aligned helmet and head models is shown in Fig. 3.

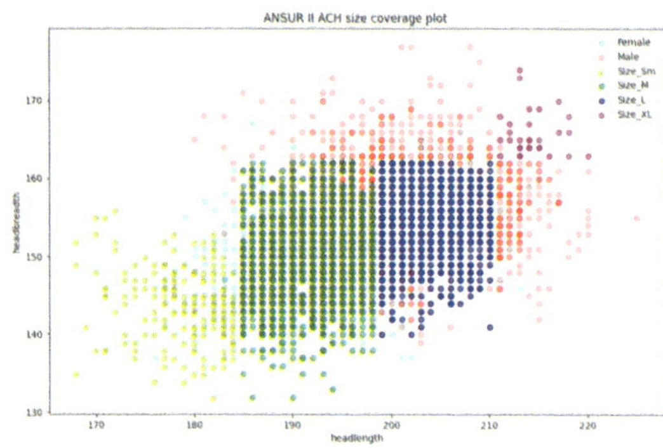

Fig. 1. Coverage of four ACH sizes on the ANSUR II population based head length (mm) and head breadth measurement (mm)

Table 2. Mean head dimensions (mm) of male within each ACH size and number of scans used to generate the average head models

	Number of 3D scans	Head length	Head breadth	Head circumference	Tragion to top of head
Small	23	180.5	147.4	531.2	124.9
Medium	1303	193.4	152.0	559.9	129.3
Large	1284	203.8	155.0	583.3	132.3
xLarge	35	214.5	166.0	614.5	138.8

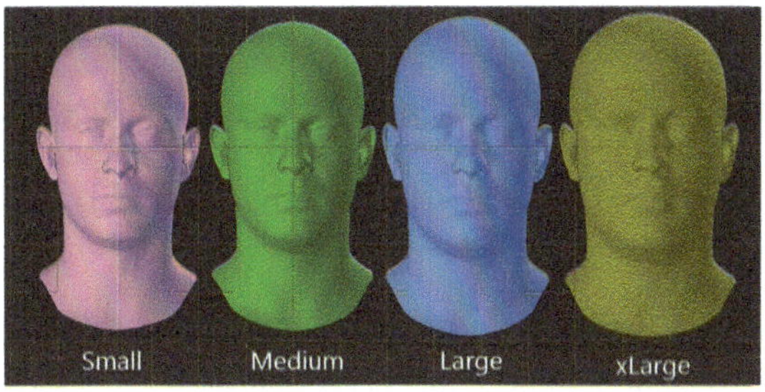

Fig. 2. ACH sized male head models.

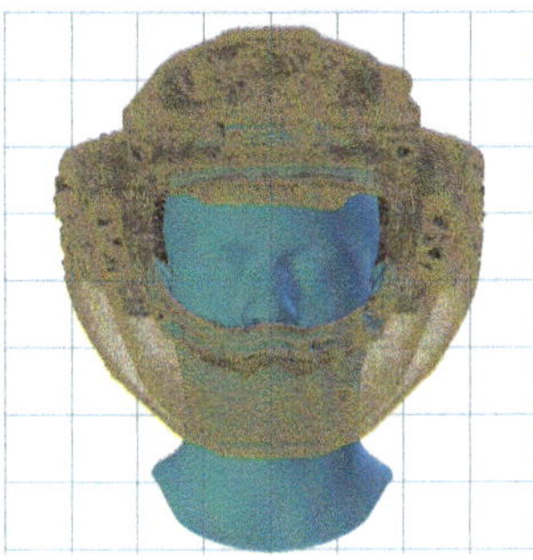

Fig. 3. Aligned helmet and head model

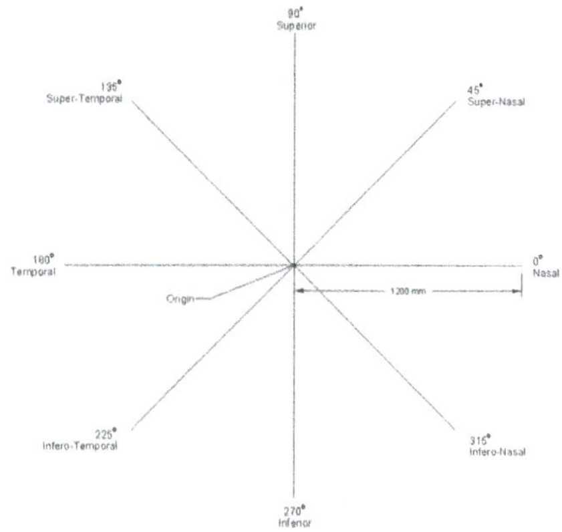

Fig. 4. NIJ static FOV axes of the left pupil

Once the digital helmet is aligned over the digital head, the four FOV planes (horizontal, vertical, 45° and 135°) for each side of the eye are assessed according to "Public Safety Bomb Suit NIJ Standard-0117.01" (2016) [8] (Fig. 4). Along each plane, the sight line rotates, as an eyeball moves, from −90° to 90° in one degree increments. When the sight line goes through the clear face shield area, there is no surface intersection taking place. The lower and upper sight line/helmet intersection angles are reported from the software and the angle interval for this space is counted as FOV range. Obstruction of the line of sight from facial structures (e.g., nose, eye socket) was not computed at this time. Figure 5 shows the FOV planes and their angle directions.

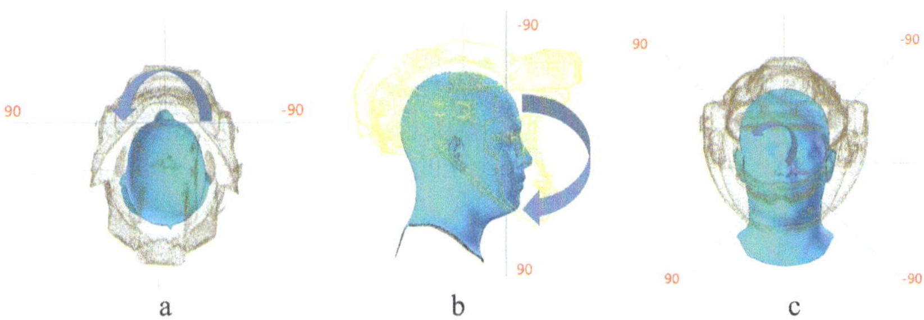

Fig. 5. FOV plane and angle direction: a) horizontal plane, b) vertical plane, c) 45° and 135° plane presume the eye freely move from −90 to 90°.

2.2 Population Test

The above computation of the FOV is limited to a single head model, which is central to the corresponding size group. However, the variation of cornea location needs be considered if we want to understand the FOV variation for the targeted population. The cornea location relative to the helmet is affected by 1) head length, 2) interpupillary breadth (IPB), 3) cornea to glabella distance in the sagittal plane (cornea depth) and 4) cornea to the top of head distance (cornea height) (these variables are defined and illustrated in Fig. 6). Table 3 presents the summary statistics of the above four measurements, by sex and within each ACH size population. It shows that while the mean head lengths of male and female are close within each size group, the differences and variations of the other three measurements are relatively large. This may indicate sex based morphological differences and the strong influence of local shape features (e.g., eye socket shape). In order to accommodate such differences, we combine the measurements from both sexes in the following analysis.

Table 3. Statistics of cornea location related measurements for each ACH size by sex (mm)

		Head length		IPB		Cornea height		Cornea depth	
		Mean	STD	Mean	STD	Mean	STD	Mean	STD
Small	Male	180.5	2.9	61.1	3.5	111.4	5.6	14.5	3.0
	Female	178.7	4.1	58.9	3.0	106.9	6.3	13.3	3.0
Medium	Male	193.4	3.4	62.9	3.3	114.7	6.3	16.0	2.8
	Female	190.6	3.5	61.4	3.3	110.6	5.9	13.7	3.2
Large	Male	203.8	3.0	64.5	3.1	117.3	6.2	16.8	2.8
	Female	201.8	2.6	63.8	3.1	114.1	6.0	13.4	3.3
X-Large	Male	214.5	2.7	67.9	3.8	123.2	7.7	13.4	3.2
	Female	–	–	–	–	–	–	–	–

Figure 7 displays the ANSUR II distribution of IPB, cornea height and cornea depth offsets of each head length interval in each size group. The offsets are computed by subtracting each variable from the corresponding values of the average head model, represented by the red dots in Fig. 7, within a size group. From this analysis, it becomes obvious the female population is concentrated in the size small, medium and large helmet size groups. Since the head models are created from male head scans, adding female data skewed the range of the above three measurements together with head length for size small and medium groups, but not significantly for size large group. There is no effective female population within the x-large group because there were no female cases in this subset.

For each measurement variable, two extreme cases were chosen for each helmet length represented by blue dots in Fig. 7. At head length 168 mm only one case of IPB and cornea depth were chosen because the maximum values of these two variables are

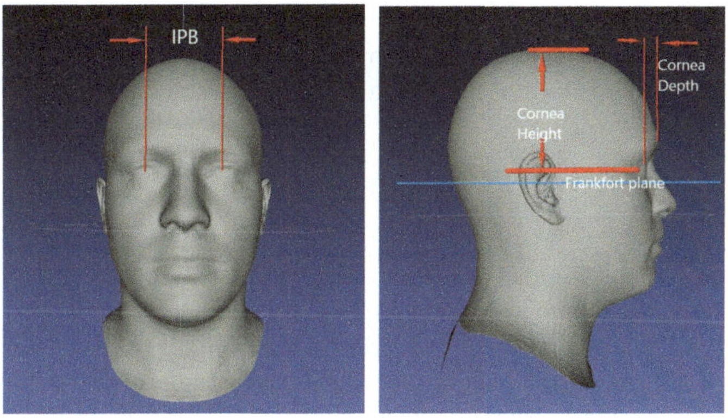

Fig. 6. Four measurements that affect cornea location

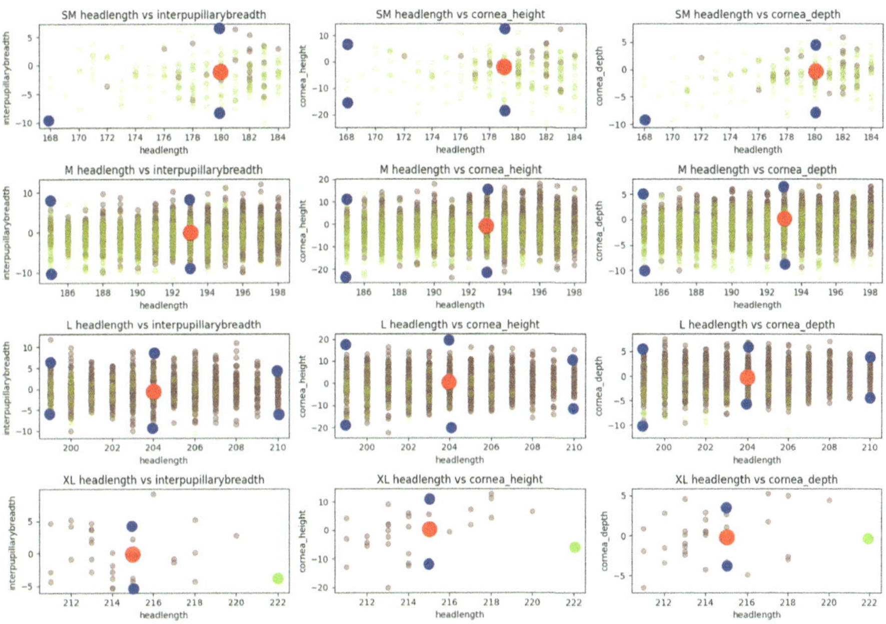

Fig. 7. The distributions of IPB, cornea height and cornea depth offsets vs. head length for the four helmet size groups. Sampled points are highlighted by big color dots, where blue dots represent individual cases in each size group; red represent the central male head model in each size group; and green represent the maximum head length in the x-large group.

close to zero. Similarly, there is only one case available at 222 mm head length. Using these case examples, we were able to obtain corresponding lower and upper amounts of offsets for IPB, cornea height and cornea depth. This resulted in 47 test cases. Running the FOV computation on those test cases provided a complete view of FOV variation, including best and worst scenarios.

It was noticed in the size small group, the head length of the central head model is 180 mm, which is not the middle value of this subpopulation. This shift is due to the large number of female heads in this size group while our central head model is generated from male scans only.

3 The Results

The results from the individual FOV test on the left eye of the head models are shown in Table 4. FOV was computed based on three repeated scans of the helmet on the headform, to simulate actual variation of cornea location due to the don-doff process. Each head model was aligned to corresponding helmet-headform scan. The mean FOV values from the three tests were presented. Note that FOV range decreases as helmet size increases for this particular item. This may be a result of increased distance from the helmet surface to the cornea due to the helmet's grading and may be unique to the design of this system.

In the population FOV test, the effect of variation of cornea location to FOV range can be summarized as:

- Cornea depth offset has the biggest effect on FOV ranges for all view directions under the same head length condition.
- Head length offset has a similar effect as cornea depth offset.
- Cornea height offset affects lower or upper view angle for vertical, 45° and 135° view directions, however FOV ranges remain relatively stable.
- IPB offset has the least impact to all FOV ranges (<3° in any direction).

Table 4. Summarized mean FOV angle values and ranges (in degree) for each helmet size (left eye) in the individual FOV test

	Horizontal			Vertical			45Degree			135Degree		
	T	N	R	S	I	R	ST	IN	R	IT	SN	R
Small	-73	57	130	-27	50	77	-34	58	92	-60	39	99
Medium	-71	57	128	-29	47	76	-38	55	92	-58	43	101
Large	-69	49	118	-32	37	69	-40	43	83	-48	45	93
X-Large	-71	43	114	-31	38	69	-38	40	78	-51	43	94

Note: T: temporal, N: nasal, R: range (=N – T, I – S, IN – ST or SN - IT), S: superior, I: inferior, ST: super-temporal, IN: infero-nasal, SN: super-nasal, IT: infero-temporal, as defined in Figure 4, 5

4 Conclusions and Discussion

The computational approach of FOV evaluation can effectively cover a large population and provide comprehensive results. However, it does not take into account the maximum limits of the visual field of the eyes. It only checks the geometric relationship between a sight line and the helmet surface. If the computed FOV range is larger than actual eye FOV, then the actual visual field of the eyes should take precedence over the computed FOV range. For the future work, we will compare the computational approach to perimetry testing of human subjects in the prototype helmet system.

The variation of FOV within a sample population is determined by the variation of cornea location relative to the helmet, primarily in three axial directions when the head is aligned to Frankfort plane. Moreover, the FOV range is mainly affected by head length, cornea height and cornea depth. That means local shape around the eyes, rather than major global measurements, has greater impact on FOV range. Even head breadth is a less important measurement because its correlation to interpupillary breadth is low (male's Pearson correlation coefficients 0.24, female's 0.16).

In summary, this computational approach to evaluate FOV has shown to be useful for the assessment of headborne equipment. However, in the future we may want to improve the approach in three aspects: a) develop female head model to reflect female facial geometry, b) obtain 3D coordinates of cornea location and use 3D point cloud of cornea location for population FOV test, and c) take obstruction of facial geometry into account when computing the ray-tracing.

References

1. Hussein, W., Nazeeh, M., Sayed, M.M.A.: Investigation of driver's FOV and related ergonomics using laser shadowgraphy from automotive interior. J Ergon. **7**, 207 (2017). https://doi.org/10.4172/2165-7556.1000207
2. Williams, H.P., Littman, E.M., Folga, R.V., Patterson, F.R.: Field of view evaluation for flight simulators used in spatial disorientation training. NAMRU-D Report Number 14–41 (2014)
3. Gaspar, J.G., et al.: Measuring the useful field of view during simulated driving with gaze-contingent displays. Hum. Factors **58**(4), 630–641 (2016). https://doi.org/10.1177/0018720816642092. Epub 2016 Apr 18 PMID: 27091370
4. Vinayak, Sen, D.: Geometric estimation of FoV for vision modeling in DHM. SAE Int. J. Passeng. Cars – Mech. Syst. **2**(1), 1614–1625 (2009). https://doi.org/10.4271/2009-01-2294
5. Karmakar, S., Pal, M.S., Majumdar, D., Majumdar, D.: Application of digital human modeling and simulation for vision analysis of pilots in a jet aircraft: a case study. Work **41**(Suppl 1), 3412–3418 (2012). https://doi.org/10.3233/WOR-2012-0617-3412. PMID: 22317240
6. Rix, J., Stork, A.: Combining ergonomic and field-of-view analysis using virtual humans. In: Proceedings of the International Conference on Design and Manufacturing (IConDM 2013), pp. 622–631 (2013)
7. Gordon, C.C., et al.: 2010–2012 anthropometric survey of U.S. army personnel: methods and summary statistics. Technical report (Natick/TR-15/007). U.S. Army Natick Research, Development and Engineering Center, Natick, MA 01760-2646 (2014)
8. https://nij.ojp.gov/library/publications/public-safety-bomb-suit-standard-nij-standard-011701

Data-Driven Personas: Expanding DHM for a Holistic Approach

Ari Kolbeinsson[1(✉)], Erik Brolin[1], and Jessica Lindblom[2]

[1] School of Engineering, University of Skövde, Skövde, Sweden
{ari.kolbeinsson,erik.brolin}@his.se
[2] School of Informatics, University of Skövde, Skövde, Sweden
jessica.lindblom@his.se

Abstract. In this paper, we detail research and development of data-driven personas in the IPS-IMMA digital human modelling system. Semi-automatically generating personas for working with user experience (UX) aspects of the Operator 4.0 side of Industry 4.0 is suggested as a viable approach for contributing to operator well-being and diversity by supporting workstation designers to take these factors into account early in the workstation design process. These data-driven personas are being developed to be generated using anthropometric data from the manikin family generation module of IPS-IMMA. Specific design suggestions are presented, what should be taken into account and how that will be implemented, and the current state of development of the data-driven personas module is discussed. Prototypes are planned under the coming year.

Keywords: Digital human modelling · DHM · Personas · Data-driven personas · Industry 4.0 · Operator 4.0 · Social sustainability

1 Introduction

Digital human modelling (DHM) has been increasingly used within industry in recent years as a tool for evaluating the design of both products and workstations. This use of DHM allows designers and production engineers to cost-effectively identify problems before a product or a workstation have been built, leading to lower overall cost through faster development with fewer prototypes having to be built, as well as less time spent on fixing after building the product or workstation. Important goals of using DHM in the development process for workstations within industry include lowering the rates of work related musculoskeletal disorders (MSD) through improved ergonomics of the workplace, as well as supporting a high level of worker satisfaction. Restrictions in the physical interaction between humans and systems, being a reason for MSDs, can often be connected to body dimensions of the users. Several methods have been developed for the consideration of body size related anthropometric diversity in design [e.g. 1, 2]. However, the physical human-machine interaction is not only affected by the size and proportions of a user but also other physical user characteristics, e.g., muscle strength and joint range of motion (ROM) [3]. To include more diverse data when generating virtual

J. L. Wright et al. (Eds.): AHFE 2021, LNNS 264, pp. 296–303, 2021.
https://doi.org/10.1007/978-3-030-79763-8_36

human models, so called manikins, is challenging since it increases the dimensionality and complexity of the data. Cluster analysis has been found to enable the generation of well distributed test cases with different body size, strength and joint range of movement, as well as other capability measures when data are available [4]. The output of a cluster analysis of diverse anthropometric data could be used to augment traditional methods for creating personas. Integrating personas into DHM systems also offers the possibility of supporting diversity goals at manufacturing plants, providing workstation designers with an intuitive tool for empathising with workers, and highlighting workers of different genders, ages, sizes, as well as their strengths and limitations. Being able to perform at least part of the work involved in creating personas automatically, while increasing the diversity of the personas, is arguably a large leap forward in working with personas to promote diversity.

An insight will be offered here into the development of a persona generation system into existing DHM software called IPS-IMMA which is being developed in collaboration with Swedish industry. IPS-IMMA (which will be referred to as IMMA throughout the remainder of this paper) is currently aimed primarily towards simulating industrial workstations, although the system can be used for a variety of ergonomics analysis, and is under active development. IMMA supports the creation of manikins or families of manikins, with these manikin families being generated from large databases of anthropometric data. The manikins are then imported into a scene where each manikin is simulated going through an activity, such as an assembly activity in manufacturing. Ergonomics evaluation results are then shown to the user, with warnings being displayed if assessed factors are seen to exceed recommendations.

Data-driven personas are seen as a natural extension of the IMMA manikin family generations system, as well as fitting neatly into evaluation tools in the future, in the end also being useful for communicating the results of ergonomics evaluations to non-experts, and supporting diversity goals in the workplace.

2 Personas

When discussing personas it is important to understand what is meant with the term. In interaction design and users experience design (UXD) personas are used as “user models”, i.e., a model of users and were first introduced in 1999 by Alan Cooper, and further developed in later work [5]. Crucially, personas are not made up based on what the designer thinks that users might be like, but rather based on user research, and should always be based on as much real-world data as possible to avoid implicit biases [5, 6]. A persona is not a real person or based off a user, but rather based on amalgamated empirical data on multiple users.

As personas are models of users it is also important to remember that models are simplifications, and may be aimed at being accurate only for a certain scenario. Likewise, personas may be created to support one or more aspects of the design process, and may be created to explore one or more user aspects, as required [6]. Most commonly, a persona has a goal and some *attributes*, and is used in conjunction with a *scenario* which contains *activities* and a *context*.

Personas have been found useful to humanise users to the design team, as well as to support different viewpoints, as personas are commonly used with a scenario and a form of role-playing to support immersion [5, 6]. That being said, personas have also been found to be a useful communication tool, within a project, to external parties, and to those on the periphery of the project such as managers. Using personas for communication can help justify why it is reasonable to spend money on a more ergonomic workstation, or why a product does not yet fulfil the design goals and requires some re-work [5]. Multiple personas are commonly used, with each persona possibly having different goals, (somewhat) different activities to perform, and even differences in contexts [5]. Each persona should contain meaningful details that have some bearing on the activity being explored.

Creating personas so that they represent the breadth of the possible group of users is time consuming, both in terms of data gathering and synthesising data. Automating parts of the process can therefore save time, especially as DHM already relies upon a lot of anthropometric data that allows the creation of diverse families of manikins. Furthermore, possibilities exist for interpreting some aspects of the anthropometric data to extract keywords that can be applied to the generated persona and can affect the ergonomic analyses, as well as manually adding keyword, increasing diversity and reducing the chance for bias when creating personas.

3 ICT as an Enabler of Sustainable Behaviour

A central basis for the sustainability movement is the United Nation report from 1987, in which the famous quote "development that meets the needs of the present without compromising the ability of future generations to meet their needs" is expressed. The Brundtland report is based on the idea of three overlapping pillars that cover *environment*, *economy*, and *society*. Social sustainability has been somewhat neglected, compared to the other two pillars [7], but is the one that is most applicable in this instance. Social sustainability covers a broad variety of issues, such as fulfilling basic human needs, equality of rights, opportunity for learning and self-development, security, health effects, sense of community attachment, and well-being [7]. The UN 2030 Sustainable Developmental Goals succeeded the Brundtland report in 2015, and aim to achieve long-lasting and immensely better development prospects within 17 prioritised areas.

A generic path to sustainability is to stimulate people to change towards more sustainable practices and behaviours. Mayordomo-Martínez et al. [8] explain that sustainability can be viewed from two complementary perspectives concerning how the usage of ICT has a positive effect on sustainability. They argue that there is a difference between on the one hand "sustainable in ICT", and on the other hand "sustainable by ICT" [8]. The former perspective refers to whether the ICT device itself is sustainable, e.g., a sustainable design process was applied to develop a mobile app. The latter perspective refers to whether using the mobile app has a positive effect on sustainability, e.g., the purpose of the app could be to support people living with cognitive impairments, or supporting the health and well-being of workers in manufacturing. This perspective on sustainability suggests that digitalisation can be a driving force for enabling sustainable development by focusing on user and societal needs and motives when developing and using ICT. Experience with these issues exists within the fields of Human-Computer interaction (HCI) and UXD [6, 9], where focusing on different groups of users in order to develop a positive user experience is a central part of common methods, promoting wellbeing @work. In this paper, the "sustainable by ICT" is considered from the wellbeing @work perspective, within which we focus on digital ergonomic assessments in the context of Industry 4.0 [9].

4 Sustainability Through Diversity via Data-Driven Personas

Industry 4.0, or the fourth industrial revolution, is characterised by the recent and rapid advances in digitalisation and intelligent autonomous technologies, e.g., internet of things, augmented and virtual reality, and collaborative robots that are entering the manufacturing domain [9]. The intended benefits of Industry 4.0 include shorter development periods, individualisation on demand for the customers, flexibility, decentralisation, and resource efficiency [9]. One consequence of this revolution is a shift in working environments throughout all aspects of the manufacturing industry towards increasingly advanced digital work spaces, leading to Operator 4.0; workers skilled in the use of intelligent tools that enhance their physical and/or cognitive capabilities, while increasing safety and well-being [10].

The changes of work roles and work practices due to more advanced forms of human-technology interaction at the shop floor will make work more demanding and flexible, and an interest in human-centred aspects has the potential to promote well-being and professional development in order to be a desired workplace. However, the envisioned digital workspace of Operator 4.0 will not materialise automatically [11].

Applying a UXD approach to the operator 4.0 paradigm has the potential to promote sustainability, and of particular interest in the DHM context is the possibility of lifting diversity. This is particularly important as the intended users of DHM tools have widely varying experience with UXD aspects of users, which means that introducing accepted and validated UXD approaches that motivations, prerequisites, and motives of primary end-user groups can be expected to result in an improved user experience, leading to a more sustainable digital work environment.

Various aspects of diversity could be: professional skills and knowledge, digital competence, prior experience of using CAD tools, age, gender, the current work practices and work culture of the actual community and/or workplace, as well as disabilities and cognitive or physical impairments [6, 12].

Moreover, in this particular context of IPS IMMA, the diversity of primary end-user groups ranges from very skilled users of the IPS IMMA software with a more engineering or manufacturing backgrounds who have been using 3D simulations tools for a long period of time that has become "trained professionals" within their company, and 'newer' users of the IPS IMMA software with various backgrounds and having prior experience of 3D simulation tools, to new users of the IPS IMMA software with limited or no experience of 3D simulation tools but having the background and competence of being professional ergonomists. On the other hand, one of the main strengths with using DHM tools is the ability to simulate a diversity of manikin families and their related attributes in the IPS IMMA software with regard to the current anthropometric module and improved bio-mechanical model generation.

Hence, the ability to simulate and evaluate various ergonomic aspects in manual assembly in manufacturing, one can identify and reduce physical problems among

Operator 4.0 and promote a sustainable (physical) ergonomic work environment. Thus, both perspectives of sustainability are contributing to improving and enhancing the functionality and interaction quality the IPS IMMA software.

5 Generating Data-Driven Personas in IMMA

Creating data-driven (or semi-automatically generated) personas requires a few components. This section will examine some of the requirements, as well as what will be shown in the first prototypes and sketches of how this will be shown (early UI sketches). More detail around components such as the manikin family generation module will be presented in following sections.

Interviews and discussions with ergonomics specialists with low DHM software experience were used to identify what personas within DHM might be useful for, and which aspects they see as being most important. In addition to this, literature on personas has been used throughout, as well as multiple sketches and mockups of both details and an overview of the system being developed. Offered here is an insight into the development process.

Interviews with ergonomics specialists identified the ability to link personas to the analyses where forces are visualised (see Fig. 1) or summarised as being potentially useful. They already use a form of personas, but are positive to getting support in generating the initial personas based on anthropometric data, and find the idea of personas that are linked to the manikins being simulated interesting.

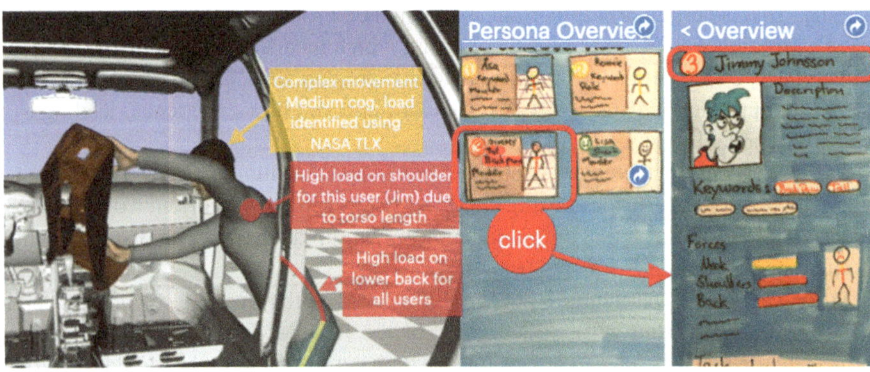

Fig. 1. A sketch of the persona supported ergonomics analysis module. This shows an overview of the personas, and on the right is a detail view of one persona.

Design limitations for the initial prototype mostly involve how the persona generator UI can be fit together with the current manikin family generator, as well as how visualisations of personas can be implemented in either the IMMA system itself or in a related ergonomics visualisation module. Figure 2 shows the initial mockup of the persona generation module within the manikin family generation/anthropometry module of IMMA. The personas module is seen as an extension of the anthropometry module, with the manikin linking together anthropometrics and persona data. As the IMMA software itself, the anthropometry module is under active development and planned to have functionality to generate manikins with specific age and body size as well as strength, as maximum joint torque, and joint range-of-motion. These resulting manikins will have data similar to the virtual test persons based on diverse anthropometric data presented by [4] which gives a basis for automatically generating input for the personas creation pane, see Table 1. The personas creation pane is located under the manikin family creation pane, and allows the user to specify the number of personas, and how the personas are generated in relation to the manikins. The simulation may be run on a larger number of manikins, while the personas can be generated from a subset of the manikins. The "spread" shown in the personas pane is not a simple percentile spread, as the manikins are generated using more advanced rules.

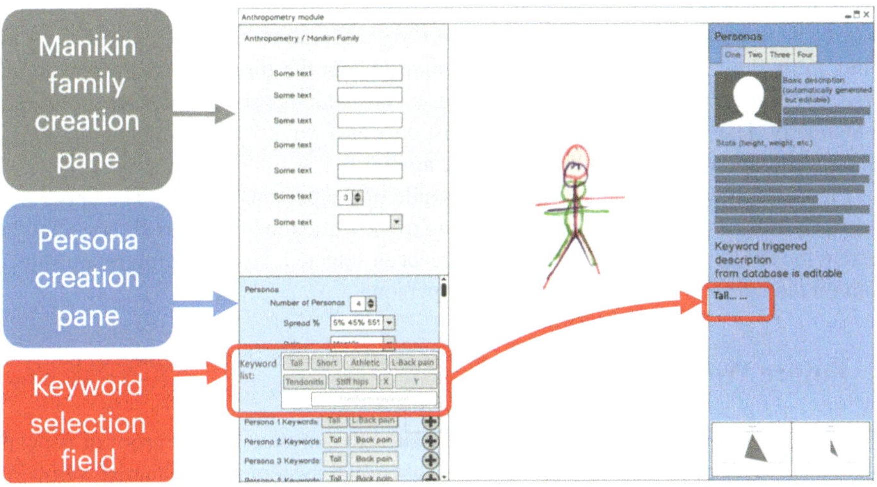

Fig. 2. A mockup of the persona generation panel, including the personas pane on the left and persona details on the right

Table 1. Summary description of cluster generated cases [4].

#Case	Age [year]	Sex	Body weight [kg]	Stature [cm]	Joint ROM	Joint torque strength
Case 1	37	Female	56	170	Med/High	Low/Med
Case 2	43	Female	64	163	Med	Low
Case 3	46	Female	88	180	Low/Med	Med
Case 4	43	Male	85	178	Med	High
Case 5	67	Female	77	154	Low/Med/High	Low
Case 6	56	Male	84	170	Low/Med	Med/High

The persona pane shows a list of keywords. These keywords fall into three main categories: anthropometrically based (automatically generated), attributes (partly automatically generated, partly manually selected), and descriptors (manually generated, possibly a random function as well). The anthropometrically based keywords are automatically generated from physical attributes in the anthropometric data. The attributes include factors based on anthropometric data, such as size and build, as well as having the possibility of manually adding attributes that can affect the analysis, as well as certain keywords affecting the simulation itself. Examples of factors that can influence the analysis are e.g. lower back pain, tendonitis, strong, etc., while keywords that can affect the manikin creation and the simulation can be e.g. long legged, broad-shouldered, etc. Finally, the descriptors are entirely manually added and could be partially randomly (or rule-based). These manual keywords mostly come from a database, and can provide descriptions of aspects of the persona that could affect the generated simulation due to diverse strategies and behaviour or may be relevant to further analyse the activity, e.g., for analysing cognitive elements or user experience of the workplace.

Below the keywords list are tools to edit or add keywords from a database that offers keywords that can be applied on the analyses. Examples of such a keyword is "Lower back pain" which would lower the threshold values for acceptable forces on joints in the lower back, or "tendonitis" which lowers the threshold for forces on the wrist. In addition to this, descriptor keywords can be added based on presents, or written in manually. These descriptor keywords exist only as a way to add characteristics to the persona and are not active in the ergonomic analyses.

Figure 1 shows sketches of the analysis side of the persona system. The upper part shows an overview of the four personas being used, while the lower part shows what the system looks like when a specific persona has been selected. Note how forces are shown both on the manikin and the corresponding persona.

6 Future Work

The development of the personas module is slated to continue, with a proof of concept prototype planned for mid-2021, and more functional system parts planned following that. These will be validated through user testing, with a particular focus on ergonomics specialists and cognitive ergonomics analysis. A future goal is to explore the effects of using data-driven personas for supporting more factors than physical ergonomics during workstation developments, in particular cognitive ergonomics, but also from a wellbeing @work perspective although the focus will be on ergonomics specialists, this will have to take into account the communication between ergonomics specialists, engineers designing the workstations, and decision makers.

Acknowledgements. This work has been realized by the Knowledge Foundation and the INFINIT research environment (KKS Dnr. 20180167) who have financially supported the Synergy Virtual Ergonomics project (SVE), as well as partner organizations.

References

1. Speyer, H.: On the definition and generation of optimal test samples for design problems. Kaiserslautern Human Solutions GmbH (1996)

2. Dainoff, M., Gordon, C., Robinette, K.M., Strauss, M.: Guidelines for using Anthropometric Data in Product Design. HFES Institute Best Practices Series. Human Factors and Ergonomics Society, Santa Monica (2004)
3. Frey Law, L., Xia, T., Laake, A.: Modeling Human Physical Capability: Joint Strength and Range of Motion. Handbook of Digital Human Modeling. CRC Press, Boca Raton (2009)
4. Brolin, E., Högberg, D., Hanson, L., Björkenstam, S.: Virtual test persons based on diverse anthropometric data for ergonomics simulations and analysis. In: NES 2017 "Joy at Work", pp. 232–239. Lund University, Lund (2017)
5. Cooper, A., Reimann, R., Cronin, D., Noessel, C.: About Face: The Essentials of Interaction Design. Wiley, Hoboken (2014)
6. Hartson, R., Pyla, P.S.: The UX Book: Agile UX Design for a Quality User Experience. Morgan Kaufmann, Burlington (2018)
7. Boström, M.: A missing pillar? Challenges in theorizing and practicing social sustainability: introduction to the special issue. Sustain.: Sci. Pract. Policy **8**(1), 3–14 (2012)
8. Mayordomo-Martínez, D., et al.: Sustainable accessibility: a mobile app for helping people with disabilities to search accessible shops. Int. J. Environ. Res. Public Health **16**(4), 620 (2019)
9. Kagermann, H., Helbig, J., Hellinger, A., Wahlster, W.: Recommendations for implementing the strategic initiative INDUSTRIE 4.0: securing the future of German manufacturing industry; final report of the Industrie 4.0 Working Group. Forschungsunion (2013)
10. Romero, D., Bernus, P., Noran, O., Stahre, J., Fast-Berglund, Å.: The operator 4.0: human cyber-physical systems & adaptive automation towards human-automation symbiosis work systems. In: IFIP International Conference on Advances in Production Management Systems, pp. 677–686. Springer, Cham (2016)
11. Rosén, J., Lindblom, J., Lamb, M., Billing, E.: Digital human modeling technology in virtual reality-studying aspects of users' experiences. In: DHM2020: Proceedings of the 6th International Digital Human Modeling Symposium, August 31–September 2, vol. 11, p. 330. IOS Press (2020)
12. Himmelsbach, J., Schwarz, S., Gerdenitsch, C., Wais-Zechmann, B., Bobeth, J., Tscheligi, M.: Do we care about diversity in human computer interaction: a comprehensive content analysis on diversity dimensions in research. In: Proceedings of the 2019 CHI Conference on Human Factors in Computing Systems, pp. 1–16 (2019)

Virtual Human Hand: Wrist Movements

Esteban Peña-Pitarch[1(✉)], Inmaculada Puig de la Bellacasa[2], Jesus Fernando Padilla Magaña[1], Anas Al Omar[1], and Iñaki Alcelay Larrión[1]

[1] Mechanical Engineering Department, Escola Politècnica Superior D'Enginyeria de Manresa (UPC), Av. Bases de Manresa, 61-73, 08242 Manresa, Spain
{esteban.pena,jesus.fernando.padilla,anas.al.omar, inaki.alcelay}@upc.edu

[2] Hand and Upper Extremity Surgery, Mútua de Terrassa Hospital Universitari, Plaça del Doctor Robert, 5, 08221 Terrassa, Spain

Abstract. Hand model with 25 Degrees of Freedom (DOF) was developed and implemented considering forward and inverse kinematics. However, the model, after some experience and use need to add new DOF. The proposed model is with 29 DOF, these new DOF serve to simulate the arc of the palm in the part of the wrist.

Once we locate a new coordinates system in the end of the radius close to scaphoid we apply Denavit-Hartenberg for all the joints. Forward and inverse kinematics will be applied. Ligaments to apply restrictions in wrist movement are included. This affect to the fingertip position.

New model of virtual human hand with more accuracy in presented and validated with a Cyberglove™ and Leap Motion.

Keywords: Human hand · Wrist · 29 DOF · Forward and inverse kinematics

1 Introduction

Author's hand with 25 Degree of Freedom (DOF) was proposed in [1]. With the forward and inverse kinematic for all the fingers, they proposed a realistic virtual simulation. However, the wrist not was included in the model. Today, several authors proposed in the literature that the wrist has a relative movement between the two rows of bones with eight bones. [2] discuss a comparison of four joint coordinates systems previously described in the literature. [3] proposed a helical movement of wrist bones in distal movements. During thumb oppositional motion, internal rotation of the first metacarpal occurred, with the palmar base rotating primarily with respect to dorsal base. This is one of conclusions achieved by [4], and it follows that first metacarpal flexes and pronates with the dorsal base as the center. However, [5] do not consider in their work a significant factor as the pisiform, which plays a minimal role in wrist kinematics.

When comparing healthy with osteoarthritic subjects, the minimal joint space was slightly higher during the neutral, adduction, extension and lateral key pinch configurations than during the abduction, flexion, power grasp and jar twist configurations is analyzed in [6].

J. L. Wright et al. (Eds.): AHFE 2021, LNNS 264, pp. 304–311, 2021.
https://doi.org/10.1007/978-3-030-79763-8_37

[7] and [8] have detail description of taxonomy for the hand. [9] studied the movement between different bones of the wrist, i.e. the movement of the radio-capitate joint during wrist flexion–extension and radio-ulnar deviation. [10] considered the motion of the scaphotrapezio–trapezoidal (STT) joint.

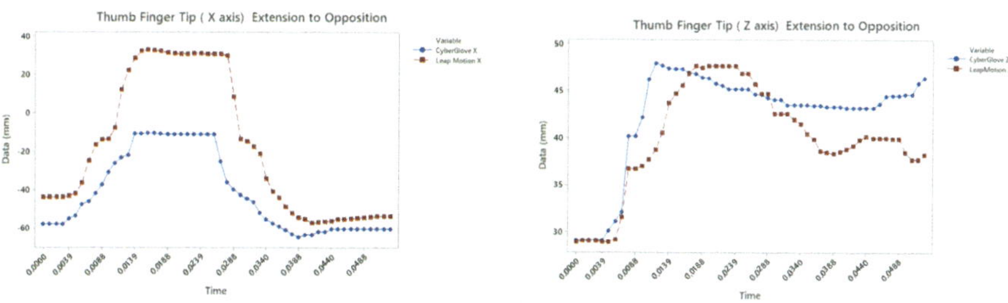

Fig. 1. For the X axis (left) we observed a big difference between the capture with Cyberglobe and Leap Motion. Similar for the Z axis (right).

Figure 1 shows a big difference in the thumb fingertip, red line (more realistic) is captured with the system of Leap Motion. The blue line is captured by Cyberglobe with eighteen sensors and adapted to model of 25 DOF. Based in this difference the need to remodeled the thumb based in the wrist movement is presented below.

From the aforementioned research, it is possible to conclude that the virtual human hand with 25 DOF will improve adding 4 DOF in the thumb chain.

Paper is organized as follow, in Sect. 2 material and methods is exposed. In Sect. 3 we show the results with a discussion. In Sect. 4 the conclusions are finally shown.

2 Material and Methods

All of joints in the hand are revolute and if we use a generalized coordinate q_i that represents one degree of freedom, then the generalized coordinate vector can be represented by $\boldsymbol{q} = \left[q_1, \ldots, q_{29}\right]$. However, the movements of fingers have some natural constraints (e.g. the middle finger cannot flex over 60°). Table 1 represents the joint limits, for the thumb, where they have units in radians. We avoid the other fingers because they are presented in [11] and [12].

Figure 2 shows a model proposed with 29 DOF.

DENAVIT-HARTENBERG (D-H) METHOD

To study forward and inverse kinematics we adapted Denavit-Hartenberg method (D-H method) [13] from robotics field to define the positions of the fingertips with respect to each local coordinate system. D-H parameters are defined as follows:

- θ_k is the angle of rotation from x_{k-1} to x_k measured about z_{k-1}.
- d_k is the distance from the origin of frame L_{k-1} to point b_k measured along z_{k-1}.
- a_k is the distance from point b_k to the origin frame L_k measured along x_k.
- α_k is the angle of rotation from z_{k-1} to z_k measured about x_k.

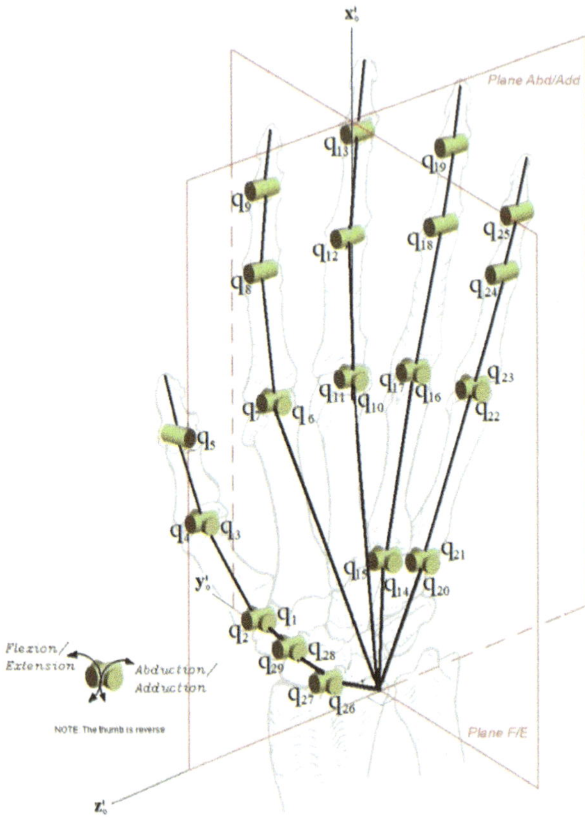

Fig. 2. Dorsal view right hand model.

Table 1. Joint limits for the thumb.

	Min.	Max.
q_1	0	$\pi/3$
q_2	$-5\pi/36$	$7\pi/36$
q_3	0	$\pi/3$
q_4	$-\pi/18$	$11\pi/36$
q_5	$-\pi/12$	$4\pi/9$

The transformation matrix ${}^{i-1}T_i$ is shown below and the subscript $i = 1 \ldots 29$

$$
{}^{i-1}T_i = \begin{pmatrix} \cos\theta_i & -\cos\alpha_i \sin\theta_i & \sin\alpha_i \sin\theta_i & a_i \cos\theta_i \\ \sin\theta_i & \cos\alpha_i \cos\theta_i & -\sin\alpha_i \cos\theta_i & a_i \sin\theta_i \\ 0 & \sin\alpha_i & \cos\alpha_i & d_i \\ 0 & 0 & 0 & 1 \end{pmatrix}
$$

D-H method is implemented in the thumb and Fig. 3 shows the sketch of D-H method and Table 2 shows the D-H parameters.

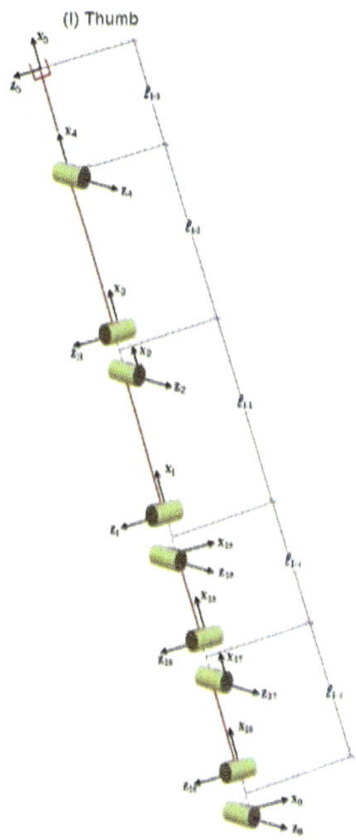

Fig. 3. Model for the thumb.

Where l_{I-1}, l_{I-2}, and l_{I-3} an found in [12]. $l_{I-4} \approx 11$ mm and $l_{I-5} \approx 22,7$ mm extracted from [14].

With these parameters we can have define the position vector, which is defined by $\boldsymbol{p}(\boldsymbol{q}^I)$ with respect to the local coordinate system by

$$
\begin{bmatrix} \boldsymbol{p}(\boldsymbol{q}^I) \\ 1 \end{bmatrix} = {}^{0}T_{26}\,{}^{26}T_{27}\,{}^{27}T_{28}\,{}^{28}T_{29}\,{}^{29}T_{1}\,{}^{1}T_{2}\,{}^{2}T_{3}\,{}^{3}T_{4}\,{}^{4}T_{5} \begin{bmatrix} 0 \\ 0 \\ 0 \\ 1 \end{bmatrix}
$$

Table 2. D-H parameters.

	θ_i	d_i	α_i	a_i
26	$q_{26}+\pi/2$	0	0	$-\pi/2$
27	q_{27}	0	l_{I-5}	$\pi/2$
28	q_{28}	0	0	$-\pi/2$
29	q_{29}	0	l_{I-4}	$\pi/2$
1	q_1	0	0	$-\pi/2$
2	q_2	0	l_{I-3}	$\pi/2$
3	q_3	0	0	$-\pi/2$
4	q_4	0	l_{I-2}	$\pi/2$
5	q_5	0	l_{I-1}	$-\pi/2$

where $\boldsymbol{q}^I = [q_{26}\ \ q_{27}\ \ q_{28}\ \ q_{29}\ \ q_1\ \ q_2\ \ q_3\ \ q_4\ \ q_5]^T$. If we want to represent the position vector with respect to the global coordinate system (wrist) there is one transformation matrix for the thumb defined by

$$ {}^0H_I = \begin{bmatrix} cos\,\gamma_I & -sin\,\gamma_I & 0 & -l_{Io}sin\,\gamma_I \\ sin\,\gamma_I & cos\,\gamma_I & 0 & -l_{Io}cos\,\gamma_I \\ 0 & 0 & 1 & 0 \\ 0 & 0 & 0 & 1 \end{bmatrix} $$

where γ_i is the angle between the ray (finger) i $(i = I, II, III, IV, V)$ and the global $y_0^{'}$ axis shown in the Fig. 2. In this case we only are talking about the thumb.

Therefore, the position vector with respect to the global frame is defined by

$$ \begin{bmatrix} \mathbf{w}(\mathbf{q}^{i}) \\ 1 \end{bmatrix} = [{}^0\mathbf{H}_i] \begin{bmatrix} \mathbf{p}(\mathbf{q}^{i}) \\ 1 \end{bmatrix} $$

The missed subindex $i = 6 \ldots 25$ are for the other joints, they do not change for the old model. For more information see [12].

3 Results

The algorithm has been implemented in a virtual hand environment. Table 3 presents neutral position of one hand to maintain sequential order, figures and their citations have been renumbered. Please check and correct if necessary.

Table 3. Angles for the neutral position of hand (in degrees).

Neutral gesture of the hand				
Thumb	Index	Middle	Ring	Little
$q_1 = 0$	$q_6 = 0$	$q_{10} = 0$	$q_{14} = 0$	$q_{20} = 0$
$q_2 = 0$	$q_7 = 30$	$q_{11} = 30$	$q_{15} = 2$	$q_{21} = 5$
$q_3 = 30$	$q_8 = 30$	$q_{12} = 30$	$q_{16} = 0$	$q_{22} = 0$
$q_4 = 0$	$q_9 = 10$	$q_{13} = 10$	$q_{17} = 30$	$q_{23} = 30$
$q_5 = 30$			$q_{18} = 30$	$q_{24} = 30$
$q_{26} = 0$			$q_{19} = 10$	$q_{25} = 10$
$q_{27} = 0$				
$q_{28} = 0$				
$q_{29} = 0$				

By the forward kinematics we can obtain the position vectors of fingertips (millimeter) corresponding to the neutral gesture of the hand shown in Table 4.

Table 4. Fingertip positions corresponding to the neutral gesture (with respect to the global coordinates).

Neutral gesture of the hand					
Position	Thumb	Index	Middle	Ring	Little
x	−109.9	3.14	16.28	36.34	55.37
y	79.3	158.25	159.71	145.73	118.74
z	−15.8	−69.5	−77.44	−79.3	−68.14

4 Discussion

The hand arches in two parts of the palm, that means, the two bones trapezium and scaphoid wrist bones move relative between them and the thumb metacarpal bone have movement with respect to the trapezium [4]. Scaphoid bone moves between the trapezium and the radius. Global coordinates system is located in the radius shown in Fig. 4. The other bones located in the wrist, trapezoid, capitate and lunate don't have movement between them and the hamate, pisiform and triquetum have relative movement between then. However, these movements are not significant for the position of the fingertip.

The action of the ligaments of the wrist are not trivial. The ligaments contribute to stabilize the bone. As a future work is interesting to study the influence of each ligament, observing which ligament is working when there is some movement of the hand, like opposing thumb to finger, palmar abduction, or retroposition.

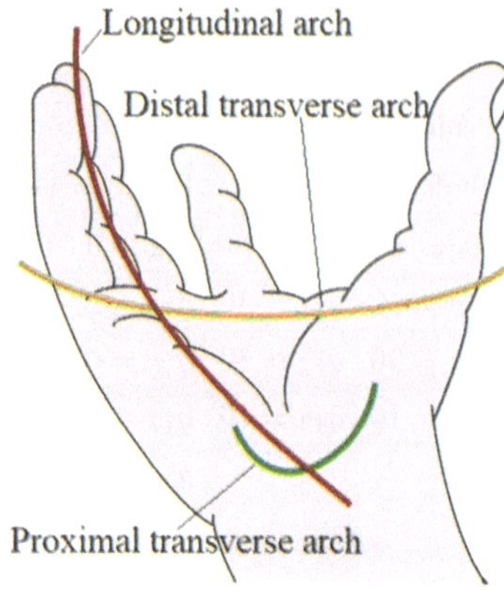

Fig. 4. Palm arch.

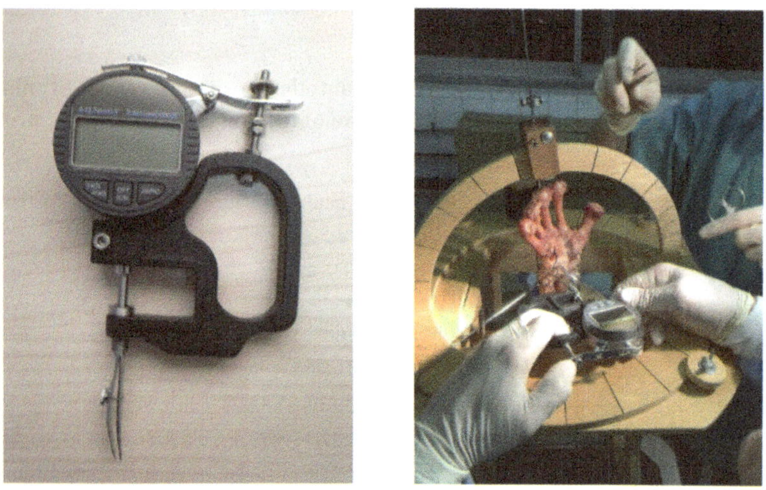

Fig. 5. Left, apparel to measure the ligament displacement. Right, jig designed by physicians of Univertitat Autonoma de Barcelona (UAB).

Figure 5 shows the testing used to investigate ligament synergies in the loaded wrist, described in [3], and adapted system shown in the right.

5 Conclusions

New hand model with 29 DOF is developed to simulate more realistic movements of the fingers, specially the thumb. Thumb is adjusted to permit movements of the palm arcs in different sections of the hand. Proposed new studies are in process, like a proposed apparel to measure the displacements induced isometrically loading the wrist or by axially distracting the metacarpal bone away from the radius.

Acknowledgments. This work was partially supported for Spanish government by the project DPI2016-80077-R.

References

1. Peña-Pitarch, E., Al Omar, A., Alcelay Larrión, J.I., Costa, J.V.: Virtual human hand: grasps and fingertip deformation. In: Advances in Additive Manufacturing, Modeling Systems and 3D Prototyping: Proceedings of the AHFE 2019 International Conference on Additive Manufacturing, Modeling Systems and 3D Prototyping, July 24–28, 2019, Washington D.C., USA, pp. 484–492. Springer, Berlin (2019)
2. Padmore, C., Langohr, G.D., Suh, N., Johnson, J.A.: The effect of coordinate system selection on wrist kinematics. J. Biomech. **109**, 1–7 (2020)
3. Gracia-Elias, M., Puig de la Bellacasa, I., Shouten, C.: Crapal ligaments: a functional classification. Hand Clin. **33**, 511–520 (2017)
4. Kawanishi, Y., Oka, K., Tanaka, H., Okada, K., Sugamoto, K., Murase, T.: In vivo 3-dimensional kinematics of the thumb carpometacarpal joint during thumb opposition. J. Hand Surg. **43**, 1–7 (2017)
5. Akhbari, B., et al.: Predicting carpal bone kinematics using an expanded digital database of wrist carpal bone anatomy and kinematics. J. Orthop. Res. **37**(12), 2661–2670 (2019)
6. D'Agostino, P., Dourthe, B., Kerkhof, F., Van Lenthe, G.H., Stockans, F., Vereecke, E.E.: In vivo biomechanical behavior of the trapeziometacarpal joint in healthy and osteoarthritic subjects. Clin. Biomech. **49**, 119–127 (2017)
7. Kapandji, I.A.: Fisiologia articular. Miembro superior, Medica Panamericana, Madrid, 5 Edición (1996)
8. Tubiana, R.: The hand, vol. I, 2nd edn. W.B. Saunders Company (1981)
9. Neu, C.P., Crisco, J.J., Wolfe, S.W.: In vivo kinematic behavior of the radio-capitate joint during wrist flexion–extension and radio-ulnar deviation. J. Biomech. **34**, 1429–1438 (2001)
10. Sonenbluma, S.E., Crisco, J.J., Kangb, L., Akelman, E.: In vivo motion of the scaphtrapzio trapezoidal (STT) joint. J. Biomech. **37**, 645–652 (2004)
11. Peña-Pitarch, E., Yang, J., Abdel-Malek, K.: Santos^{TM} hand: a 25 Degree of Freedom Model. In: Digital Human Modeling for Design and Engineering Symposium, Iowa (USA) (2005)
12. Peña-Pitarch. E.: Virtual human hand: grasping strategy and simulation. Ph.D. UPC. ISBN 9788469356678 (2008)
13. Denavit, J., Hartenberg, R.S.: A kinematic notation for lower-pair mechanisms based on matrices. J. Appl. Mech. ASME **22**, 215–221 (1955)
14. Elatta, M.A., Elgaind, S.M., Talat, E., Alqaseer, A.M., Basheer, H.M.: Scapho-capitate ratio for estimation of scaphoid length. J. Hand Surg. (Asian-Pac. Vol.) **14**(2), 1–6 (2019)

A Framework to Model the Use of Exoskeletons in DHM Tools

Francisco Garcia Rivera(✉), Anna Brolin, Estela Perez Luque, and Dan Högberg

School of Engineering Science, University of Skövde, Skövde, Sweden
{francisco.garcia.rivera,anna.brolin,estela.perez.luque, dan.hogberg}@his.se

Abstract. Work-related musculoskeletal disorders (WMSDs) constitute a large part of work absences among industry workers, together with all the health and economic problems that it carries. Exoskeletons developed for overhead operations can potentially be a solution to reduce risks for WMSDs. However, some companies are still hesitant to implement exoskeletons in their workplace, since the effects of using exoskeletons are still not fully proved. Digital human modeling (DHM) could help with this dilemma by facilitating studies of the viability of the exoskeletons for specific work tasks. This paper proposes a DHM based framework to implement the study of upper body exoskeletons focused on overhead assembly operations. The framework emphasizes the kinematics and forces interaction between the human and the exoskeleton.

Keywords: Exoskeleton · DHM · Digital human modelling · Simulation

1 Introduction

Simulation technologies are widely used in the industry since they enable efficient design and testing of products and production systems. The usage of simulation tools has increased in recent years, which highlights the usefulness and cost-effectiveness of these tools [1]. Simulations also enable the assessment of product and production ergonomics, typically made by the use of digital human modeling (DHM) software [2].

Narrowing down the scene to the use of exoskeletons, there is commonly a lack of functionality in DHM tools to simulate the physical interaction between an exoskeleton and a human. Moreover, there is a lack of ergonomic evaluation methods specifically designed to assess the use of exoskeletons [3].

An exoskeleton is a wearable device that assists its user by applying forces, torques, or structural support. It seems like a promising solution to avoid work-related musculoskeletal disorders (WMSDs) [4, 5], although the consequences of using exoskeletons during work have not been fully proved yet, especially the long-term effects. An advantage of exoskeletons is that very little to no modification of the workplace is required for its implementation [6].

Most studies on the effects of using exoskeletons during work use electromyography (EMG) or motion capture (mocap) equipment, on real humans, doing real or representative work tasks in the real world [7–10]. There are fewer studies published regarding the

J. L. Wright et al. (Eds.): AHFE 2021, LNNS 264, pp. 312–319, 2021.
https://doi.org/10.1007/978-3-030-79763-8_38

use of DHM tools and simulated work tasks to study the effects of using exoskeletons during work, but examples of such studies are [4, 11]. To be able to simulate the effects of wearing an exoskeleton, the DHM tools should consider certain parameters of the exoskeleton such as the force that it applies and the weight of the device.

In this work, we propose a framework for a DHM tool, in this case, IPS IMMA [12], to simulate the effects of wearing an exoskeleton for overhead operations. The objective is that the framework will describe the functions needed, and will guide the DHM development to facilitate proactive assessments of the effects of using exoskeletons during work.

2 DHM Based Framework

To define the framework, assumptions were made that relate to the requirements that the DHM tool, as well as the virtual exoskeleton, has to fulfill.

1. An automated manikin that can perform instructions with realistic and representative postures and motions. A manikin in a DHM tool is considered automated if it can perform tasks without any help from the user (except the given instructions in simple commands, i.e. grasp, move to, etc.). Besides, these tasks should be performed with ergonomically sound, and representative postures.
2. An instruction language that can be interpreted by all the objects in the scene including manikins [12].
3. The behaviour of the exoskeleton is known (since the purpose is not the design of the exoskeleton, but rather the assessment of its use). It is possible to model this behavior as a function that uses postural data of the human as input, and it gives the output as the force that the exoskeleton applies on the human. The output force applied to the digital manikin by the virtual exoskeleton is, therefore, the predicted force between the exoskeleton and the human, and can be used by the DHM tool to perform simulations. Furthermore, these predicted forces and simulations can be used to evaluate ergonomics in virtual environments.
4. To make a study of the ergonomic effect of wearing an exoskeleton, the digital manikin should have a realistic musculoskeletal model. The main biomechanical loads can be calculated and saved along the timeline of the simulation to be used as input in ergonomic evaluation or be analyzed by ergonomists.

According to [13], human-product interaction systems can be seen as consisting of four elements: the users, the product, the task, and the environment. Furthermore, [14] presents a schematic user-product model with the aforementioned four elements, based on a literature review on existing interaction modeling approaches in DHM tools. Figure 1 shows a modified version of the model presented in [14], adapted to illustrate the human-exoskeleton interaction. In this approach, we consider the exoskeleton as the virtual product, the virtual manikins as the users, and the virtual tasks as well as the virtual environment defined by the assembly tasks that the virtual manikins have to perform.

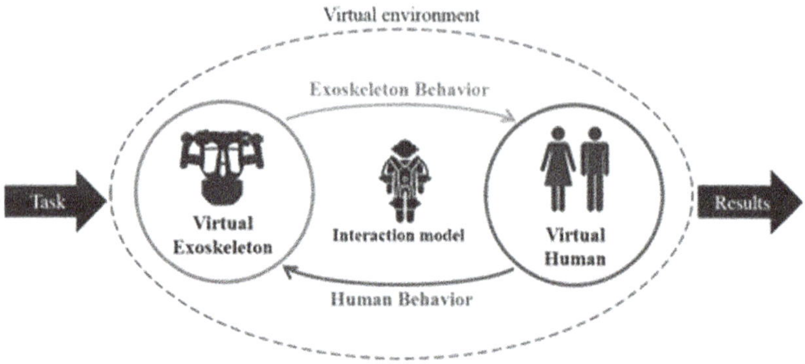

Fig. 1. Illustration of human-exoskeleton interaction, adapted from [14].

The authors of [14] state that there are three types of interaction models. Predicting human behavior as a function of product behavior, predicting product behavior as a function of predefined human behavior, or predicting human and product behavior as a mutually dependent feedback loop.

There are many different aspects involved in the human-exoskeleton interaction that can be simulated: forces, range of motion, muscle activity, etc. From a physical ergonomics point of view, the interaction can be simplified as an exchange of forces between the exoskeleton and its users. In this study, the objective is to predict human behavior as a function of predefined product behavior. The behavior of the human will depend on the product's behavior, which is predefined by the modeling function.

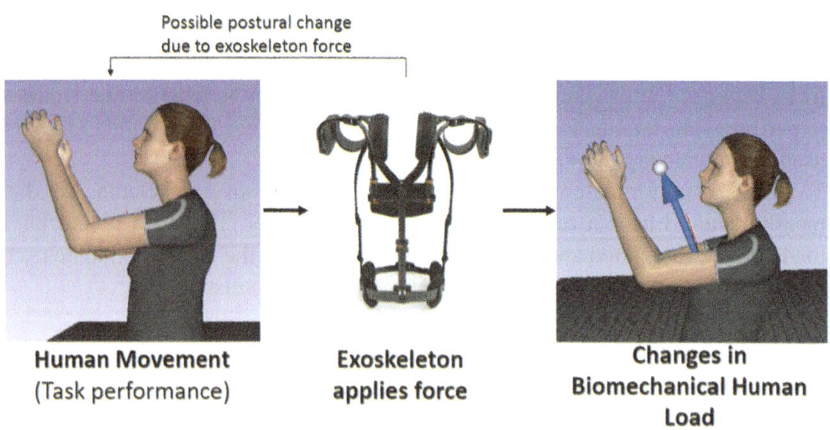

Fig. 2. Model of task chain during simulation of human-exoskeleton interaction.

The simulation starts with the definition of the human and the exoskeleton. Then, the task is defined. The definition of the task is done by instructing the manikin on how to perform the task. After the manikin is instructed, the software generates the human motions necessary to perform that task. The manikin's movements will then move the exoskeleton, and the exoskeleton will apply forces on the manikin as a function of the movements of the exoskeleton. The application of this force might influence the posture of the virtual manikin, as described with the loop going back in Fig. 2. The applied force from the exoskeleton will have a biomechanical effect on the virtual manikin and that value is then the result that the upcoming interaction study of human-exoskeleton aims to analyze.

3 Method for Generating Interaction Effects of Human-Exoskeleton in IPS IMMA

A proof of concept including an upper-body exoskeleton has been performed to test and validate the proposed framework and to model the use of exoskeletons in DHM tools, here refereeing to the IPS IMMA software [12]. An exoskeleton for overhead operations is a dynamic system that applies a force in the upper limbs to assist and improve the user's capabilities in this type of awkward operation. The exoskeleton used here and intended for the upcoming interaction study is Paexo Shoulder from Ottobock [16]. He initial version of the framework, presented in this paper, is not able to simulate dynamics, since IPS IMMA is a quasi-static system, i.e. the manikin is in static equilibrium in each frame, but without the consideration of inertia effects. Dynamics will be implemented in future versions of IPS IMMA, enabling dynamic simulations of human exoskeleton interaction.

To simulate the effect of the exoskeleton, the force that the real exoskeleton applies to the real human has to be applied similarly to the virtual manikin. The weight of the device has also to be considered. To model this human-exoskeleton interaction a script has been developed which uses the position of the manikin as input and returns the force applied by the exoskeleton (in magnitude and direction).

4 Generating Interaction Effects from Human-Exoskeleton

In this paper, three key aspects were identified for further use in upcoming studies and simulations of the effects of an exoskeleton in DHM tools.

- Force Applying Points
- Force Magnitude and Force Direction

4.1 Force Applying Points

Exoskeletons for overhead work apply force on the upper arm to relieve the stress from the shoulder. While the weight of the exoskeletons applies forces on the shoulders, similar to a backpack, in most cases the low back holds the majority of the weight when the operator is standing. In Fig. 3, the contact points between the exoskeleton and the human are shown as red dots. In this case, the exoskeleton applies forces in the two red points in the upper arm, and its weight in the contact point of the lower back.

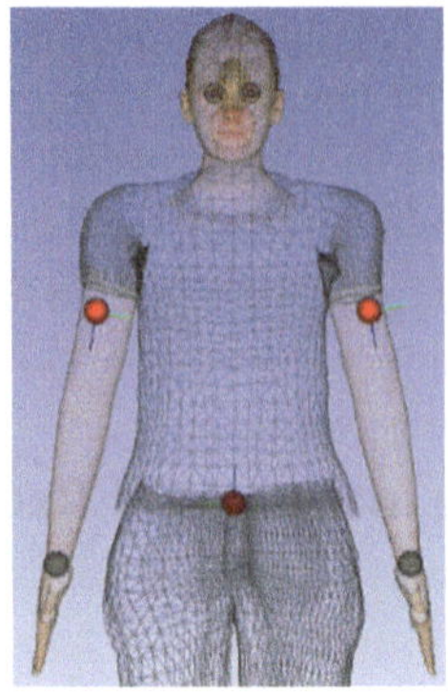

Fig. 3. Attachment points between the digital manikin and the exoskeleton.

4.2 Force Magnitude and Force Direction

The forces applied on the upper limbs from the exoskeleton involve magnitude as well as direction. This can vary from one exoskeleton model to another, and even the same model might have different configurations of where on the human body the forces are applied. Most of the exoskeletons do not apply the same forces in all the range of motion of the arm, but rather, the applied force is higher when the posture is more awkward (60 and higher degrees of flexion) and lower when the arm is in a more neutral position (less than 40°) [10].

To solve the issue of having different magnitude and direction of the forces on the exoskeleton, due to different configurations and exoskeleton models, an investigation of mathematical relations was conducted. There is a mathematical relationship between the angle of the shoulder and the force applied in the form of Eq. 1, where α is the angle of the Glenohumeral joint.

$$F(\alpha) = \text{Force applied} \quad (1)$$

This mathematical relation can be measured manually for each exoskeleton or provided by the manufacturer [10]. Figure 4 shows the force-angle relation of the exoskeleton used in this study, based on manual measurements.

Regarding the direction of the force, it is assumed that the exoskeleton applies the force perpendicular to the vector coincident to the upper arm, in the direction of the flexion of the upper arm (always in the direction of lifting the arm) as shown in Fig. 5. The weight of the exoskeleton is applied in the direction of gravity.

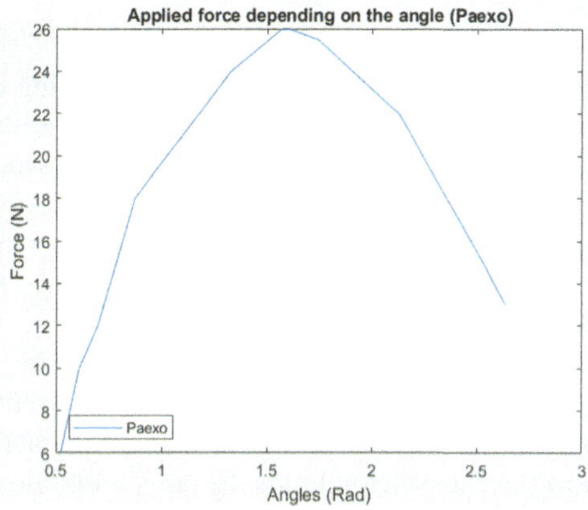

Fig. 4. Illustration of the force-angle relation of the exoskeleton.

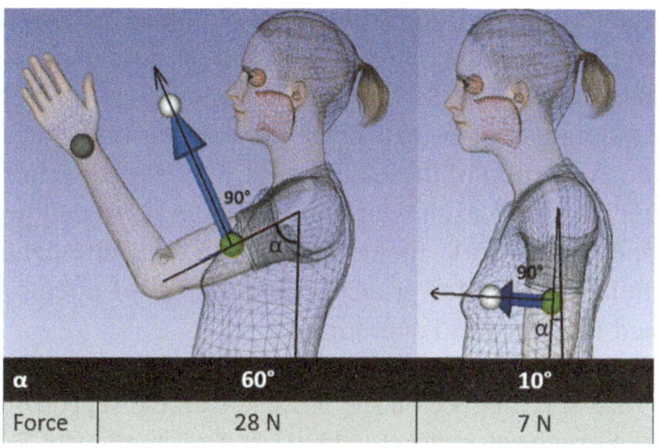

Fig. 5. Example of the direction and magnitude of the force applied to the digital manikin.

Table 1. Example of Glenohumeral joint torque in two different angles, with and without estimated forces applied from the exoskeleton.

Glenohumeral joint torque	Alpha 60°	Alpha 10°
With exoskeleton	3.4 Nm	2.7 Nm
Without exoskeleton	6.6 Nm	2.7 Nm

Through the process of the developed framework, the estimated force of the exoskeleton, applied to the manikin through an attachment point, affects the manikin's biomechanical model. Functionality in the DHM system IPS IMMA could then be used to assess differences in joint torques, for example in the Glenohumeral joint in the shoulder, for different body postures (Table 1).

5 Discussion

The use of DHM tools to study the ergonomic effects of wearing exoskeletons during work can facilitate proactive predictions of potential effects on the operators, before implementing esoskeletons in real work situations. Hence, DHM tools could potentially reduce the time, cost, and resources that a company has to invest in deciding whether or not an exoskeleton can be used in its production system or not. The framework presented in this paper is assumed to be possible to apply to any DHM tool that meets the requirements stated in Sect. 2.

This paper presents research and development work in progress, and there are limitations of the developed framework and proof of concept in its current state. For example, it is not possible to study how the exoskeleton affects muscular activity, inertia effects, long term effects, or how the exoskeleton limits the range of motion of the user, among others. Another possible issue with the suggested framework is that it can create an infinite feedback loop as the exoskeleton model both affects the posture of the manikin, and uses the posture as input data, i.e. the estimated force from the exoskeleton affects the posture of the manikin, which in turn takes a new posture due to the force, which new posture then causes the exoskeleton to give a new estimated force. This possible loop effect needs to be handled in the forthcoming research and development.

Although the development of the exoskeletons has come quite far, common and standard methods for the evaluation of these devices (from data collection to data analysis) do not exist [6], which is one reason why more research is needed in this field. In conclusion, the developed framework, including the stated requirements and limitations, supports DHM simulation used for proactive assessment of effects while wearing an exoskeleton designed for over-head operations.

Acknowledgment. This work has been made possible with the support from the Knowledge Foundation supported research environment INFINIT at the University of Skövde, in the project Synergy Virtual Ergonomics (SVE), and with support from VINNOVA in the project VIVA - the Virtual Vehicle Assembler, and by the participating organizations. This support is gratefully acknowledged.

References

1. Rüßmann, M., et al.: Industry 4.0: The future of productivity and growth in manufacturing industries. Company report. Boston Consult. Group, **9**(1), 54–89 (2015)
2. Scataglini, S., Paul, G.: DHM and Posturography. Academic Press, Cambridge (2019)
3. Gull, M.A., Bai, S., Bak, T.: A review on design of upper limb exoskeletons. Robotics **9**, 16 (2020)
4. Tröster, M., Schneider, U., Bauernhansl, T., Rasmussen, J., Andersen, M.S.: Simulation framework for active upper limb exoskeleton design optimization based on musculoskeletal modeling. tech. Unterstützungssysteme, 11 (2018)
5. Maurice, P., et al.: Objective and subjective effects of a passive exoskeleton on overhead work. IEEE Trans. Neural Syst. Rehabil. Eng. **28**, 152–164 (2020)
6. Bornmann, J., Schirrmeister, B., Parth, T., Gonzalez-Vargas, J.: Comprehensive development, implementation and evaluation of industrial exoskeletons. Curr. Dir. Biomed. Eng. **6** (2020)

7. Schmalz, T., et al.: Biomechanical and metabolic effectiveness of an industrial exoskeleton for overhead work. Int. J. Environ. Res. Public. Health. **16**, 4792 (2019)
8. de Vries, A., Murphy, M., Könemann, R., Kingma, I., de Looze, M.: The amount of support provided by a passive arm support exoskeleton in a range of elevated arm postures. IISE Trans. Occup. Ergon. Hum. Factors. **7**, 311–321 (2019)
9. Huysamen, K., Bosch, T., de Looze, M., Stadler, K.S., Graf, E., O'Sullivan, L.W.: Evaluation of a passive exoskeleton for static upper limb activities. Appl. Ergon. **70**, 148–155 (2018)
10. Perez Luque, E., Högberg, D., Iriondo Pascual, A., Lämkull, D., Garcia Rivera, F.: Motion behavior and range of motion when using exoskeletons in manual assembly tasks. In: Presented at the 9th Swedish Production Symposium, 7–8 October 2020, Jönköping, Sweden (2020)
11. Zhou, X.: Predictive human-in-the-loop simulations for assistive exoskeletons. In: ASME 2020 International Design Engineering Technical Conferences and Computers and Information in Engineering Conference (2020)
12. Högberg, D., Hanson, L., Bohlin, R., Carlson, J.S.: Creating and shaping the DHM tool IMMA for ergonomic product and production design. Int. J. Digit. Hum. **1**, 132–152 (2016)
13. Högberg, D., et al.: DHM based test procedure concept for proactive ergonomics assessments in the vehicle interior design process. In: Bagnara, S., Tartaglia, R., Albolino, S., Alexander, T., Fujita, Y. (eds.) IEA 2018. AISC, vol. 822, pp. 314–323. Springer, Cham (2019). https://doi.org/10.1007/978-3-319-96077-7_33
14. Wolf, A., Miehling, J., Wartzack, S.: Challenges in interaction modelling with digital human models – a systematic literature review of interaction modelling approaches. Ergonomics **63**, 1442–1458 (2020)
15. Reinhard, R., et al.: The use and usage of virtual reality technologies in planning and implementing new workstations. In: DHM2020: Proceedings of the 6th International Digital Human Modeling Symposium (2020)
16. Paexo Shoulder. https://paexo.com/paexo-shoulder/?lang=en

Design Specifications for Eyewear with an Integrated Headborne System

Hyegjoo E. Choi-Rokas(✉), Peng Li, and Todd N. Garlie

U.S. Army Combat Capabilities Development Command (DEVCOM) Soldier Center, Natick, MA 01760, USA
{Hyegjoo.E.Choi-Rokas.civ,Peng.Li.civ, Todd.N.Garlie.civ}@mail.mil

Abstract. The current study investigates eye location relative to adjacent facial landmarks in support of the design of eyewear and Integrated Headborne systems. The primary objective is to define, summarize, and visualize the distribution of eye (pupil) location in three-dimensional (3D) space relative to the forehead (Glabella). The variability of pupil location relative to vertical and horizontal distances from the Glabella, as well as Interpupillary Breadth were summarized to better inform the development of adjustment ranges for an eyewear designed to integrate with a headborne system.

Keywords: 3D landmarks · Integrated headborne system · Military anthropometry · Eye wear · Facial structure · 3D eye volume · Interpupillary Breadth (IPB) · Product design

1 Introduction

This study follows two previous investigations [1, 2] related to head and face shape intended to inform supplementary design guidance of headborne systems for U.S. Army Soldiers. In the previous study [1], selected neck-up dimensions from the U.S. Army Anthropometric Survey (ANSUR II) [3] were compared between males and females and DoD population subgroups to document similarities and differences. A follow-up study [2], investigated eye region utilizing three-dimensional (3D) coordinates extracted from 3D head scans to quantitatively define the eye location (pupil location) relative to the Glabella. On average, it was calculated that pupils are located at 27 mm below the Glabella landmark for both male and female Soldiers, and on average, pupils are located at 17 mm and 10 mm posterior relative to the Glabella landmark for male and female Soldiers, respectively.

Knowing the eye location relative to specific facial landmarks is critical to design and evaluate eyewear for a headborne system. Given the average horizontal and vertical distance from the Glabella to pupil, the variability of those dimensions from the target population is additional critical information for computing optimal accommodation ranges. Specifically, the amount of eyewear adjustment, once mounted to a headborne system, is the key for the Soldiers to properly locate the eyewear to achieve optimal

J. L. Wright et al. (Eds.): AHFE 2021, LNNS 264, pp. 320–327, 2021.
https://doi.org/10.1007/978-3-030-79763-8_39

functions (i.e. proper vision through the system, maximum field of view, overlaying of information over the desired real world objects) that the headborne system can offer.

The current study investigates variability of pupil location in 3D space by expanding on the findings from the earlier studies. The magnitude of the variability is then applied to develop adjustment ranges for eyewear intended to be incorporated into headborne systems. The findings from the current study will aid in developing the design specifications of future headborne systems.

2 Method

2.1 Scan Databases and Standardization Process

Three-Dimensional head scan images collected during the U.S. Army Anthropometric Survey [3] were processed. First, the noise, gap and outliers were removed from the scans, then all void data and holes were filled using a template matching approach [4]. Once cleaned, all the scans were aligned to the Frankfort Horizontal Plane (FHP) (Fig. 1). 3D coordinates for the aligned scans represent the front and back, the up and down, and the left and right directions relative to the FHP. Figure 2 represents the scan captures before and after the processing for ANSUR II male scan database.

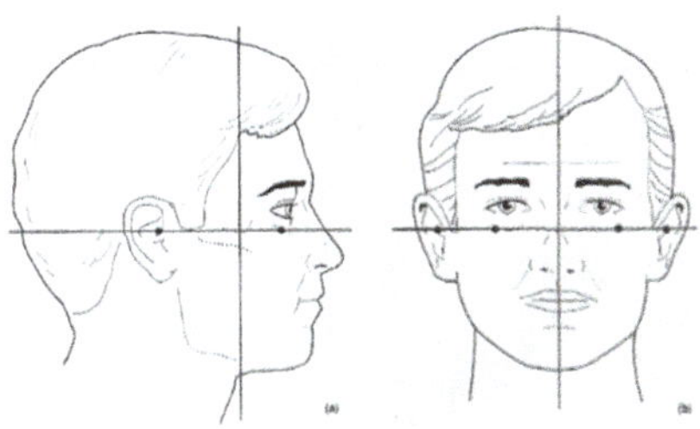

Fig. 1. Frankfort Horizontal Plane (Photo courtesy [5]), the coordinate system for this project is Front = +X, Up = +Y, Left = +Z and the Tragion-Tragion line is coincident to Z axis

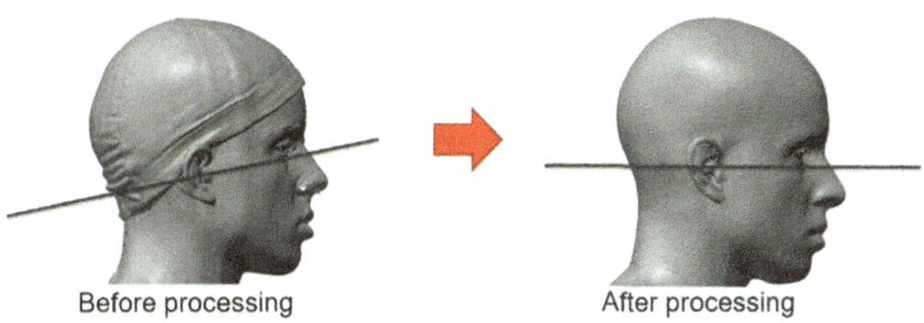

Fig. 2. Screen capture of before and after template matched scan

Currently, all post-processed scans in ANSUR II are aligned by FHP and standardized by template matching. The final standardized dataset includes 5509 head scans representing 1903 female heads and 3606 male heads.

2.2 Derived Eye Dimensions

After aligning and standardizing the head scans, facial landmarks were extracted. For this study, five facial landmarks including Cornea (Left and Right), Glabella, and Tragion (Left and Right) were used to compute four distances: 1) Cornea Depth relative to the Tragion; 2) Glabella Depth relative to Tragion; 3) Cornea Height relative to FHP; and 4) Glabella Height relative to FHP. The Cornea to Glabella horizontal distance was derived by subtracting the Cornea Depth (averaged between right and left Cornea Depth) from the Glabella Depth relative to the Tragion. The Cornea to Glabella vertical distance was then derived by subtracting the Cornea Height (averaged between right and left Cornea Height) from the Glabella Height relative to FHP. Those derived distances, Cornea to Glabella horizontal distance and Cornea to Glabella vertical distance were then named Pupil Depth and Pupil Height relative to Glabella, respectively (Fig. 3). For this study, those two derived dimensions along with IPB were further analyzed to produce the variabilities of eye location.

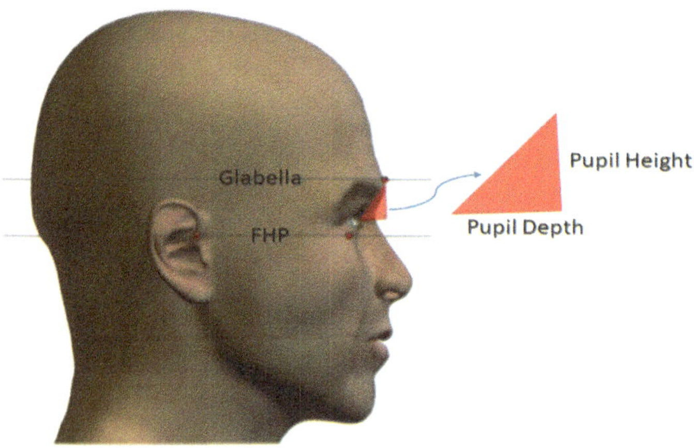

Fig. 3. Illustration of two derived dimensions (Pupil Height and Pupil Depth), Glabella level and FHP

2.3 3D Eye Volume

The 3D coordinates of left and right Corneas and Glabella along with IPB were used to construct and visualize the 3D pupil space. Due to the standardization process, the X-coordinates for facial landmarks represent anterior (positive) and posterior (negative) direction relative to Tragion, and Y-coordinates represent the vertical deviation, superior (positive) and inferior (negative) relative to FHP. To be consistent with the derived eye dimensions, Pupil Depth and Pupil Height relative to Glabella, 3D coordinates for both Corneas were transferred relative to Glabella. By doing this, Glabella becomes the origin of the coordinate system when 3D coordinates of the left and right Corneas are plotted.

When the 3D coordinates of the Cornea landmarks were plotted with Glabella, the location of the Corneas on the Z-axes (left and right) was systematically deviated toward one side relative to Glabella. For females, it shifted right by 12 mm, on average, and for males, it shifted left by 5 mm, on average. Since there is no strong evidence of asymmetric distribution between right and left Cornea, Z-coordinates for the left and right Cornea were adjusted by IPB. Therefore, the Z-coordinate for Cornea became half of IPB in the positive and negative position. Adjusted 3D landmarks of left and right Cornea were then plotted with Glabella as the origin of the coordinate system to visualize the 3D pupil space, named 3D eye volume.

2.4 Data Processing

Two derived dimensions, Pupil Height and Pupil Depth, along with Interpupillary Breadth were processed to calculate and quantify the variability of those three dimensions. To control for the extreme outliers, only data points within two standard deviations from the mean values were included for additional analyses.

Means and standard deviations for those three selected dimensions were computed for males and females. Percentile values at 0.5^{th}, 2nd, 5th, 95th, 98th and 99.5^{th} for the dimensions were computed for population subgroups[1] and for males and females. Then, within each sex, the most extreme number at each percentile value out of all population subgroups was selected to construct 90%, 96% and 99% accommodation ranges so that the variability between population subgroups are accounted for by the accommodation ranges [6].

3D eye volume was produced to visualize the eye location in 3D space. The central 90%, 96% and 99% accommodation ranges of the 3D eye volume on the X, Y, and Z axes was computed to construct design specifications on the adjustment ranges for the eyewear with a headborne system.

3 Results

3.1 3D Eye Variation: Ranges for the Eyewear Adjustment

Descriptive statistics on three selected dimensions, Pupil Depth, Pupil Height, and IPB are presented in Table 1. For males, pupils are located, on average, 27 mm inferior and 17 mm posterior from Glabella [2], and the distance between the left and right pupils is, on average, 64 mm. For 90% of the U.S. Amy male population, pupils are located between 19.9 mm and 33.9 mm inferior and between 10.1 mm and 23.1 mm posterior from Glabella, and the distance between the left and right pupils is between 56.5 mm and 71.5 mm. Obviously, the accommodation ranges get wider to capture 99% of the population: pupils are between 17.9 mm and 35.9 mm inferior and between 8.6 mm and 24.6 mm posterior relative to Glabella, and the distance between pupils is as narrow as 53 mm, and as wide as 75 mm.

[1] For valid statistical comparisons in the current analyses, only race based population subgroups with more than 100 personnel were selected, and these included: White, Black, and Hispanic.

For females, pupils are located, on average, 27 mm inferior and 10 mm posterior from Glabella [2], and the distance between the left and right pupils is, on average, 61.7 mm. The calculated accommodation ranges for females are relatively narrower than those of males, except for Pupil Depth. For 90% of the U.S. Amy female population, pupils are location between 20.5 mm and 33.5 mm inferior and between 1.7 mm and 18.7 mm posterior from Glabella, and the distance between left and right pupils is between 54.2 mm and 69.2 mm. Like with males, a wider range is needed to capture 99% of the population in that pupils are between 19 mm and 35 mm inferior and between −0.3 mm and 20.7 mm posterior relative to Glabella, and the distance between pupils is as narrow as 51.7 mm, and as wide as 71.7 mm.

Table 1. Accommodation ranges for the eye location relative to Glabella

Dimensions (Unit: mm)	Sex	Mean	Std. Dev.	Accommodation range (±)		
				90%	96%	99%
Pupil depth	Male	16.6	3.4	6.5	7.0	8.0
	Female	10.2	4.7	8.5	9.5	10.5
Pupil height	Male	26.9	3.8	7.0	8.0	9.0
	Female	27.0	3.4	6.5	7.5	8.0
Interpupillary breadth	Male	64.0	3.4	7.5	9.5	11.0
	Female	61.7	3.6	7.5	8.5	10.0

3.2 3D Eye Volume

Differences on the facial structure between males and females are better visualized by plotting the pupil location with U.S. Army male average head form as a background. Figure 4 illustrates a front and side profile of Pupil Depth, Pupil Height and Interpupillary Breadth with 90% accommodation ellipses. In both views, the origin of the X, Y, Z coordinates is the Glabella location. The blue ellipses represent male data and the red ellipses represent female data.

The front view represents Pupil Height on the y-axes, and both pupils are plotted together on the X-axes depicting the Interpupillary Breadth. Ellipses for both male and female Soldiers are closely aligned indicating that the accommodation ranges for Pupil Height and Interpupillary Breadth are similar between male and female populations. The side view represents Pupil Height on the y-axes and Pupil Depth on the x-axes. The differences on the side view depict the Pupil Depth difference between males and females and illustrates that females have a wider accommodation range than males.

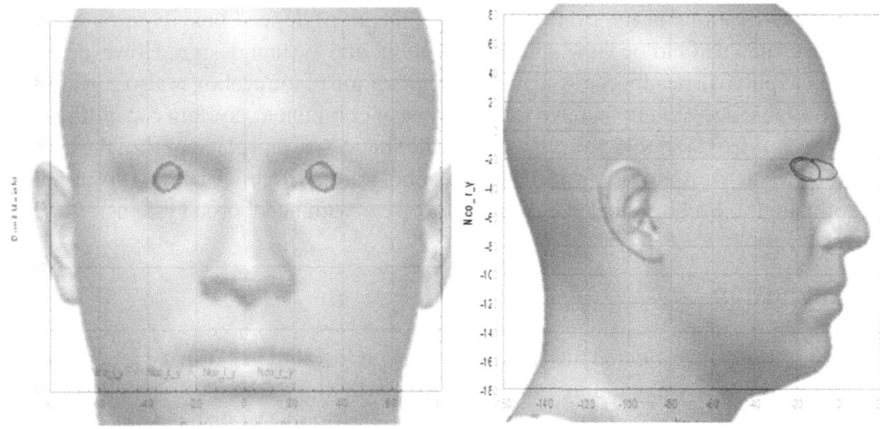

Fig. 4. 3D Eye Volume, front (Left) and side (Right) profile, where blue represents male data and red represents female data

4 Discussion

4.1 Sex Differences Relative to Accommodation Ranges

The accommodation ranges on Pupil Height and Interpupillary Breadth for both males and females tend to be similar to each other. The accommodation ranges for female Soldiers on Pupil Height are completely inscribed within the bounds of the male Soldiers' data. The accommodation ranges of Interpupillary Breadth for both males and females can be combined by expanding the ranges for the male population on the lower end by maximum 2 mm.

However, the larger variability seen relative to Pupil Depth represent facial structure differences between males and females. The differences on Pupil Depth for males and females ($M = 6.4$ mm) is due to the larger protrusion on brow ridge for males where the Glabella landmark is located; The difference ($M = 6.7$ mm) on average of Glabella Depth relative to the Tragion for males ($M = 99.2$ mm) and females ($M = 92.5$ mm) is almost identical to the difference in Pupil Depth between males and females. Given the large mean differences ($M = 9.7$ mm) for Head Length between males ($M = 199.4$ mm) and females ($M = 189.7$ mm), it can be inferred that two thirds of the size difference in Head Length is related to structural differences of the forehead region between males and females.

4.2 Eyebox

One of the critical concepts to design and evaluate of eyewear for headborne systems is the "eyebox" that refers to an optimal 3D eye space of the visor and displays when the headborne system is properly donned. If Soldiers can properly place their eyes within this eyebox, they should be able to fully achieve "an acceptable view of the image with respect to a set of criteria and thresholds [7]" or in other words, they can obtain the optimal resolution of the system as well as the maximum field of view the system offers.

The specifications of the eyebox (i.e., size, location) depend heavily upon external constraints such as field of view, image resolution, and system design. However, if the eyewear is a part of a headborne system, the specification of the eyebox is also constrained by the available adjustment features for the eyewear component. System capabilities and user anthropometrics should be used to define the shape, size and placement of the eyebox for the eyewear with a headborne system. Authors are currently planning to visualize and compare the eyeboxes between various eyewear with headborne systems.

5 Conclusion

The current study investigated the variability of pupil location in 3D space relative to the Glabella landmark by expanding on the significant findings from earlier studies. Table 2 represents the adjustment ranges for Pupil Depth, Pupil Height, and Interpupillary Breadth per target accommodation rate. The lower and upper bounds for the combined ranges are the closest integers within which the adjustment ranges produced for males and females could be inscribed.

To align the eyewear with a headborne system to the eye location of 99% of the US Army population, Soldiers should be able to lower the central focal point of eyewear components of the system from the Glabella by at least 17 mm and at most by 36 mm, and to slide the eyewear closer to the eyes by up to 25 mm relative to the Glabella. In addition, to accommodate 99% of both male and female Soldiers, the eyewear should accommodate IPB of between 51 mm and 75mm.

The magnitude of the adjustability (Table 2) of the eyewear relative to Glabella produced in this study should be applied to develop overall adjustment ranges for eyewear intended to be incorporated into headborne systems. The findings from this current study will aid in developing design specifications for such systems.

Table 2. Adjustment Ranges per target accommodation rate

Dimensions (Unit: mm)	Adjustment ranges	90%		96%		99%	
		Lower	Upper	Lower	Upper	Lower	Upper
Pupil depth (from Glabella)	Male	10.1	23.1	9.6	23.6	8.6	24.6
	Female	1.7	18.7	0.7	19.7	-0.3	20.7
	Combined	1	24	0	24	-1	25
Pupil height (From Glabella)	Male	19.9	33.9	18.9	34.9	17.9	35.9
	Female	20.5	33.5	19.5	34.5	19.0	35.0
	Combined	19	34	18	35	17	36
Interpupillary Breadth	Male	56.5	71.5	54.5	73.5	53.0	75.0
	Female	54.2	69.2	53.2	70.2	51.7	71.7
	Combined	54	72	53	74	51	75

Acknowledgements. This research was funded by the US Army DEVCOM Soldier Center research program 19–117. The authors would like to thank those Soldiers who participated in ANSUR II for their time and effort, as well as for the sacrifices they make every day for our country.

1. References

1. Choi, H.J., Li, P., Garlie, T.N.: Facial shape analyses using 3D landmark coordinates for eye wear design with an Integrated Headborne system. Paper presented at AHFE 2020, Virtual conference, July 2020
2. Choi, H.J., Garlie, T.N., Parham, J.L.: Anthropometric analyses of head and face shape to design protective headgear for U.S. army personnel. In: Di Nicolantonio, M., Rossi, E., Alexander, T. (eds.) Advances in Additive Manufacturing, Modeling Systems and 3D Prototyping. AHFE 2019. Advances in Intelligent Systems and Computing, vol. 975. Springer, Cham (2020). https://doi.org/10.1007/978-3-030-20216-3_50
3. Gordon, C.C., et al.: 2010–2012 Anthropometric Survey of U.S. Army Personnel: Methods and Summary Statistics. Technical Report (Natick/TR-15/007). U.S. Army Natick Research, Development and Engineering Center, Natick, MA 01760–2646 (2014)
4. Hirshberg, D.A., Loper, M., Rachlin, E., Black, M.J.: Coregistration: simultaneous alignment and modeling of articulated 3D shape. In: Fitzgibbon, A., Lazebnik, S., Perona, P., Sato, Y., Schmid, C. (eds.) ECCV 2012. LNCS, vol. 7577, pp. 242–255. Springer, Heidelberg (2012). https://doi.org/10.1007/978-3-642-33783-3_18
5. Vetter, J.P.: Biomedical Photography. Focal Press, London (1992)
6. Garlie, T., Choi, H.: Anthropometric difference among population group and the impact on military clothing and equipment design. 7th International conference on Applied Human Factors and Ergonomics, Orlando, Florida, July 2016
7. Cholewiak, S.A., Basgoze, Z., Cakmakci, O., Hoffman, D.M., Cooper, E.A.: A perceptual eyebox for near-eye displays. Opt. Express **28**(25), 38008–38028 (2020)

The Relationship Between the Position of Bicycle Saddle and Human Comfort Based on Jack

Yu Yalan[1], Liu Litao[2], and RongRong Fu[1](✉)

[1] College of Art Design and Media, East China University of Science and Technology, Shanghai, China

[2] College of Mechanical Engineering, An Yang Institute of Technology, An yang, China

Abstract. In order to explore the relationship between the position of the saddle and the comfort of various parts of the body in the riding state, this paper establishes a simulation system model of human riding a bicycle based on ergonomics theory, human physiology, and jack simulation experiment technology, adjust the height of the saddle to obtain the joint angle and comfort analysis in different postures. The analysis results show that when the position of the saddle is adjusted, the joint angle changes, and the comfort of each part changes. At the same time, the experimental results can obtain the bicycle saddle position data that are more suitable for 5%, 50%, 95%, and 99% of digital people. The analysis method and results provide reference and basis for the personalized position design of bicycle saddles.

Keyword: Jack · Simulation system · Bicycle saddle · Digital human · Comfort · Ergonomics

1 Introduction

As of 2019, China has nearly 400 million bicycles and nearly 300 million e-bikes, ranking first in the world. The application range of bicycles is so wide that people have higher and higher requirements for bicycles in addition to their aesthetic requirements for appearance and comfort. Whether they conform to ergonomics is also an important consideration in the bicycle manufacturing process. People riding on a bicycle, as a whole, different riding postures will have different effects on the body [1]. The position of the saddle relative to the handle affects the upper body posture during riding, while the position of the saddle relative to the pedal axis mainly affects the lower body posture of the human body [2]. This article mainly discusses the influence of the height of the saddle relative to the pedal axis on the comfort of various parts of the human body.

Jack software, as a very representative human body simulation and ergonomic analysis software, has been widely used for human-machine analysis to judge whether the product meets user needs and human factors. Its analysis tools and data have also been proven to be true and effective, and have a high similarity to the results of physical experiments [3]. This paper uses Jack analysis technology to establish digital humans with different percentiles, import three-dimensional bicycle models, and adjust the data

J. L. Wright et al. (Eds.): AHFE 2021, LNNS 264, pp. 328–335, 2021.
https://doi.org/10.1007/978-3-030-79763-8_40

to complete the simulation model of the riding state. In the experiment, the height of the saddle was adjusted to obtain the joint angle and comfort analysis under different postures, and the position data of the bicycle saddle more matching with people of different height and weight were obtained through the experimental results. This analysis method can obtain real and effective human riding comfort data in a short period of time, which can be applied to the bicycle design and manufacturing process. At the same time, the data conclusions in this article also provide references and basis for the personalized position design of bicycle saddles.

2 Jack Simulation Platform

Jack is a human body modeling and simulation system platform developed by the University of Pennsylvania. As a software that can verify and evaluate efficacy, it has been widely used in industrial fields for many years [4]. The simulation environment built by Jack can import the CAD model required by the user, introduce the three-dimensional model of the human body with bio-mechanical characteristics, and can give the digital human specific tasks, analyze the behavior and environment of the digital human and obtain valuable information. Jack has detailed three-dimensional human body model data from multiple countries and can adjust the posture of the human body model. Jack's human body model includes 69 segments and 68 joints, it uses the inverse kinematics control method of the extremity to control the movement of the human body [5]. The Dreyfuss 3D model in Jack can conduct a comprehensive analysis of human comfort, and the comfort corresponds to the corresponding range of motion of joints [6]. The human factor analysis tool in Jack performs human factor analysis on the human body in different postures. This tool can be used to judge whether a certain job or state conforms to human characteristics, so jack is often used as a common analysis software in ergonomics. There are mainly four analysis tools in the Jack software. This experiment uses the commonly used comfort analysis in OPT tools.

3 Establish the Model of Man Riding Bicycle System

3.1 Parameter Setting and Creation of Digital Human

In the process of human factor analysis using JACK, creating a digital human is the first step. According to product characteristics and target users, selecting a digital human model with the appropriate region, gender, and percentile is the basis for obtaining accurate human factor analysis results. The picture shows the human body size of a Chinese adult male aged 18–25 in the national standard GB10000–88. The six measurement items of human body height, weight, upper arm length, forearm length, thigh length, and calf length are considered to be highly related to bicycle saddles. In order to obtain the use of most people and make the experimental data more reliable, this experiment selects 18–25 year-old adult male digital people with percentiles of P5, P50, P95, and P99 as the main body of the analysis. Four different percentiles of digital humans represent cyclists of four sizes: large, medium, relatively small, and small. The parameter adjustment of the digital human and model effect is shown in Fig. 2.

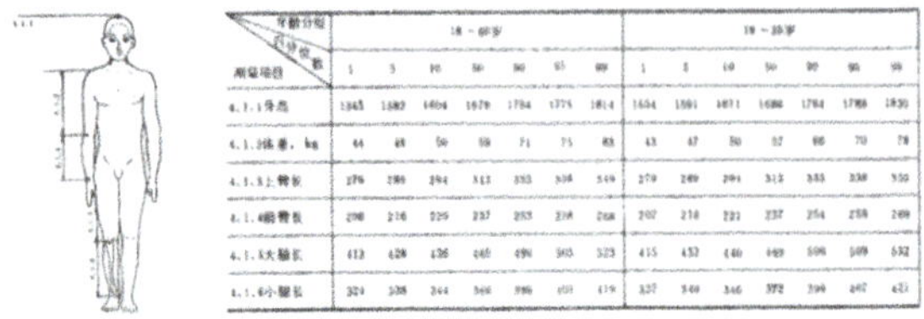

Fig. 1. The main dimensions of human body and The main dimensions of human body (male)

Fig. 2. From left to right are P5 digital human, P50 digital human, P95 digital human, and P99 digital human

3.2 Selection and Creation of Bicycle Model

At present, there are as many as twelve types of mainstream bicycles on the market, and many brands are also dazzling, but the overall structure of the bicycle is basically the same. The difference between different brands of bicycles is mostly in the design of the appearance of the product. In this study, a relatively popular bicycle on the market was selected as the simulation object, and the Rhino modeling software was used to establish a 1:1 three-dimensional model of the bicycle according to product size parameters. The bicycle selected for this experiment and its size parameters 99 are shown in the Fig. 3. The figure shows that the height of the pedal axis from the ground is 30 cm.

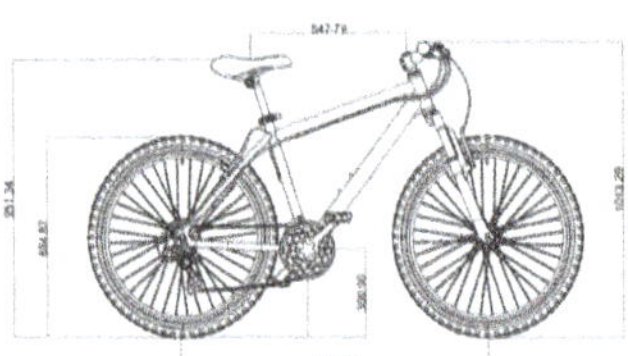

Fig. 3. Bicycle related dimensions (unit: mm)

3.3 Adjust Various Parameters to Complete the Static Cycling Simulation Model

Import the three-dimensional model of the bicycle into the Jack simulation environment, set and adjust the posture of the digital human and the position relationship with the bicycle, and complete the static simulation model of riding. The posture adjustment of the digital human first selects the digital human posture that comes with the jack system that is closest to the target posture, and then adjusts the parameters of each joint to make it conform to the best riding posture of the bicycle.

Fig. 4. Adjustments of human joint parameters in jack and Simulation model of a person riding a bicycle.

4 Experimental Process/Method

Figure 5 shows the human-bicycle-vehicle interface relationship and system composition diagram. In order to study the joint comfort of humans during riding, humans and bicycles cannot be considered separately, whereas they should be regarded as a whole system. The two systems cooperate with each other to complete the driving task. It can be seen from Fig. 5 that the human and the bicycle are coupled through the hand-handle, hip-saddle and foot-pedal. The support part and power-receiving part of the human-bicycle system determine the human comfort. For this reason, on the jack simulation software platform, the posture of the human body is adjusted so that the hands, buttocks and feet of the human body are located in the corresponding positions of the bicycle, so as to obtain the riding state of the human being on the bicycle.

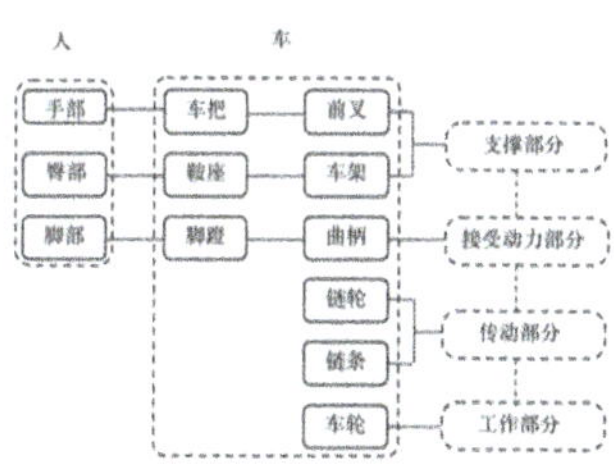

Fig. 5. The composition of the human-bicycle system [7]

The change of the saddle position in this paper takes the pedal axis as the reference base point. The reason is that the lowest point of a person's posture during riding is the pedal, and the human body is in a suspended state, so the relative position of the saddle and the pedal axis is more valuable than the position relative to the ground.

According to the difference of digital people, the experiment is divided into 4 groups, namely P5, P50, P95, and P99. The contents of each group are: Keep the horizontal distance of the saddle relative to the axis of the pedal at 30 cm unchanged, change the vertical distance between the saddle and the axis of the pedal, adjust the body posture to adapt to the position of the saddle. In the process of adjusting the posture, it should be noted that the position and posture of the hands and feet of the digital person should be kept unchanged, that is, the parameters of the hands and feet should remain unchanged.

5 Experimental Data Statistics and Processing

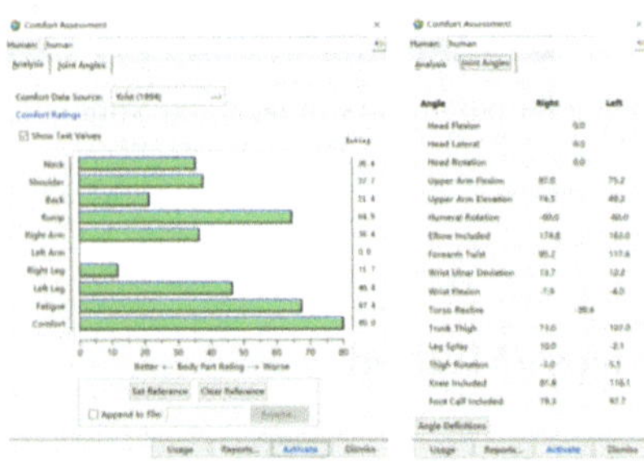

Fig. 6. Comfort and joint angles of various body parts of the P5 digital human body in the riding state

Keep the horizontal distance of 20 cm (the saddle relative to the axis of the pedal) unchanged, that is, the horizontal distance between the saddle and the handle does not change, set the saddle to the ground at 85 cm, 90 cm, 95 cm, 100 cm, 105 cm, the height difference from the pedal axis is 55 cm, 60 cm, 65 cm, 70 cm, 75 cm. The comfort experience of each part of the body when the P5 digital human body is riding is shown in Fig. 6:

Simulating the comfort conditions when the height difference between the saddle and the pedal axis h is 55 cm, 60 cm, 65 cm, 70 cm, 75 cm, respectively, and obtain the comfort test results of different body parts and count them in Table. As shown below, the smaller the comfort value is, the better the comfort level will be. When the height difference h reaches 70 cm and 75 cm, the human body cannot make the hips reach this height, and it is difficult to complete the normal riding posture, so there is no corresponding comfort data. In order to more intuitively see the influence of the different height difference between the saddle and the pedal axis on the human comfort, the Four sets of data is displayed in a line graph, as shown in Fig. 7–10.

Analysis of the experimental results of the first group of P5 digital humans:

The height of the P5 figure is 158.3 cm, the weight is 48 kg, and the sum of thigh length and calf length is 76.6 cm. When the height difference reaches 70 cm and 75 cm, it is difficult for the human body to complete the normal riding posture, so there is no corresponding comfort data. It can be seen from Table 3 that the joint angles on the right: humerus twist, elbow flexion, forearm twist, wrist tilt, wrist flexion angle unchanged, including upper arm elevation, upper arm flexion angle increased, thigh flexion, knee flexion angle reduced; the joint angle on the left: humerus twist, elbow flexion, forearm

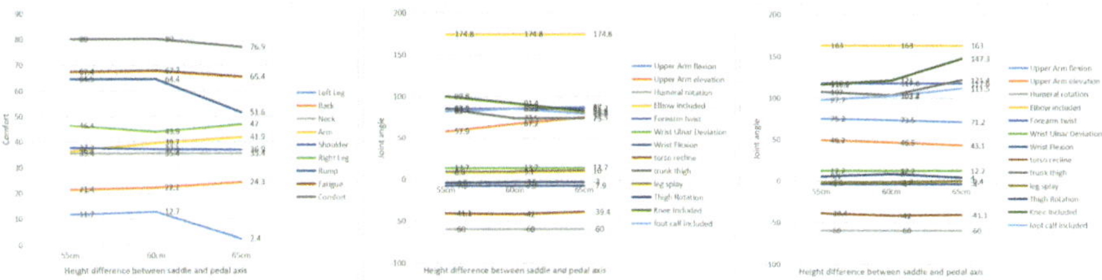

Fig. 7. From left to right is the comfort analysis of P5 digital human, right joint angle analysis, left joint angle analysis

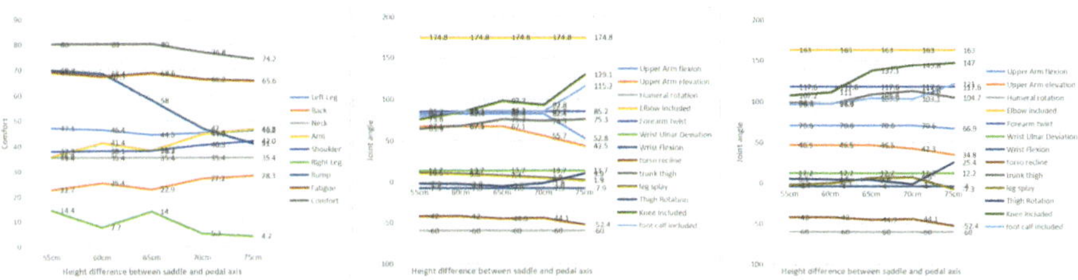

Fig. 8. From left to right is the comfort analysis of P50 digital human, right joint angle analysis, left joint angle analysis

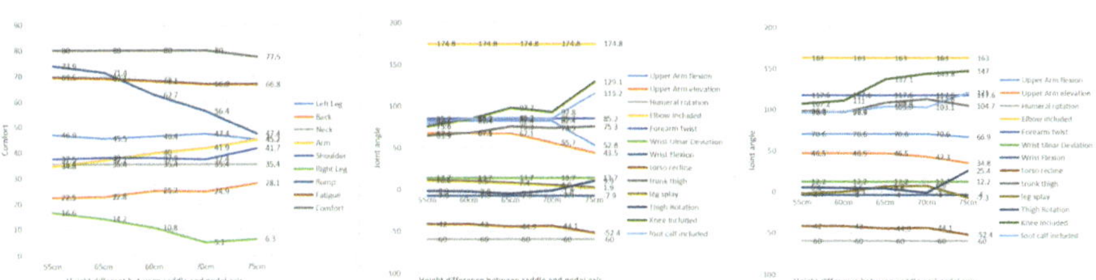

Fig. 9. From left to right is the comfort analysis of P95 digital human, right joint angle analysis, left joint angle analysis

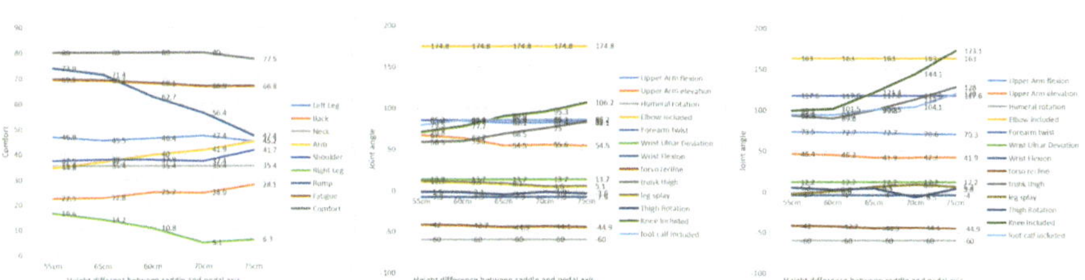

Fig. 10. From left to right is the comfort analysis of P99 digital human, right joint angle analysis, left joint angle analysis

twist, wrist tilt, wrist flexion angle unchanged, including upper arm elevation, upper arm flexion angle decreased, thigh flexion, knee flexion angle increased. It can be seen from Table 2 that with the increase in height, in addition to the worsening of the comfort experience of the arms and back, the better the comfort experience including the hips and the left leg, among which the comfort of the right leg first deteriorates and then becomes better, when the height difference is 60 cm, the comfort of the right leg is the best. The overall comfort becomes better as the height increases, and fatigue decreases.

Analysis of the experimental results of the second group of P50 digital humans:

The height of P50 digital person is 167.8 cm and the weight is 59 kg.As can be seen from Tables 9 and 10, The change trend of joint angle is the same as the first group. It can be seen from Table 7 that although the data fluctuates with the increase of height, the overall change trend is close to the experimental results of the first group. Except for the arm, shoulder and back, the comfort experience is worse, and the other comfort experience including hip and right leg is better. The comfort of the left leg first becomes worse and then better. When the height difference is 65 cm, the comfort of the left leg is the best. The overall comfort becomes better with the increase of height, and the fatigue decreases.

Analysis of the experimental results of the third group of P95 digital humans:

As can be seen from Table 14 and Table 15, The change trend of joint angle is the same as the first group. As can be seen from Table 12, with the increase of height, except for the arm, shoulder and back, the comfort experience becomes worse, and the other comfort experience including hip and left leg becomes better. The comfort of right leg becomes worse first and then better. When the height difference is 70 cm, the comfort of left leg is the best. The overall comfort becomes better with the increase of height, and the fatigue decreases.

Analysis of the experimental results of the forth group of P99 digital humans:

As can be seen from Table 19 and Table 20, The change trend of joint angle is the same as the first group. It can be seen from Table 17 that with the increase of height, except for the arm, the comfort experience of shoulder and back becomes worse, and the comfort experience of hip and left leg becomes better. When the height difference is 75 cm, the comfort of left leg is the best. The overall comfort becomes better with the increase of height, and the fatigue decreases.

In general, when the height of the saddle changes, the human body will adjust its posture to ensure the riding state and trigger changes in the angle of each joint. There is a suitable saddle height. When the rider is at this saddle height, the angle of motion of his joints is most suitable for muscle contraction [8]. When the joint angle changes, the muscle contraction also changes, and the muscle contraction will affect the comfort of the human body. In the process of increasing saddle height, we can see that the data changes of different digital people also have the same thing, that is, the joint angle of the wrist is basically unaffected, the upper arm flexion angle decreases, the thigh flexion and knee flexion increase, the comfort of the shoulders, backs, and legs of the corresponding arms also increase. When the height reaches a certain level, the comfort will reach a peak as the joint flexion angle increases.

6 Conclusion

Experimental results show that during riding, the position of the saddle will indeed affect the comfort of various parts of the human body, and for people of different heights and proportions, the optimal height difference between the saddle and the pedal axis is also different. Among them, the best height difference for a P5 digital human with a height of about 158 cm is 60 cm, the best height difference for a P50 digital human with a height of about 167 cm is 65, and the best saddle height difference for a P95 digital human with a height of about 177 cm is 70 and a P99 digital human with a height of 181 cm is suitable for 75 cm. This experiment only considers the position of the saddle relative to the pedals, but has not studied the influence of the distance between the saddle and the handle on the comfort, so this experiment has certain limitations.

References

1. Wei, Q., Liu, H.: The effect of bicycle saddle height on injury prevention and sports performance. In: Sports Biomechanics Branch of Chinese Sports Science Association. Collection of abstracts of the 19th National Sports Biomechanics Academic Exchange Conference. Sports Biomechanics Branch of Chinese Sports Science Society: Sports Biomechanics Branch of Chinese Sports Science Society, pp. 249–250 (2017)
2. Gong, C., Feng, C.: On the principle of bicycle dynamics. China Bicycle **09**, 68–69 (2018)
3. Lan, S.: Human factors and ergonomics analysis based on Jack virtual simulation technology. Ind. Eng. **20**(06), 96–100 (2017)
4. Su, S., Wang, H.: The main points of man-machine engineering simulation analysis based on JACK software. Ind. Des. (11), 115–116 (2017)
5. Shuang, L.: Ergonomics analysis based on Jack virtual simulation technology. Ind. Eng. **20**(06), 96–100 (2017)
6. Ma, Z., Xue, H., Rune, S.: Human modeling and ergonomics analysis based on Jack. Aeronaut. Comput. Technol. **01**, 97–100 (2008)
7. Chen, G., Feng, C.: On the principle of bicycle dynamics. China Bicycle **09**, 68–69 (2018)
8. Zhou, S., Li, M., Zhang, X.: Research on the design of bicycle saddle based on ergonomics. In: Chinese Sports Science Association. Collection of abstracts of the 11th National Sports Science Conference. Chinese Sports Science Association: Chinese Sports Science Association: 2 (2019)

Digital Human Modeling by Women in Human Factors

Mixed Reality Simulation for Medical Training: How It Affects Learners' Cognitive State

Agnese Brunzini[1(✉)], Alessandra Papetti[1], Michele Germani[1], Pamela Barbadoro[2], Daniele Messi[3], and Erica Adrario[4]

[1] Department of Industrial Engineering and Mathematical Sciences, Università Politecnica delle Marche, Via Brecce Bianche 12, 60131 Ancona, Italy
a.brunzini@pm.univpm.it, {a.papetti,m.germani}@staff.univpm.it

[2] Department of Biomedical Sciences and Public Health, Section of Hygiene, Preventive Medicine and Public Health, Università Politecnica delle Marche, Via Tronto, 10/a, 60126 Ancona, AN, Italy
p.barbadoro@staff.univpm.it

[3] Università Politecnica delle Marche, Via Tronto, 10/a, 60126 Ancona, AN, Italy
d.messi@pm.univpm.it

[4] Department of Biomedical Sciences and Public Health, Section of Anaesthesia and Intensive Care Unit, Università Politecnica delle Marche, Via Tronto, 10/a, 60126 Ancona, AN, Italy
e.adrario@staff.univpm.it

Abstract. A mixed reality (MR) system, by providing visual, auditory, and haptic feedback to the learner, can offer a high level of immersion and realism, especially in the healthcare context. In medical training through MR simulations, it is particularly important to avoid mental overload, discomfort, fatigue, and stress, to guarantee productive learning. The present work proposes a systematic assessment of stress, cognitive load, and performance (through subjective and objective measures) of students during an MR simulation for the rachicentesis procedure. A specific application has been developed to enhance the sense of realism, by showing, over the skill trainer, a digital patient that responds with auditory and visual feedback, based on the learner's interaction. A sample of 18 students has been enrolled in the pilot study. Preliminary results suggest the effectiveness of the proposed MR application using Hololens: high performances are achieved, and the cognitive conditions are well balanced.

Keywords: Medical simulation · Mixed reality · Cognitive load · Stress · Performance

1 Introduction

The simulation in clinical education is considered an excellent means of training healthcare professionals as it allows improving technical and non-technical skills (e.g., clinical decision-making, situation awareness, team working) in controlled environments, without putting real patients at risk. Typical and rare medical conditions can be simulated so

J. L. Wright et al. (Eds.): AHFE 2021, LNNS 264, pp. 339–347, 2021.
https://doi.org/10.1007/978-3-030-79763-8_41

that learners can experience different clinical situations, which they may encounter in their careers, in a practical, realistic, and immersive way.

In recent years, a lot of studies have tried to integrate manikins and augmented reality (AR), obtaining simulators in mixed reality (MR). Several review studies tried to collect and categorize AR applications in healthcare and showed how they mainly focused on high-risk, invasive skills such as endoscopy and surgery [1]. Beyond surgery, anatomy, and anesthesia are the most frequent subjects of AR/MR applications in healthcare education [2]. Many studies applied extended reality for helping learners in needle insertion [3–5]. Even if the medical procedures are different, most of them use AR to overlay the internal human anatomy on the skill trainer to help the user having a visual of the patient's internal body state, interactively [6]. The technology is then used to provide additional visual material, which is not available in real practice. In this way, the simulation moves away from reality. Instead, an MR system, by sensorial feedback, should offer a high level of immersion and realism. It also allows students to experience stress levels similar to those in real practice. However, medical simulations could generate excessive levels of acute stress or emotional states of anxiety and worry, that can compromise students' performance and health [7]. For this reason, it is important to monitor and manage stress levels during simulations. Indeed, stress management education may enhance technical performance [8]. It is worth also considering the influence that the use of AR/MR applications has on the emotions of students, assessing their effect on performance and learning.

The multimodal dimension of stress makes the research field very broad; however, according to ISO 10075–3, 2004, four main criteria can be distinguished in detecting stress: psychological, physiological, behavioral, and biochemical. The most common analyses typically include the subjective assessment based on self-report and physiological assessment based on ECG and skin conductivity. Another key aspect is related to cognitive load (CL). Indeed, in the education field, it is particularly important to avoid mental overload to guarantee productive learning. Although performance measures are strongly used in the field of medical education to evaluate the skills of trainees, the assessment of their cognitive state is relatively "uncommon". This fact is disadvantageous especially if technologies, such as advanced simulators and AR/MR devices, are introduced. They can support learning and teaching by improving the students' immersion in the simulated scenario or, conversely, may represent a potential risk of information overload [9]. Indeed, some studies reported that AR decreases cognitive load, others stated that it causes cognitive overload [10]. Moreover, learners with limited clinical experience are more exposed to the risk of cognitive overload, with a resultant decline in learning [11]. Therefore, the CL study and the related AR implications need to be further investigated.

Concerning the CL assessment methods, researchers agree in classifying them into three main broad categories (also according to the ISO 10075–3, 2004): performance assessment method, self-assessment method, and physiological measurements method. From a last more specific review about the assessment of students' cognitive state in medical simulation training, 78% of the analyzed papers applied only one method, while the remaining 22% used a couple of them. The self-assessment method was the most applied (72%), with the NASA-TLX as the most used questionnaire (56%). The method of task performance measure was applied in 39% of papers, while the psychophysiological measures were applied only in 11% of studies [12]. Current findings suggest that physiological parameters are the most sensitive means for detecting variations in CL levels during simulation training. However, it is suggested to use multiple concurrent measurements to increase the validity of CL assessment and help to improve the rigor of studies in medical education and simulation training [13].

More recent studies still confirm that AR needs to be investigated more robustly [14], since rigorous, objective measurements of clinical procedural skills, and human performance metrics are limited or absent [15]. A significant shortcoming is also the wide heterogeneity among research designs and outcome measurements. Establishing guidelines and standard methodologies to analyze and assess outcomes in medical education due to the use of AR technology would lead to higher-quality studies [2]. For this aim, this work proposes a systematic assessment of stress, cognitive load, and performance of students during an MR simulation for the rachicentesis procedure.

2 Development of a Mixed Reality Prototype for Rachicentesis

An MR prototype for the training of the rachicentesis has been designed and developed, based on the guidelines proposed in [16]. Since providing the student with information that he/she would not have in real practice could be disadvantageous, the objective of the MR system (manikin + AR) was not to make the procedure easier, but to keep the student in a more immersive and realistic environment and to put him/her under pressure during the exercise. Indeed, the low fidelity partial manikin consisted only in the patient's abdomen, keeping the learners distant from reality. For this reason, the AR application overlays the 3D model of a patient, positioned as in the real procedure (left lateral decubitus or sitting position), on the real manikin, thus allowing also the visualization of the patient's head, arms, and legs. Moreover, it has been developed to provide visual and sound feedback based on the action performed (i.e., patient's spasms and complaints concurrently with the needle insertion). In this way, the AR application should have an impact on students' cognitive and emotional conditions, since they should feel more involved in the simulation and have the impression of being immersed in a real context.

For the development of the AR application, Vuforia was used to track the position of two target images related to the skill trainer and the needle. Blender 3D was used to modify the position of the 3D human model (downloaded from an anthropometric database (as suggested in [17]) in the upright, open arms position) and to create the animation related to the spasms. Indeed, Blender 3D allows using statistical body shape modeling, that permits to apply the same motion to a wide range of different body shapes [18]. The skill trainer was then scanned by the non-contact Konica Minolta Range 7 3D Scanner, to guarantee the optimal superimposition of the digital human model over the manikin. The graphic engine Unity was used to realize the main AR scene, putting together all the previous steps.

The used skill trainer was the Lumbar Puncture Trainer by Gaumard®, while Hololens by Microsoft® was selected to provide the AR experience. This technologically very advanced device consists of transparent glasses that populate the environment with holograms. The user has the sensation of seeing a life-size hologram in front of him/her, thus feeling immerse in a mixed real and digital environment. However, one limit is the field of view (FOV) for the holograms' visualization, which is only 30°.

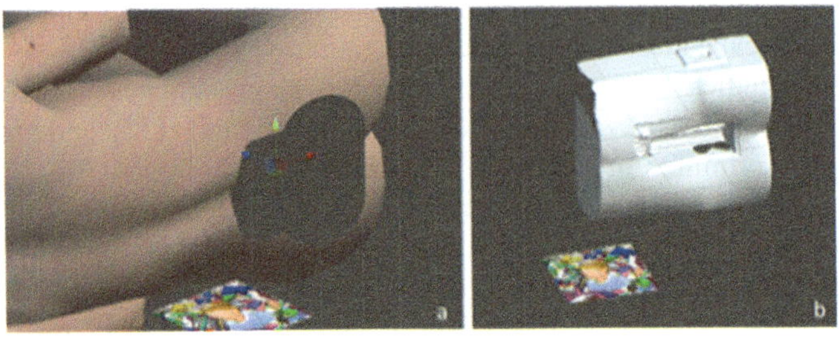

Fig. 1. (a) Portion of the 3D human model and (b) 3D scan of the skill trainer in Unity.

3 Simulation Analysis - Experimental Procedure

3.1 Participants

Eighteen students of the 6th year of Medicine and Surgery were randomly selected for the trial. Ten of them were males and eight females, with a mean age of 25.6 (±0.6) years. 83% of them was working on a degree thesis pertinent to the rachicentesis procedure (e.g., anesthesia and intensive care, cardiology, surgery, emergency, pneumology) and 38% has already had previous experience with invasive procedures.

After the explanation of the trial and the eventual contraindications in the use of Hololens and wearable devices for biometric signals monitoring, all participants signed the informed consent and the form for the processing of personal data. Then they were divided into three subgroups of six students each. For each simulation session, in turns, one student performed the rachicentesis while the other five participants looked at him/her. The execution sequence was randomly established and not revealed, to avoid eventual influences on students' anxiety, stress, and cognitive states.

3.2 Data Acquisition Workflow

The data acquisition workflow has been defined to analyze the effect that simulations in medical training have on students' cognitive conditions (especially on anxiety, stress, and cognitive load), and consequently on their performance. It includes, for each participant, self-assessment (subjective) analysis, physiological measurements (objective analysis), and performance assessment. It consists of the following steps:

- Administration of the Numerical Analogue Scale (10-points Likert scale) at the arrival in the classroom, to record the basal level of perceived stress (NAS 0);
- State-Trait Anxiety Inventory administration to assess the perceived level of anxiety in general life (STAI Trait) and before the simulation (STAI State PRE);

- NAS administration ten minutes after the first acquisition, to record the level of perceived stress before the simulation (NAS 1);
- Equipment with wearable devices (Zephyr BioHarness chest band and Empatica E4 bracelet) for the collection of physiological parameters;
- Theory explanation, briefing, and demonstration by the teacher;
- Rachicentesis simulation performed by the student;
- State-Trait Anxiety Inventory administration to assess the perceived level of anxiety after the simulation (STAI State POST);
- NASA-TLX administration to assess the perceived workload (i.e., physical demand, mental demand, temporal demand, effort, performance, and frustration);
- NAS administration ten minutes after the end of the simulation (NAS 2);
- NAS administration twenty minutes after the end of the simulation (NAS 3);
- NAS administration thirty minutes after the end of the simulation (NAS 4);
- Removal of the wearable devices (Zephyr BioHarness and Empatica E4).

The performance was assessed by a checklist that evaluates each task of the rachicentesis procedure as 'correctly performed', 'non correctly performed', or 'not performed'. Times (i.e., patient preparation time, time to succeed, total time), number of errors (i.e., touching the needle in the wrong place, do not re-insert the stylet, coming in-and-out with the needle in the skill trainer), number of attempts to succeed, and teacher interference were also recorded and assessed. Moreover, a 5-items questionnaire was administered before and after the simulation to evaluate students' skills.

Figure 2 shows one participant who is equipped with Hololens, Zephyr BioHarness chest band, and Empatica E4 bracelet while performing the rachicentesis.

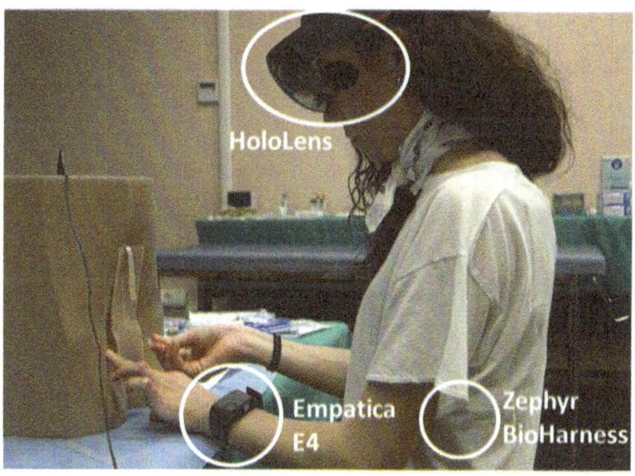

Fig. 2. Trial setup (Hololens + E4 + BioHarness)

4 Results and Discussion

Results refer to 15 participants. Three have been discarded because they did not answer all the questions in the subjective surveys and/or their physiological data were affected by errors.

4.1 Psychophysiological Analysis

Results related to the subjective self-assessment about the perceived anxiety, stress, and cognitive conditions are summarized in Table 1.

Table 1. Psychological self-assessment before and after the MR simulation

	MEAN	STD. DEVIATION
STAI TRAIT	45.99	8.84
STAI STATE PRE	41.65	9.96
STAI STATE POST	37.07	9.88
NAS 0	2.24	1.30
NAS 1	2.41	1.23
NAS 2	3.41	1.84
NAS 3	3.06	1.92
NAS 4	2.71	1.93
NASA-TLX	50.41	19.52
Mental demand	188.50	126.82
Physical demand	59.61	68.51
Temporal demand	108.28	116.26
Performance	203.50	120.56
Effort	136.67	88.30
Frustration	59.67	103.81

On average, participants are more anxious in their life than in the classroom before the simulation (STAI Trait > STAI State PRE). Perceived anxiety decreases after the simulation (mean STAI State Post < mean STAI State Pre). Moreover, the mean STAI State Post agrees with the value expected for college students [19], while the mean STAI State Pre is few points higher, confirming the increment of anxiety felt before the simulation.

From NAS 0 and NAS 1 it is possible to note that, before the MR simulation, students do not feel stressed. The peak in NAS 2 corresponds to the perceived stress during the rachicentesis training. However, even if the feeling of stress constantly decreases from NAS 2 to NAS 4, thirty minutes after the end of the simulation the perception of stress is still higher than the baseline NAS 0. A debriefing session should be expected, to keep the perceived stress under the basal level.

The NASA-TLX shows the perceived workload related to the MR rachicentesis, and the weight assigned to the six indexes. The mean perceived workload for the simulation in MR is on the inferior boundary of "high-range" (high for 50–79 total score), according to the literature's score interpretation [20]. The performance index received the greatest weight, followed by the mental demand and the perceived effort.

Stress and CL levels, which range between 0 and 6, have been measured through the proprietary Phasya's algorithm for the analysis of physiological signals. They were computed for simulation phases (Table 2). The mean stress and CL levels, computed over the entire MR rachicentesis, are respectively equal to 2.24 (±2.06) and 2.20 (±1.28). Thus, the stress is on average a bit higher than the CL. However, the CL is higher than the stress during the puncture execution. It is worth to note that the CL increases from preparation to puncture phase, and then decreases toward the end of the simulation until the rest after the training. Conversely, the stress increases from the basal level before the simulation to the preparation phase, and then constantly decreases during the following phases. The stress and CL levels after the training on average turn back to the basal levels (even if debriefing is not performed). On average, stress, and CL during the execution of the procedure in MR can be considered medium, not high (on the 6 levels scale).

Table 2. Stress and CL levels during the MR simulation

SIMULATION PHASES	STRESS	COGNITIVE LOAD
Rest	2.21 (±1.83)	2.12 (±1.35)
Patient preparation	2.27 (±2.12)	2.17 (±1.28)
Lumbar puncture	2.23 (±2.06)	2.33 (±1.46)
Fluid collection and end of the procedure	2.20 (±2.26)	2.10 (±1.51)
Rest	2.17 (±2.18)	2.08 (±1.67)

4.2 Performance Analysis

The questionnaire for the skills assessment was answered before and after the MR training, and, while the mean number of correct responses before the simulation was equal to 3.61 (±1.04), after the simulation is increased to 4.72 (±0.57).

Regarding the performance, success was reached in 94.45% of cases (liquor collection). Thus, most participants were able to correctly perform the puncture and make the liquor pouring out. The percentage of incorrectly performed or not executed tasks is under 6%. The mean number of attempts to reach success was 2.21 (±2.03), and the mean number of errors per subject was 0.33 (±0.49). 27.78% of participants came in-and-out with the needle during the procedure, and only 5.56% touched the needle where it is not allowed (both actions should be avoided for hygienic reasons). No one extracted the needle without re-inserting the stylet.

Mean execution times were calculated for students who reached success:

- Preparation Time: 1.16 ± 0.50 min
- Time to success: 3.15 ± 2.87 min
- Time to finish: 0.92 ± 1.79 min
- Simulation Total Time: 5.92 ± 3.09 min.

5 Conclusions

The paper investigates the AR use in medical education to increase the student's level of immersion during the simulation of the rachicentesis procedure. Satisfactory levels of performance and improvement of skills have been achieved. Moreover, the monitoring of stress and cognitive load showed the achievement of the right compromise between realism and overload. The results also highlight their correct trend during the simulation.

Future works will include the analysis and the comparison with traditional simulation training, enrolling a larger sample of students.

References

1. Zhu, E., Hadadgar, A., Masiello, I., Zary, N.: Augmented reality in healthcare education: an integrative review. Peer J. **2**, e469 (2014)
2. Gerup, J., Soerensen, C.B., Dieckmann, P.: Augmented reality and mixed reality for healthcare education beyond surgery: an integrative review. Int. J. Med. Educ. **11**, 1–18 (2020)
3. Kobayashi, L., Zhang, X.C., Collins, S.A., Karim, N., Merck, D.L.: Exploratory application of augmented reality/mixed reality devices for acute care procedure training. Educ. Adv. 158–164 (2017)
4. Rochlen, L.R., Levine, R., Tait, A.R.: First-person point-of-view-augmented reality for central line insertion training: a usability and feasibility study. Simul. Healthcare **12**, 57–62 (2017)
5. Margarido Mendes, H.C., Costa, C.I.A.B., da Silva, N.A., Leite, F.P., Esteves, A., Lopes, D.S.: PIÑATA: pinpoint insertion of intravenous needles via augmented reality training assistance. Comput. Med. Imaging Graph. **82**, 101731 (2020)
6. Sherstyuk, A., Vincent, D., Berg, B., Treskunov, A.: Mixed reality manikins for medical education. In: Furht B. (eds) Handbook of Augmented Reality. Springer, New York, NY (2011). https://doi.org/10.1007/978-1-4614-0064-6_23
7. Dias, R.D., Ngo-Howard, M.C., Boskovski, M.T., Zenati, M.A., Yule, S.J.: Systematic review of measurement tools to assess surgeons' intraoperative cognitive workload. Br. J. Surg. **105**, 491–501 (2018)
8. Goldberg, M.B., et al.: Optimizing performance through stress training - an educational strategy for surgical residents. Am. J. Surg. **216**, 618–623 (2018)
9. Atalay, K.D., Can, G.F., Erdem, S.R., Muderrisoglu, I.H.: Assessment of mental workload and academic motivation in medical students. J. Pak. Med. Assoc. 66(5), 574-578 (2016)
10. Akçayır, M., Akçayır, G.: Advantages and challenges associated with augmented reality for education: a systematic review of the literature. Educ. Res. Rev. **20**, 1–11 (2017)
11. Fraser, K., Ma, I., Teteris, E., Baxter, H., Wright, B., McLaughlin, K.: Emotion, cognitive load and learning outcomes during simulation training. Med. Educ. **46**, 1055–1062 (2012)
12. Scafà, M., Serrani, E.B., Papetti, A., Brunzini, A., Germani, M.: Assessment of students' cognitive conditions in medical simulation training: a review study. In: Cassenti D. (eds) Advances in Human Factors and Simulation. AHFE 2019. Advances in Intelligent Systems and Computing, vol. 958, pp. 224–233 (2019)
13. Naismith, L.M., Cavalcanti, R.B.: Validity of cognitive load measures in simulation-based training: a systematic review. Acad. Med. **90**, S24–S35 (2015)
14. Munzer, B.W., Khan, M.M., Shipman, B., Mahajan, P.: Augmented reality in emergency medicine: a scoping review. J. Med. Internet Res. **21**(4), e12368 (2019)
15. Linde, A.S., Geoffrey, T.M.: Applications of future technologies to detect skill decay and improve procedural performance. Mil. Med. **184**, 72–77 (2019)

16. Brunzini, A., Papetti, A., Serrani, E.B., Scafà, M., Germani, M.: How to improve medical simulation training: a new methodology based on ergonomic evaluation. In: Karwowski, W., Ahram, T., Nazir, S. (eds.) AHFE 2019. AISC, vol. 963, pp. 145–155. Springer, Cham (2020). https://doi.org/10.1007/978-3-030-20135-7_14
17. Paul, G., Scataglini, S.: Open-source software to create a kinematic model in digital human modeling. In: DHM and Posturography, pp. 201–213 (2019)
18. Scataglini, S., Danckaers, F., Haelterman, R., Huysmans, T., Sijbers, J.: Moving statistical body shape models using blender. In: Bagnara, S., Tartaglia, R., Albolino, S., Alexander, T., Fujita, Y. (eds.) IEA 2018. AISC, vol. 822, pp. 28–38. Springer, Cham (2019). https://doi.org/10.1007/978-3-319-96077-7_4
19. Spielberger, C.D., Gorsuch, R.L.: State-trait anxiety inventory for adults: Sampler set: manual, test, scoring key. Mind Garden, Redwood City, California (1983)
20. Sugarindra, M., Suryoputro, M.R., Permana, A.I.: Mental workload measurement in operator control room using NASA-TLX. In: IOP Conference Series: Materials Science and Engineering, vol. 277, no. 1, p. 012022 (2017)

User-Centered Design of Cues with Smart Glasses for Gait Rehabilitation in People with Parkinson's Disease: A Methodology for the Analysis of Human Requirements and Cues Effectiveness

Silvia Imbesi[1(✉)], Mattia Corzani[2], Filippo Petrocchi[1], Giovanna Lopane[3], Lorenzo Chiari[2], and Giuseppe Mincolelli[1]

[1] Department of Architecture, University of Ferrara, Via della Ghiara 36, 44123 Ferrara, Italy
{silvia.imbesi,filippo.petrocchi,giuseppe.mincolelli}@unife.it
[2] Department of Electrical, Electronic, and Information Engineering, University of Bologna, Bologna, Italy
{mattia.corzani,lorenzo.chiari}@unibo.it
[3] IRCCS Istituto delle Scienze Neurologiche di Bologna, Bologna, Italy
giovanna.lopane@isnb.it

Abstract. As an alternative to the technical industrial mindset, User-Centered Design has proven to be an effective tool to realize products and services for the Healthcare sector. In this scenario, this project aims to develop an innovative wearable gait rehabilitation solution obtained by integrating the Smart Glasses Vuzix Blade into the smartphone-based CuPiD-system. The specific aims of the testing phase described in this paper were to observe how submitted visual, auditory, and vibratory cues influence the user's gait; and which were the most impacting and efficient typologies of cues. To achieve these results, we analyze the users' qualitative and quantitative feedback, respectively obtained from an interview and a specific gait analysis protocol.

Keywords: User-Centered Design · Tele-Rehabilitation · Smart glasses · Visual cues · Auditory cues · Vibratory cues · Ageing users · Gait Monitoring System · Parkinson's disease · Wearable sensors

1 Introduction

It is well known that our world is getting older and older. While the ageing society can be considered an achievement in terms of societal progress, it also brings an economic burden [1]. With social, economic, cultural, and political changes, governments are exploring different strategies to face the rapid ageing of their societies, including how to provide effective aged care services [2].

At the same time, the world is also getting more digital, technological, and connected. The Internet and the Internet of Things (IoT) are tools that have revolutionized almost any sector, and healthcare is one of them.

J. L. Wright et al. (Eds.): AHFE 2021, LNNS 264, pp. 348–358, 2021.
https://doi.org/10.1007/978-3-030-79763-8_42

Healthcare Systems have recognized, for several years, the advantages of using Information and Communication Technologies (ICT) to improve the quality of care, and they are now working, although at a different pace worldwide, to turn traditional into smart healthcare [3]. To meet the increasing demands of an aging population with chronic diseases and comorbidities, IoT technology seems an effective tool to shift from clinic-centric to patient-centric healthcare [4].

Nevertheless, to accelerate the shift toward the brave new world of e-health, it is necessary that technology, IoT included, is appropriately designed with the aid of end-users. Many e-health technologies have failed to innovate the current clinical practice because they ignored the interaction between technology, human characteristics, and socio-economic environment [5].

As an alternative to the technical industrial mindset, User-Centered Design has proven to be an effective tool to realize products and services for the Healthcare sector. User-Centered approach has to be included in the design process since the starting phases to develop a product or system that is effective due to the close relationship with the users' requirements and the high capacity of satisfaction of their needs [6].

1.1 Gait Rehabilitation with Smart Glasses in Persons with PD

This project is developed to provide gait training to people with Parkinson's disease (PD), a disease that markedly impacts their functional independence, well-being, and health-related quality of life [7]. In particular, parkinsonian gait is generally expressed as reduced step length and gait speed, increased asymmetry and double support time, and increased variability in step/stride time and length [8]. It is known that the use of external sensory cues (e.g., auditory, visual) to reinforce attention toward the task [9] is an effective gait-rehabilitation strategy for persons with PD; the cues stimulate the voluntary executive component of action [10, 11] by activating the attentional-executive motor control system and bypassing the dysfunctional, habitual, sensorimotor basal ganglia network [10, 12].

In the past, auditory cueing during gait has typically been provided continuously in an open loop (regardless of gait performance) [13, 14]. In the therapeutic context, people are instructed to match foot strike with each beat of the auditory rhythm: these so-called 'cues' [15] facilitate the movement's initiation or continuation.

Recent advances in wearables and mobile technology offer promising new ways to provide closed-loop feedback to people with PD during gait [16], which can overcome traditional open-loop cue, providing customized cueing: stimuli are triggered in real-time when gait deviates from normal, thus providing patients with immediate feedback on their performance.

Casamassima et al. [17] developed a unique wearable and smartphone-based system that provides verbal feedback to improve the dynamic balance and gait performance of people with PD [18, 19]. During the former EU-funded CuPiD-project, Ginis et al. [18] tested this system's feasibility (CuPiD-system) in a real-life context: they discovered that the wearable system was well-accepted and seemed to be a practical approach to promote gait training in PD subjects.

In the field of sensory feedback, it is known that proprioceptive feedback (such as vibrational stimuli) requires little or no cognitive processing or attention [20]. Besides, a multi-sensory approach enhances perceptual processing [21], known to be reduced in PD subjects [22]. Therefore, using a different combination of sensory inputs may significantly change the motor response's effectiveness and efficiency to the feedback.

Thanks to advances in technologies, visual feedback is possible through Smart Glasses (SG) [23]. SG represents an ideal modality to provide personalized feedback and assistance to people with PD in daily living situations. Indeed, McNaney et al. [24] reported that participants with PD were generally positive about SG as an everyday assistive device; however, usability issues and social stigma still hinder its general acceptance.

The aim of this project is to develop an innovative wearable gait rehabilitation solution obtained by integrating the Vuzix Blade SG [25] into the smartphone-based CuPiD-system described above [17]. Thus, the resulting gait rehabilitation solution allows therapists to design personally tailored rehabilitation programs based on individual patient sensory preferences, cognitive aspects, and medical knowledge.

Blade SG can augment reality by overlaying pertinent information on top of the user's visual field: projecting a bidimensional image on the right lens. Besides, it also can trigger dual vibratory feedback on its right and left temples.

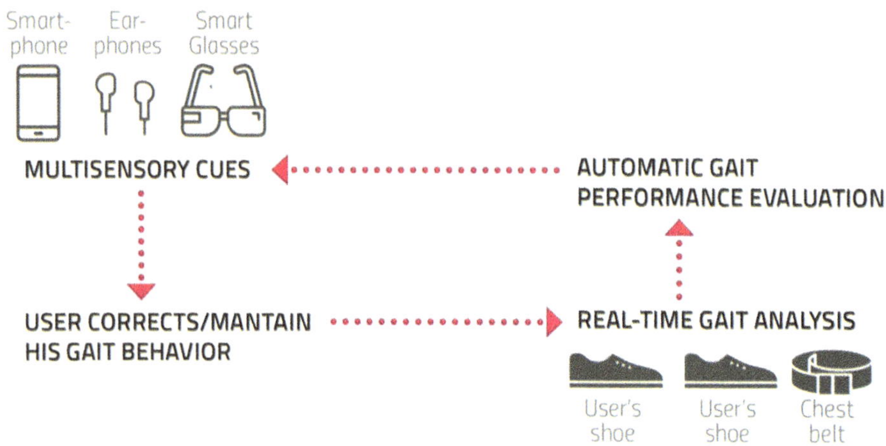

Fig. 1. Schematic representation of the innovative gait rehabilitation system. Adapted from Casamassima et al. [17].

In the light of the previous consideration, it was decided to follow a User-Centered approach due to the necessity to gain a high degree of usability of the system, whose effectiveness could be compromised by a low fitting to this particular users' necessities.

The aim of the specific testing phase described in this paper was to observe how submitted visual, auditory, and vibratory cues influence the user's gait; which were the most impacting and efficient typologies of feedback, and on which spatio-temporal gait parameters.

2 Methods

The authors used the mGait system, a CE-marked medical device from mHealth Technologies [26], to evaluate the spatio-temporal gait parameters. In our walking paradigm, the mGait system was used with two inertial sensors applied to the feet, and data is transmitted via Bluetooth to the mGait smartphone app.

In designing significant and effective cues to be tested with people with PD, it was first decided to plan a pre-trial on several people not belonging to a particular demanding category of users. Thus, we test these preliminary cues developed following a Human-Centered approach to guide a selection of the most effective ones for further experimentations with persons with PD.

It was decided to propose three types of sensory stimuli: auditory by wireless earphones, visual and vibratory by the Vuzix Blade SG [25].

2.1 Designing Sensory Cues

The first decision was to design signals suggesting a rhythm to be followed instead of suggestions on incrementing and decrementing aspects of the gait [27, 28]. This rhythm was identified with walking cadence and set at the average value of 100 steps/min as suggested by Tudor-Locke et al. [29].

Thus, it was decided to elaborate signals requiring a low effort for their comprehension and interpretation, for not compromising the attention on the gait path.

The scientific literature on visual cues developed for people with PD reveals that static and dynamic cues significantly contribute to body balance, standing, and walking when conditions of equilibrium are compromised [30, 31]. Considering the Vuzix Blade's possibilities, it was decided to elaborate static patterns composed of two figures and dynamic patterns with six figures (in both cases played in a loop following the established cadence). For all patterns, to avoid complex icons and written parts, it was preferred to use simple figures that do not involve interpretation in reading. Among analyzed ones, that of the circle was considered particularly immediate and re-proposed in several variations. The authors developed several groups of visual patterns, suggesting different ways of swing and alternation, intuitively imitating the gait movements (Fig. 2).

Also, vibratory cues impact gait parameters, especially if associated with the cadence [32]. In this experiment, vibrations are provided by the SG temples. The only setting possibility is related to the stimulus duration but not on position or intensity. It is also possible to set differently if it comes from the right or left temple. Following these constraints, the authors elaborated several vibratory patterns to be submitted to the tested users.

The field of auditory cues is the one that has undoubtedly been more detailed in the scientific literature [14, 15]. As previously described, there have been many experiments testing the effect of auditory cues on persons with PD, most of them using rhythmic pathways [33]. Moore et al. suggest the use of musical cues [34] that have been shown to have an impact on the human brain, including areas that are responsible for movement, language, attention, memory, executive function, and emotion [35]. The authors created a basic melody, then reproduced it in various versions using one or more musical instruments or electronic sounds. Those pathways were registered at 100 bpm, corresponding to 100 steps/min.

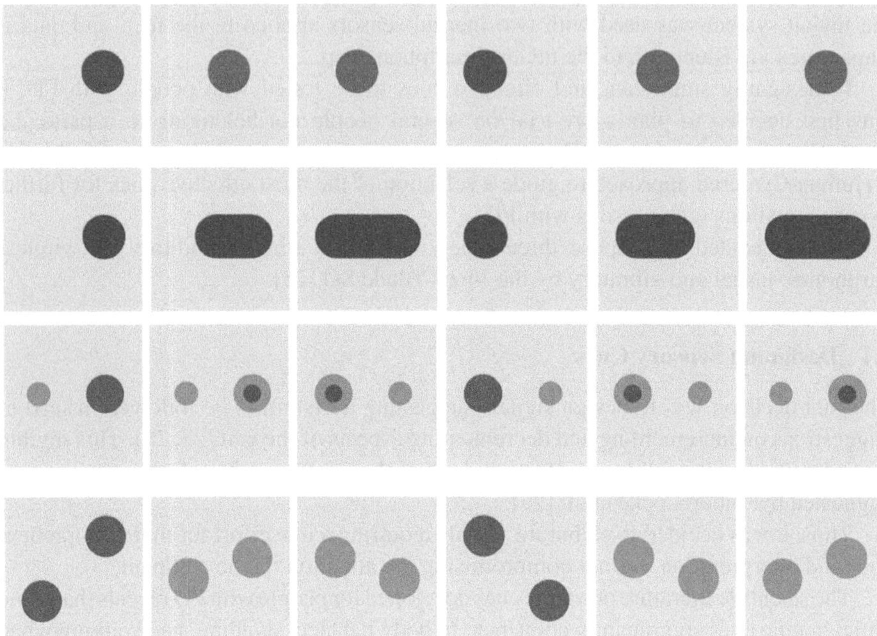

Fig. 2. In this image are some examples of the elaborated dynamic patterns, based on the circle, suggesting the rhythm to be followed while walking.

2.2 Protocol

The path has a total length of 25 m, is linear and free of physical and perceptive obstacles (unevenness, carpets, floor changes, sudden changes of lighting). The environment has a minimum width of 2 m without impediments (furniture, walls, etc.) immediately before the start or immediately after the end.

The user wears IMUs on his shoes connected to the mGait system, and he starts with a walk that will be the reference point of comparison for the following ones. Hence, he wears SG and earphones and starts the cues' tests where between 5 and 9 m from the start of the walk, the stimulus is transmitted and will only be interrupted at the end of the path (Fig. 3).

Each subject, therefore, performs 21 tests following this scheme:

- 1 baseline trial without stimuli as a reference;
- 3 unisensory tests (visual, vibratory, auditory) with randomized administration order (Z1, Z2, Z3);
- 14 unisensory tests (6 visual, 2 vibratory, 6 auditory) and 3 multisensory tests (1 visual + vibratory, 1 auditory + vibratory, 1 visual + auditory + vibratory) with randomized order (A1: A6, B1: B6, C1: C2, D1: D3).

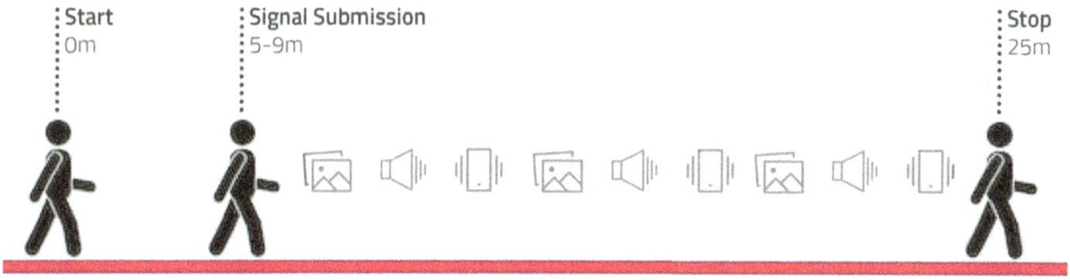

Fig. 3. Schematic representation of the path for the test.

The first three tests are carried out without providing information to the user either on the type of signal that will be received or on the behavior that he/she will have to adapt to verify the perceptual impact of mono-sensory stimuli with the least possible conditioning. Subsequently, in the second part of the test, the user is invited to align with the cadence suggested by the mono- or multi-sensory stimulus that will be transmitted (Fig. 4).

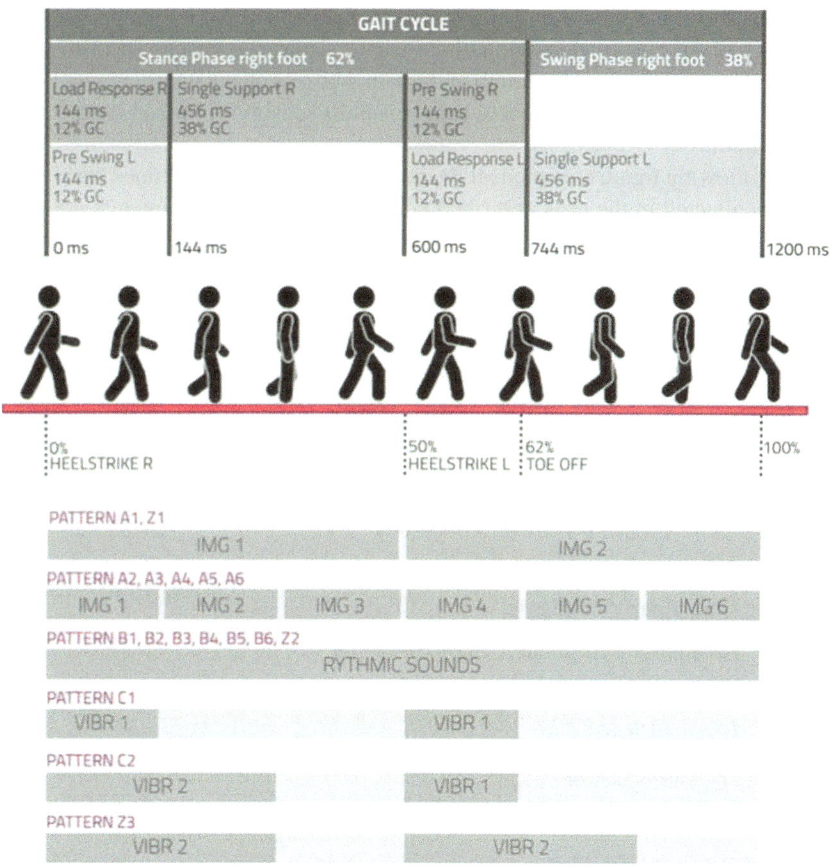

Fig. 4. The scheme represents the different phases of the gait cycle and the correspondence between the phases and the submitted cues' timing. Adapted from Iosa et al. [36].

After each test, the user answers a few questions on the just submitted cue to express his perception of the signal and its effectiveness, giving a score from 1 to 5. At the end of all tests, the user has a short interview on his perception of the SG, their wearability, ease of use, etc.

3 Results

The testing phase aimed to detect qualitative and quantitative feedback from the users, respectively obtained from interview and gait analysis.

Regarding the qualitative feedback, it was asked to the person to answer questions on signals' evocative, perceptive and emotive properties, comparing their feelings with their physical reactions. From this information, it is possible to collect users' requirements to improve both the multisensory cues and the Smart Glasses' shape.

Concerning the quantitative feedback, it was detected and analyzed several values related to the gait performance for the different typologies of signals to analyze and compare the relationship between the cue submission and the performance improvement. In particular, from the spatio-temporal gait parameters provided by the mGait system, the most significant parameters the authors want to quantify are impact of the signal; differences between the various stimuli of the same sensory channel; differences between the three sensory channels; differences between multi-sensory stimuli and related mono-sensory stimuli.

In the following figure are listed all the parameters that are determined starting from the values collected in the tests described before. The obtained parameters are used to guide a conscious choice on the typology of cues that will be selected to be used in future tests (Fig. 5).

01	Gait Speed variation
02	Stride Lenght variation
03	Cadence variation
04	Variation between final cadence and suggested cadence
05	Intra-test gait parameters variability
06	Response time
07	Duration of gait phases
08	Parameter most influenced by the signal
09	Parameter less influenced by the signal
10	Usability level
11	Usneasiness level
12	Synchronization difficulty perceived by the user

Fig. 5. List of the findings that this phase of the research wants to obtain.

Results of this trial are available at the link provided in the section "supplementary materials". Methodologies and results of the forthcoming experiments and developments will be shared in other scientific publications.

4 Conclusions

The results of this analysis of human requirements and cues effectiveness are supposed to be the starting point for the selection of several numbers of tested cues for the submission to persons with PD in ambulatory context, in order not to stress them with a high number of training tests, but to use cues that have already proven to be suitable for the established project's objectives.

From this first testing phase, according to detected parameters, it was possible to identify critical points as the synchronization between the submission of the signal and the person's gait phase, and the gap between the user's cadence and the one proposed by the system; and draw conclusions on which cues were most appreciated for their clarity and non-invasiveness, and on which ones have an appreciable impact on gait parameters.

Another limitation of our approach is certainly about the extent that each subject is visually-, auditory-, or somatosensory-dependent. The nervous system uses augmented sensory information differently depending primarily on individual proclivities to rely on a sensory channel over another to control motor behavior [37].

To overcome some of these limitations, in the innovative wearable gait rehabilitation solution indicated in Fig. 1, the target cadence will be subject-specific. In particular, during gait training with the CuPiD-system, the target cadence can be based on the one detected during the user's reference gait of the CuPiD-system [17]. Alternatively, to comply with the therapeutic needs, the target cadence should be a modified version of that recorded during the user's reference gait.

Importantly, even SG must be personalizable in their physical dimension according to the user's requirements: most people with PD are over 50 and have vision problems, thus SG must be compatible with prescription lenses. This feature is possible with the SG used [25].

Gait and balance impairments afflict most persons with PD and impact their quality of life and independence [38]. With their sensors and feedback capabilities, SG may help alert the patient to correct this behavior and provide support via visual, auditory, or vibratory cues, instructional videos, or step by step verbal instructions.

Although the potential of real-time feedback systems in gait rehabilitation through wearable devices is underexploited [39], these new real-time systems seem to increase adherence to treatment, self-management, and quality of life [40], allowing also personalized and tailored rehabilitation on the individual patients' needs [18].

In conclusion, SG has the potential to empower persons with PD. However, more research is necessary to ensure safety, efficacy, usability, and social acceptance.

Supplementary Materials. At this link, there are all the cues described in this paper and the tables of the parameters obtained from the tests.

https://drive.google.com/drive/folders/1qQU0cwB9wR8wnMF0–5fyy0CXpRBa-81?usp=sharing

References

1. Jämsén, A., Kukkonen, T.: Age-friendly society supports active ageing. In: Publications of Karelia University of Applied Sciences B, Handbooks and Article collections: 63, Karelia University of Applied Sciences (2020)
2. Pakulski, J.: Facing the challenges of an ageing society. In: Kendig, H., McDonald, P., Piggott, J. (eds.) Population Ageing and Australia's Future, pp. 111–134. Anu Press, Australia (2016)
3. ActiveAdvice: What is Smart Health and How do People Benefit?. https://www.activeadvice.eu/news/concept-projects/what-is-smart-health-and-how-do-people-benefit/
4. Farahani, B., Firouzi, F., Chang, V., Badaroglu, M., Constant, N., Mankodiya, K.: Towards fog-driven IoT eHealth: promises and challenges of IoT in medicine and healthcare. Future Gen. Comput. Syst. **78**, 659–676 (2018)
5. Van Gemert-Pijnen, J.E., et al.: A holistic framework to improve the uptake and impact of eHealth technologies. J. Med. Internet Res. **13**(4), e111 (2011)
6. Imbesi, S., Mincolelli, G.: Design of smart devices for older people: a user centered approach for the collection of users' needs. In: Ahram, T., Karwowski, W., Vergnano, A., Leali, F., Taiar, R. (eds.) IHSI 2020. AISC, vol. 1131, pp. 860–864. Springer, Cham (2020). https://doi.org/10.1007/978-3-030-39512-4_131
7. Peterson, D.S., Horak, F.B.: Neural control of walking in people with parkinsonism. Physiology, **31**(2), 95–107 (2015)
8. Hobert, M.A., Nussbaum, S., Heger, T., Berg, D., Maetzler, W., Heinzel, S.: Progressive gait deficits in parkinson's disease: a wearable-based biannual 5-year prospective study. Front. Aging Neurosci. **11**, 22 (2019)
9. Lee, S.J., Yoo, J.Y., Ryu, J.S., Park, H.K., Park, H.K., Chung, S.J.: The effects of visual and auditory cues on freezing of gait in patients with Parkinson disease. Am. J. Phys. Med. Rehabil. **91**(1), 2–11 (2012)
10. Morris, M.E.: Locomotor training in people with Parkinson disease. Phys. Ther. **86**(10), 1426–1435 (2006)
11. Ferrazzoli, D., Ortelli, P., Madeo, G., Giladi, N., Petzinger, G.M., Frazzitta, G.: Basal ganglia and beyond: the interplay between motor and cognitive aspects in Parkinson's disease rehabilitation'. Neurosci. Biobehav. Rev. **90**, 294–308 (2018)
12. Redgrave, P., et al.: Goal-directed and habitual control in the basal ganglia: implications for Parkinson's disease. Nat. Rev. Neurosci. **11**(11), 760–772 (2010)
13. Rubinstein, T.C., Giladi, N., Hausdorff, J.M.: The power of cueing to circumvent dopamine deficits: a review of physical therapy treatment of gait disturbances in Parkinson's disease. Mov. Disord. **17**(6), 1148–1160 (2002)
14. Rocha, P.A., Porfírio, G.M., Ferraz, H.B., Trevisani, H.B.: Effects of external cues on gait parameters of Parkinson's disease patients: a systematic review. Clin. Neurol. Neurosurg. **124**, 127–134 (2014)
15. Lim, I., et al.: Effects of external rhythmical cueing on gait in patients with Parkinson's disease: a systematic review. Clin. Rehabil. **19** (7), 695–713 (2005)
16. Sánchez-Ferro, Á., Maetzler, W.: Advances in sensor and wearable technologies for Parkinson's disease. Mov. Disord. **31**(9), 1257 (2016)
17. Casamassima, F., Ferrari, A., Milosevic, B., Ginis, P., Farella, E., Rocchi, L.: A wearable system for gait training in subjects with Parkinson's disease. Sensors **14**(4), 6229–6246 (2014)
18. Ginis, P., et al.: Feasibility and effects of home-based smartphone-delivered automated feedback training for gait in people with Parkinson's disease: a pilot randomized controlled trial'. Parkinsonism Rel. Disord. **22**, 28–34 (2016)

19. Ferrari, A., Ginis, P., Nieuwboer, A., Greenlaw, R., Muddiman, A., Chiari, L.: Handling gait impairments of persons with parkinson's disease by means of real-time biofeedback in a daily life environment. In: Chang, C.K., Chiari, L., Cao, Y., Jin, H., Mokhtari, M., Aloulou, H. (eds.) ICOST 2016. LNCS, vol. 9677, pp. 250–261. Springer, Cham (2016). https://doi.org/10.1007/978-3-319-39601-9_22
20. Peterson, D.S., Smulders, K.: Cues and attention in parkinsonian gait: potential mechanisms and future directions. Front. Neurol. **6**, 255 (2015)
21. Shams, L., Seitz, A.R.: Benefits of multisensory learning. Trends Cogn. Sci. **12**(11), 411–417 (2008)
22. Davidsdottir, S., Cronin-Golomb, A., Lee, A.: Visual and spatial symptoms in Parkinson's disease. Vis. Res. **45**(10), 1285–1296 (2005)
23. Wiederhold, B.K.: Time to port augmented reality health apps to smart glasses? Cyberpsychol. Behav. Soc. Netw. **16**(3), 157–158 (2013)
24. McNaney, R., et al.: Exploring the acceptability of google glass as an everyday assistive device for people with parkinson's. In: Proceedings of the 32nd annual ACM conference on Human factors in computing systems - CHI 2014, Toronto, Ontario, Canada, pp. 2551–2554 (2014)
25. 'Vuzix Blade®'. https://www.vuzix.com/products/blade-smart-glasses. Accessed 03 Dec 2020
26. Assessment - mTEST3, mHealth Technologies. http://mhealthtechnologies.it/products-wearable-sensors-smartphone-apps/assessment-mtest3/. Accessed 05 Feb 2021
27. Ginis, P., Heremans, E., Ferrari, A., Dockx, K., Canning, C.G., Nieuwboer, A.: Prolonged walking with a wearable system providing intelligent auditory input in people with parkinson's disease. Front. Neurol. **8**, 128 (2017)
28. Ginis, P., et al.: Feasibility and effects of home-based smartphone-delivered automated feedback training for gait in people with Parkinson's disease: a pilot randomized controlled trial. Parkinsonism Rel. Disord. **22**, 28–34 (2016)
29. Tudor-Locke, C., Han, H., Aguiar, E.J., et al.: How fast is fast enough? walking cadence (steps/min) as a practical estimate of intensity in adults: a narrative review Br. J. Sports Med. **52**, 776–788 (2018)
30. Amblard, B., Cremieux, J., Marchand, A.R., Carblanc, A.: Lateral orientation and stabilization of human stance: static versus dynamic visual cues. Exp. Brain Res. **61**, 21–37 (1985)
31. Prokop, T., Schubert, M., Berger, M.: Visual influence on human locomotion modulation to changes in optic flow. Exp. Brain Res. **114**, 63–70 (1997)
32. Winfree, K.N., Pretzer-Aboff, I., Hilgart, D., et al.: The effect of step-synchronized vibration on patients with Parkinson's disease: case studies on subjects with freezing of gait or an implanted deep brain stimulator. In: IEEE Transactions on Neural Systems and Rehabilitation Engineering: a Publication of the IEEE Engineering in Medicine and Biology Society, vol. 21, no. 5, pp. 806–811 (2013)
33. Woerd, E.S., Oostenveld, R., Bloem, B.R., Lange, F.P., Praamstra, P.: Effects of rhythmic stimulus presentation on oscillatory brain activity: the physiology of cueing in Parkinson's disease. NeuroImage: Clin. **9**, 300–309 (2015)
34. Moore, E., Schaefer, R.S., Bastin, M.E., Roberts, N., Overy, K.: Diffusion tensor MRI tractography reveals increased fractional anisotropy (FA) in arcuate fasciculus following music-cued motor training. Brain Cogn. **116**, 40–46 (2017)
35. Thaut, M.H.: Neurologic music therapy in cognitive rehabilitation. Music Percept. **27**(4), 281–285 (2010)
36. Iosa, M., et al.: The golden ratio of gait harmony: repetitive proportions of repetitive gait phases. BioMed Res. Int. 918642, 7 (2013)
37. Dozza, M., Horak, F.B., Chiari, L.: Auditory biofeedback substitutes for loss of sensory information in maintaining stance. Exper. Brain Res. **178**(1), 37–48 (2007)

38. Maetzler, W., Nieuwhof, F., Hasmann, S.E., Bloem, B.R.: Emerging therapies for gait disability and balance impairment: promises and pitfalls. Mov. Disord. **28**, 1576–1586 (2013)
39. Monje, M.H.G., Foffani, G., Obeso, J., Sánchez-Ferro, Á.: New sensor and wearable technologies to aid in the diagnosis and treatment monitoring of parkinson's disease. Ann. Rev. Biomed. Eng. **21**(1), 111–143 (2019)
40. Dorsey, E.R., et al.: Moving Parkinson care to the home, Mov. Disord. **31**(9), 1258–1262 (2016)

RGB-D Sensors as Marker-Less MOCAP Systems: A Comparison Between Microsoft Kinect V2 and the New Microsoft Kinect Azure

Benedetta Rosa, Filippo Colombo Zefinetti, Andrea Vitali(✉), and Daniele Regazzoni

University of Bergamo, Bergamo, Italy
b.rosa1@studenti.unibg.it, {filippo.colombozefinetti, andrea.vitali1,daniele.regazzoni}@unibg.it

Abstract. Marker-less motion capture (MOCAP) systems based on consumer technology simplify the analysis of movements in several research fields such as industry, healthcare and sports. Even if the marker-less MOCAP systems have performances with precision and accuracy lower than the marker-based MOCAP solutions, their low cost and ease of use make them the most suitable tools for full-body movements analysis. The most interesting category is relative to the use of RGB-D devices. This research work aims to compare the performances of the last two generations of Kinect devices as marker-less MOCAP systems: Microsoft Kinect v2 and Azure devices. To conduct the tests, a list of specific movements is acquired and evaluated. This work measures the improvements of the Azure in tracking human body movements. The gathered results are presented and discussed by evaluating performances and limitations of both marker-less MOCAP systems. Conclusions and future developments are shown and discussed.

Keywords: Kinect V2 · Kinect Azure · Marker-less MOCAP systems · Accuracy

1 Introduction

A motion capture (MOCAP) system is able to detect one (or multiple) human body shapes used to reconstruct the corresponding virtual avatar that mimics the real movements. The animated virtual avatar can be recreated by means of different innovative technologies [1]. The market offers a wide range of technologies with different technical specifications, purposes, costs, limits and advantages. It is important to understand which device better fits with any specific purposes and budget. Among the main MOCAP solutions, the marker-less MOCAP systems based on RGB-D sensors are playing an important role to evaluate human motions in several research fields, such as industry, healthcare and sports.

Even if this family of MOCAP systems does not guarantee as accurate and precise acquisitions as the most powerful marker-based MOCAP systems, the low-cost and the ease of use of RGB-D sensors is preferred in all those contexts in which an error of measurement of several millimeters can be considered negligible. These are the cases of the motion evaluation of actions in which the analysis is usually based on wide

J. L. Wright et al. (Eds.): AHFE 2021, LNNS 264, pp. 359–367, 2021.
https://doi.org/10.1007/978-3-030-79763-8_43

movements of limbs or part of them. The most important RGB-D devices are relative to the family of Microsoft Kinect. In the last decade, Microsoft has released 2 devices (i.e., Kinect v1 and Kinect v2) for gaming, which have been exploited by scientific researchers to develop marker-less MOCAP solutions in several research fields.

In 2020, Microsoft released the Kinect Azure that has been totally designed as RGB-D device for research and development. The new Kinect Azure has to be investigated about its accuracy and precision. The aim of this paper is the comparison of the Kinect Azure with the well-known Microsoft Kinect v2 to understand which are the real potentialities as RGB-D device for a marker-less MOCAP system. The comparison is performed by the definition of a specific list of movements. The movements known a-priori allow a person to reach predefined positions useful to evaluate the accuracy of the MOCAP systems exploited. This approach is useful to avoid the introduction of a more precise and costly marker based Mocap system for validating the measurements relative to the tracked movements.

First, the paper presents the scientific background relative to the Kinect Azure device and the Kinect v2 device with particular emphasis on their use as low-cost marker-less MOCAP systems. Then, the reached results are compared and discussed to evaluate performances and limitations. Finally, conclusions are discussed to understand which features of the Kinect Azure will really increase the accuracy of a future marker-less MOCAP systems.

2 Scientific Background

MOCAP devices have been developed since the 1970s, when they were first created for military use, and have been developed for the entertainment industry since the mid-1980s [2]. Concerning the most used optical systems, the academic literature defines the accuracy of a MOCAP system as the comparison between the positions tracked from the investigated MOCAP system with the same motions simultaneously tracked by a gold standard solution, which is usually a marker-based mocap system, such as Vicon, Optitrack and Qualisys [3–7]. A maker-based MOCAP system is composed by retro-reflective markers and is considered the solution with the higher accuracy, even though costly, available in the market. For example, the accuracy of Vicon system is between 0,15 mm and 2 mm [8]. Several research works have been done using Vicon as reference for evaluating marker-less system based on one or multiple RGB-D devices. Scano et al. [9] evaluated the accuracy and reliability of the Microsoft sensor by means of a gold standard marker-based MOCAP system. Each subject performs two upper-limbs movements in three different orientation. Results confirm that RGB-D sensors can track upper-limbs movement for rehabilitation, but only in specific devices orientations. Cai et al. [10] quantify the accuracy and the reliability of Kinect V2 by means recording and assessing certain upper limbs movements. In particular, the Microsoft sensors are compared to Vicon MOCAP system as gold standard. The four performed movements lead to confirm that the Kinect V2 has potential in upper limbs assessment in patients' rehabilitation analysis.

The accuracy has been also evaluated by comparing movements and positions measured in the real world and compered with the virtual joints of the avatar generated by a marker-less MOCAP system. Vitali et al. [11] have already assessed the accuracy performance of Kinect V2 under a double depth sensor configuration and they compared the empirical values with ground-truth values known a priori. This work evaluates and compares the performances of the two systems used to track the position of human articulations of both upper and lower limbs during the execution of predefined movements. They found that the 8 GoPro system is more accurate than the Kinect V2 double configuration for joints of both the upper and lower body.

The Microsoft Kinect Azure device was released in March 2020 and, differently from its predecessors, this version of the Kinect is less focused on gaming and more oriented towards logistics, robotics, healthcare and retail [12]. The main study conducted until now about Kinect Azure performances led by Albert and his research group [13]. The aim of the study is to assess motion tracking performance of a single Kinect Azure device compared to a single Kinect V2 device as MOCAP systems by evaluating a treadmill walking. The tracked movements have been compared with the gold standard Vicon multi-camera motion capturing. The main results show a higher accuracy of Kinect Azure with respect to its predecessor Kinect V2 regarding the spatial parameter, while no significant improvements were detected for the temporal parameters. A better tracking accuracy of Kinect Azure was found for the foot marker trajectories while the Kinect V2 gave more positive results for the mid and upper body region.

In this paper, the well-known Kinect V2 sensor is compared with the new Kinect Azure. Both RGB-D sensor performances are assessed by means a series of specific movements. In particular, the exercises involve not only upper limbs but also legs in vertical and horizontal directions. The acquisition campaign permits to evaluate the accuracy of the innovated Microsoft sensors compared to the previous version.

3 Data Acquisition Planning

The methodology of the research has been chosen by taking as reference approach the research work by Vitali et al. [11]. The best plan for the acquisition has been evaluated in order to fit the purpose of the research and to limit systematic error for both systems. In order to maintain continuity with the previous research and given the similarity of the devices under consideration the same movements and layout configurations are considered.

3.1 Scene Set-Up

The system layouts have been designed to optimize the acquisition of lower and upper limbs by considering the presence of a staircase and the subject position during the established movements (Fig. 1). Both MOCAP markerless systems are composed by two RGB-D devices. All the sensors are mounted on tripods at 120 cm of height. Both the Kinect V2 and Azure devices have been placed to guarantee a frontal view of the person as well as an opposite lateral perspective. For each MOCAP system, RGB-D has been placed in front of the first step of the staircase, while the other sensors are in the direction of the upper left corner of the staircase. The frontal Kinects are both 200 cm far from the centre of the staircase while their counterparts are 230 cm $\times$ 170 cm distant.

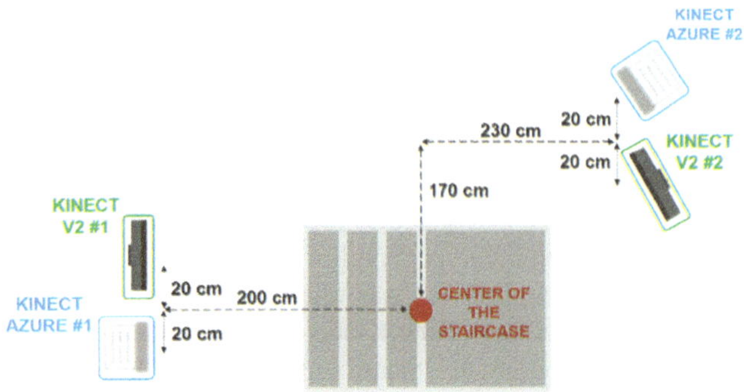

Fig. 1. Lay-out configuration of the two marker-less MOCAP systems.

3.2 Identification of Movements

Three specific movements have been identified to estimate the accuracy of the marker-less devices: climb and descent the staircase (Fig. 2a), free vertical movement of arms (Fig. 2b) and vertical movement of a hand following a predefined trajectory (Fig. 2c) [11]. In order to follow a common guideline and to make a consistent comparison, the subject is asked to reach the specified positions whose ground truth value is known a priori.

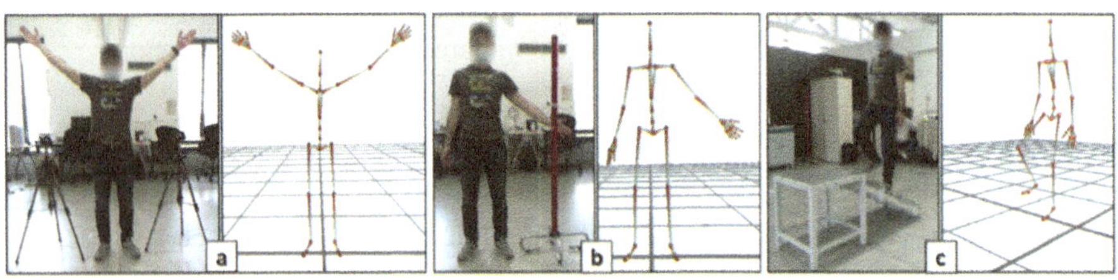

Fig. 2. Performed movement and the relative virtual avatar: climbing the staircase (a), putting hands up (b) and Vertical movement of hand by following a predefined trajectory (c).

Climb and Descent a Staircase. A rehabilitation staircase composed by 3 steps is used. The steps are 16 cm height except for the first one that is 17 cm height. Users have to climb and descent all the 3 steps. The subject climbs the first step with the right foot then he/she starts the descent with the left foot on the third step. The virtual joint evaluated are the Left and the Right Toe that are relative to the metatarsal of subject's feet. The measured parameter is the height of Left and Right Toe when the foot touches each step of the staircases.

Free Vertical Movement of Arms. A strand is horizontally placed at a height of 177 cm. Users abducts their arms until they touch the strand (i.e., the maximum height reached). The strand has to be touched for 5 s using the wrists before lowering the arms and placing wrists on the upper parts of two tripods placed under the strand. Both tripods have a height of 105 cm (i.e., the minimum height reached). The virtual joints under consideration are the Left Hand and Right Hand, while the measured parameters are the maximum and minimum height of the wrist human joints.

Vertical Movement of Hand by Following a Predefined Trajectory. A vertical stick is used as aid to the users who move the left hand along the pipe, top down. Users place their wrists on a marker on the fixed stick at a height of 165 cm (i.e., the maximum height reached) and then they lower the hand keeping the contact with the stick since it reaches the second marker at the height of 100 cm (i.e., the minimum height reached). The virtual joints under consideration are the Left Hand and Right Hand and the measured parameter is the height of the virtual joint considered.

4 Acquisition Campaign

The acquisition campaign involved 11 subjects, 8 males and 3 females. A subject has been randomly chosen to perform the vertical movement of hand by following a predefined vertical trajectory 14 times. This task of acquisition permits to assess the influence of the measurement of different testers to the accuracy of the MOCAP systems. The translation from RGB-D sensors output data to virtual human avatar and the kinematic data are provided by iPiSoft MOCAP suite. The accuracy is evaluated by the Mean Absolute Error (MAE) and its standard deviation (SD), that is the average value of the differences between the virtual joint measurement and the real values.

4.1 Climb and Descent a Staircase

Table 1 is relative to the MAE and its SD for each step of the staircase. Microsoft Kinect V2 has an accuracy for the feet tracking on the ground (i.e., 0.96 cm) and a higher MAE for the other steps. The MAE reaches a value of 1.50 cm for the first step and the second step with the latter also having the higher SD of 1.31 cm. Microsoft Kinect Azure has a different behaviour: it shows the higher MAE for the ground position (i.e., 1.46 cm) and reaches its lower value in correspondence of the third step (i.e., 0.32 cm). With the exception of the ground value, the Kinect Azure has significantly lower MAE for each of the staircase step.

Table 1. MAE and the standard deviation for each step of the staircase using Kinect Azure and Kinect V2.

Step	MAE Kinect Azure [cm]	Std. Dev. Kinect Azure [cm]
Ground	1.46	0.84
1st Step	0.74	0.83
2nd Step	0.47	0.56
3rd Step	0.32	0.53

Step	MAE Kinect V2 [cm]	Std. Dev. Kinect V2 [cm]
Ground	0.96	0.75
1st Step	1.45	0.90
2nd Step	1.48	1.31
3rd Step	1.37	0.93

4.2 Free Vertical Movement of Arms

Table 2 shows the values for MAE and SD for the maximum and the minimum height, whose ground truth values are 177 and 105 cm respectively. The MAE of V2 highlights the lower accuracy in the higher position of the arms (i.e., 2.12 cm). It is meanly caused by the proximity to the end of the vertical field of view of the device [11]. This feature has been improved with the Kinect Azure that has vertical FOV (V-FOV) of 65°. The MAE in the maximum height for Kinect Azure is lower than of the V2 one (i.e., 0.98 cm). The minimum height is strongly improved with the Kinect Azure with a MAE value of 0.77 cm that is lower than Kinect v2 (i.e., 1.41 cm). The maximum height with the Kinect V2 also has a high standard deviation (i.e., 1.68 cm).

Table 2. MAE and SD for the free vertical movement of arms with both MOCAP system.

Microsoft Kinect Azure			
Max. Height [cm]		Min. Height [cm]	
MAE	Std. Dev.	MAE	Std. Dev.
0.98	0.95	0.77	0.58

Microsoft Kinect V2			
Max. Height [cm]		Min. Height [cm]	
MAE	Std. Dev.	MAE	Std. Dev.
2.12	1.68	1.41	0.87

The left wrist values are better than the right wrist ones even though the configuration of the Kinects leave the left side of the body at the extremity of the FoV. Kinect Azure and Kinect V2 have a quite stable MAEs for both arms. Also in this case, the Kinect V2 shows limitations during the movement tracking near the limit of the V-FOV. Its MAE in the maximum position reaches 2.00 cm for the right hand and 2.25 cm for the left hand. The MAE of the upper position for Kinect Azure has a value of 1.29 cm, with a SD of 1.09 cm.

4.3 Vertical Movement of Hand by Following a Predefined Trajectory

Table 3. MAE and SD for the movement of the left arm following a predefined trajectory using Microsoft Kinect Azure and Microsoft Kinect V2.

Microsoft Kinect Azure			
Max. Height [cm]		Min. Height [cm]	
MAE	Std. Dev.	MAE	Std. Dev.
0.99	0.79	1.58	1.55

Microsoft Kinect V2			
Max. Height [cm]		Min. Height [cm]	
MAE	Std. Dev.	MAE	Std. Dev.
2.59	1.70	1.47	1.27

The left-hand joint has been measured with respect to a minimum and a maximum height (i.e., 100 cm and 165 cm respectively). Table 3 shows the MAE and the SD for the two devices in both positions. The Kinect V2 has the higher MAE for the upper position of the hand. The Kinect Azure has a MAE of 0.99 cm for the maximum position and 1.58 cm for the minimum, with a standard deviation having a high value for the minimum height (1.55 cm) and 0.79 cm for the maximum. In this case, the accuracy of Kinect V2 at the minimum position is better (MAE value of 1.47 cm) with respect to the one of the Kinect Azure (MAE value of 1.58 cm).

5 Repeatability Analysis

A repeatability study has been planned in order to evaluate the impact of the users' actions, that affect the measurements of motions. Hence, a single tester has been chosen to repeat the same movement for 14 times. Table 4 shows the MAE and the SD for both devices. The Kinect Azure has a MAE of 0.86 cm and a SD of 0.56 cm and the Kinect V2 has a MAE of 0.92 cm and a SD of 0.69 cm. These values demonstrate a very low difference of accuracy among the Kinect Azure and Kinect v2.

Table 4. MAE and SD for the repeatability analysis using Microsoft Kinect Azure.

Microsoft Kinect Azure				**Microsoft Kinect V2**			
Max. Height [cm]		Min. Height [cm]		Max. Height [cm]		Min. Height [cm]	
MAE	Std. Dev.	MAE	Std. Dev.	MAE	Std. Dev.	MAE	Std. Dev.
0.80	0.61	0.92	0.54	0.42	0.40	1.43	0.53

The repeatability test gives positive result: the variations found between different testers are also due to the subject performing the exercise and not only to the sensors.

6 Discussion and Conclusions

The paper presents a research work about the evaluation of the accuracy of a marker-less Mocap system based on two Microsoft Kinect Azure device compared with a Mocap system based on two Microsoft Kinect V2. The main aim is evaluating if Kinect Azure really represents a step forward in the realization of competitive and reliable RGB-D sensors. The analysis was carried out through the execution of three different predefined movements as well as a repeatability analysis of a specific movement to evaluate intra-operator errors. The comparison concerned the evaluation of the virtual joints position compared to ground truth values. According to the results, the Kinect Azure presents generally better tracking capabilities than the older Kinect v2 device. The MAEs of Kinect Azure are generally lower with a clear worst case about the tracking of the lower human body (Fig. 3). The chosen layout is meant to create challenging conditions for the Mocap systems (e.g. acquisition close to the border of the field of view or occlusions) to better highlight performances and differences between the two. Occlusions penalized more Kinect V2 acquisitions among which the MAEs are over 1.7 cm, while the wider V-FoV of Kinect Azure allowed avoid tracking problems. As a matter of fact, during the vertical movement of arms we reached a height of 177 cm with no issue for the Kinect Azure devices, that has a MAE lower than 1 cm. Kinect v2 presents a MAE of 1.77 cm that cannot be accepted for wide movement performed with arms. At present, the main lack of the Kinect azure is the tracking of the feet on the ground. This feature, on the other hand, is accurately calculated by the Kinect V2. As far as it concerns the limits of the study, we considered a sample of only 11 subjects. Despite of the aforementioned open issues, the presented results allow us to conclude that the Kinect Azure presents several relevant features, which will be exploited to further improve the quality of acquisitions.

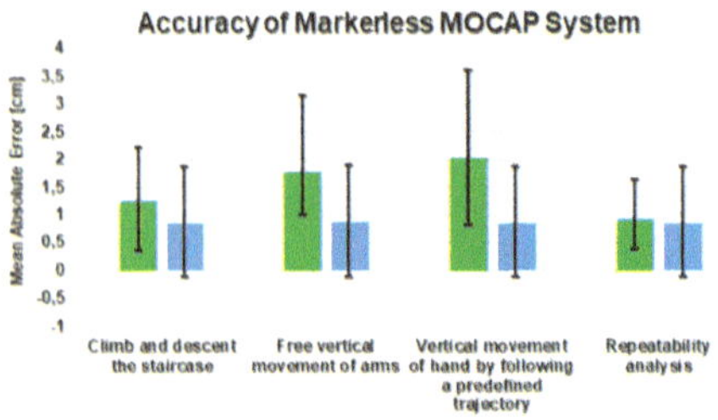

	Microsoft Kinect V2	Microsoft Kinect AZURE
Climb and descent the staircase	1.25	0.86
Free vertical movement of arms	1.77	0.88
Vertical movement of hand by following a predefined trajectory	2.03	0.86
Repeatability analysis	0.92	0.86

Fig. 3. Accuracy measured for each movement tracked by means both MOCAP systems.

References

1. Zhang, Z.: Microsoft kinect sensor and its effect. IEEE Multimedia **19**(2), 4–10 (2012)
2. Furniss, M.: Motion Capture (1999). http://web.mit.edu/m-i-t/articles/index_furniss.html
3. Fernandez-Baena, A., Susın, A., Lligadas, X.: Biomechanical validation of upper- body and lower-body joint movements of kinect motion capture data for rehabilitation treatments. In: Proceedings of the 4th International Conference on Intelligent Networking and Collaborative Systems, INCoS 2012, pp. 656–661 (2012)
4. Chen, C., Jafari, R., Kehtarnavaz, N.: Improving human action recognition using fusion of depth camera and inertial sensors. IEEE Trans. Hum.-Mach. Syst. **45**(1), 51–61 (2015)
5. Kotsifaki, A., Whiteley, R., Hansen, C.: Dual Kinect V2 system can capture lower limb kinematics reasonably well in a clinical setting: concurrent validity of a dual camera markerless motion capture system in professional football players. BMJ Open Sport Exerc. Med. **4**(1) (2018)
6. Stone, E.E., Butler, M., McRuer, A., Gray, A., Marks, J., Skubic, M.: Evaluation of the microsoft kinect for screening ACL injury. In: Proceedings of the Annual International Conference of the IEEE Engineering in Medicine and Biology Society, EMBS, vol. 2013, pp. 4152–4155 (2013)
7. Bernardina, G.R.D., Monnet, T., Pinto, H.T., De Barros, R.M.L., Cerveri, P., Silvatti, A.P.: Are action sport cameras accurate enough for 3D motion analysis? A comparison with a commercial motion capture system. J. Appl. Biomech. **35**(1), 80–86 (2019)
8. Vicon: Vicon Award Winning Motion Capture Systems. https://www.vicon.com/
9. Scano, A., Mira, R.M., Cerveri, P., Molinari Tosatti, L., Sacco, M.: Analysis of upper-limb and trunk kinematic variability: accuracy and reliability of an RGB-D sensor. Multimodal Technol. Interact. **4**(2), 14 (2020)

10. Cai, L., Ma, Y., Xiong, S., Zhang, Y.: Validity and reliability of upper limb functional assessment using the Microsoft Kinect V2 sensor. Appl. Bionics Biomech. (2019)
11. Vitali, A., Regazzoni, D., Rizzi, C., Lupi, G.: Low cost markerless motion capture systems: a comparison between RGB cameras and RGB-D sensors. In: Proceedings of the ASME IMECE 2020. American Society of Mechanical Engineers (ASME) (2020)
12. Microsoft: Azure Kinect DK. https://azure.microsoft.com/it-it/services/kinect-dk/
13. Albert, J.A., Owolabi, V., Gebel, A., Brahms, C.M., Granacher, U., Arnrich, B.: Evaluation of the pose tracking performance of the azure Kinect and Kinect V2 for gait analysis in comparison with a gold standard: a pilot study. Sensors **20**(18), 5104 (2020)

Towards Wearable Technology Assisting in Training and Risk Assessment Against Musculoskeletal Disorders for Maritime Workers

Alexander Van Gastel[1], Sofia Scataglini[1], Sander Zelck[1], Lenie Denteneer[2], Hanne Vanden Bossche[1], and Stijn Verwulgen[1(✉)]

[1] Department of Product Development, University of Antwerp, Antwerp, Belgium
alexander.vangastel@student.uantwerpen.be, {sofia.scataglini, sander.zelck,hanne.vandenbossche,stijn.verwulgen}@uantwerpen.be
[2] Rehabilitation Sciences and Physiotherapy, University of Antwerp, Antwerp, Belgium
lenie.denteneer@uantwerpen.be

Abstract. Nowadays, wearable technologies are capable of monitoring human activities and performances in different environments. In this paper, we present an investigation of their usage as tools for risk assessment and training in a maritime environment. In particular, bend sensors, stretch sensors and wearable inertial measurement units (IMUs) were investigated in regard to monitoring container lashers' movement in a maritime environment. Design and technological issues will be discussed. In this context, we also consider human system integration and interaction of those technologies. Notably, flex and stretch sensors show a low position repeatability precision as measurements were strongly affected by unwanted repositioning of the sensor. As a result, they introduce accuracy and reliability issues in uncontrolled environments. Assessment of each sensor type lead to the conclusion that contemporary IMU's are a suitable solution.

Keywords: Wearable · Motion capture system · Musculoskeletal disorder · Marine workers

1 Introduction

Impact, pinching and fall injuries have been reported to contribute to the poor occupational health of container lashers [1, 2]. The objective of this research is to define specifications for a product, capable of monitoring dockworkers health, and risk assessment taking in account both technical and design issues. Existing guidelines for risk assessment of workers are not exhaustive due to high variability, receptiveness and a low standardization of tasks [3–5]. For example, physiotherapists will often assess range of motions of dockworkers and work towards flexibility goals and judge flexibility based on their own observations. More general, biomechanical risk assessments remain largely based on observational and subjective approaches without use of human motion capture

J. L. Wright et al. (Eds.): AHFE 2021, LNNS 264, pp. 368–376, 2021.
https://doi.org/10.1007/978-3-030-79763-8_44

tools [6]. However, monitoring devices have been proposed, accepted and used in the health sector as risk assessment tools [6]. Using motion analysis, measuring kinematic and dynamic data acquired through wearable motion capture devices, risk assessments can be performed with quantitative and evolving data, which allows for more representative assessments in real working scenarios. Moreover, a multitude of sensors and systems exist, each with different sets of specifications, measured variables and restrictions.

Wearable human motion monitoring devices are also used in sports like golf e.g. Hackmotion and Golfsense, where they directly help improve technique. Training with motion capture devices help enhance worker's awareness of flawed technique. This is especially relevant since a gap was discovered between the learned practices at the training facility and the behavior applied on the work floor [7].

In addition, wearables could assist rehabilitation, a dynamic process, which allows patients to restore their functional capabilities to normal. To reach this target, patients' activities need to be continuously monitored and subsequently corrected [8]. In fact, workers with reoccurring or recovering injuries can be monitored closely during their rehabilitation. Maritime organizations could do so through quantitative and evolving biomechanical data, before, during and after shifts.

In addition, the issue of work-related musculoskeletal injuries (WMSDs) extends beyond the maritime sector. Between 1992–2010 MSDs accounts for 29–35% of all occupational injuries and illnesses involving days away from work in the United States (US). In 2007 this resulted in $1.5 billion direct costs to US businesses [9]. This shows the partial financial burden caused by absenteeism as a result of WMSDs. They bring financial burdens to workers, their families and society, through loss of income, productivity and increasing medical expenses. A report in 2020 on MSDs in the construction industry [10] shows that 46% of construction workers self-reported they had one or more MSDs related symptom(s). Lastly, over 27% of construction workers aged 55 and older reported arthritis or joint pain symptoms limited their activities.

By estimation, each individual lashes and de-lashes up to 75 and 250 containers, each shift. Workers are trained to work in pairs, however often work alone, increasing the likelihood of injury. Two lifting operations are estimated per lashing and de-lashing procedure, giving a total of 650 lifting operations (turn buckle + lashing bars). The average weight of one lifting operations is estimated at 22 kg. An estimation is provided as the weight of a lashingbar varies between two sizes: 15 kg ($\pm$2/3 lashings) and 25 kg ($\pm$1/3 lashings). The weight of turn buckles remains constant at around 25 kg (each lashing). Improving training, technique and increased accuracy in measurements used for risk assessments allow for potential prevention of maritime worker's health deterioration.

2 Method

In this study, three different wearable sensors (dielectric elastomer sensors, bend and IMU wearables), capable of monitoring human body motion were selected to monitor hand, wrist and elbow movements in a maritime environment. Technological and design issues were evaluated together with aesthetical ones. Technological issues relate to the sensing and processing, safety, durability, reliability, and monitoring duration

[11, 12], while design issues relate to anthropometry, adherence, unobtrusiveness, body positioning, sustainability, intuitiveness and user-friendliness [11–13].

A maritime environment is humid, corrosive and harsh. For bend and stretch sensors humidity and temperature influences their behavior, lowering the reliability and accuracy over a sustained period [14]. Additionally, the degree to which they are influenced is not reliably enough understood to account for in post processing.

In case of IMU's, containers, equipment made out of ferromagnetic materials, magnets or strong currents will cause magnetic field disturbances, disrupting their measurements. Workers are often challenged with temperature fluctuations and extremes during various seasons. Currently they maintain a strategy of layering their clothes during cold periods. This has to be taken into close consideration as bend sensors and stretch sensors are sensitive to repeatability errors caused by warping or moving garments. Taking off upper layers could interfere, cause inaccurate sensor readouts and ruin a calibration.

2.1 Wearable Sensors Evaluation

Assessment of the dielectric elastomer sensors (DESs) is based on a motion monitoring system (Fig. 1) proposed by Huang et al. 2017 [15]. It showed good lab accuracy, repeatability (0.81%), high linearity (0.14%) and sensitivity (0.0082 V/mm). High linearity refers to the measured change in voltage scaling linearly with the change in angle/bend. These characteristics of the DES meet the potential requirements of wrist motion measurements.

Additionally, DES based systems, fully integrated in garments show excellent unobtrusiveness, sensing, safety and durability in a controlled environment [16]. Nevertheless, they suffer from multiple key disadvantages when deployed in uncontrolled situations. The first of which is measurement errors caused by moving and warping garments. Often considered to solve this are tight, stretchable garments. However, this solution is impaired when the tight garment stretches alongside the sensors instead of only the sensors stretching.

Additionally, in real world scenarios where individuals get pulled on their sleeves, take off or otherwise manipulate their clothing, DESs based monitoring devices' low tolerance to displacement as a result of a 0.5 cm position repeatability precision, will be a critical hinderance.

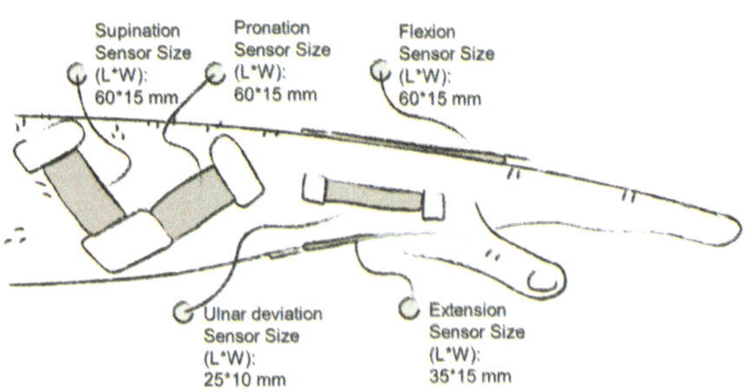

Fig. 1. A dielectric elastomer sensor based motion monitoring system [15]

Secondly, high levels of personalization would be required for each individual as a result of different forearm and wrist dimensions. After the setup of a personalized configuration, in all following use cases each sensor needs to be placed in the same spot with a tolerance of 5 mm.

Lastly, in case of direct attachment to the skin, workers would need to attach twelve separate sensors to their wrists and forearms, as each component motion, F/E, U/R, P/S requires one sensor, for both arms. This is something they cannot be expected to do, even more so if the system is to expand into a full or upper body monitoring system. As a result, personalization poses hindrances to wide scale deployment and day to day use.

Bend sensors manufactured by Flexpoint Sensor Systems (FLXT) were studied, producing four different observations focused on elbow flexion and extension monitoring. In this paper the last observation is presented where a Flexpoint Sensor was included in a smart textile prototype that consists of a Flexpoint Sensor fixed at proximal side (Fig. 2). This observation was tested using Kinovea software for studying the reliability of joint angle measurements from flexion and extension at the elbow.

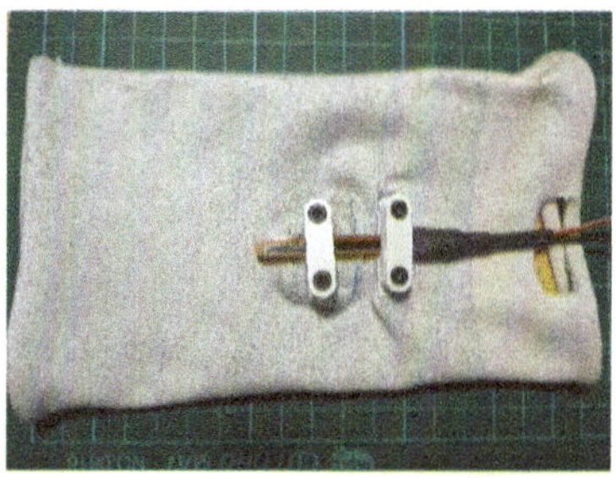

Fig. 2. Flexpoint 2 Inch - Fixed at the proximal side – 3D printed PLA prototype of mounting fixtures. The distal mount functions as the guide through which the bend sensor can slide while being bend. The proximal side keeps the sensor in place.

Assessment of flex/bend sensors was based on reliability, accuracy and robustness in a maritime environment, where direct impacts, humidity and rough usage are common occurrences. One advantage to using bidirectional bend sensors compared to stretch sensors would be lower cost. This is caused by a lower average cost per sensor and that one sensor could be implemented for both flexion and extension monitoring. However, the final hand monitoring system would need to be combined with stretch sensors, as bend sensors are not capable of monitoring ulnar, radial, supination and pronation angles. This is due to those biomechanical movements not creating enough measurable internal resistance. Another key advantage is that they are unobtrusive to users compared to IMUs based motion capture systems.

The key disadvantage to bend sensors are their low reliability. Manufacturers and multiple other sources state their bend sensor to be sensitive to external pressure forces. This was confirmed during observation of the experiments. This behavior would negatively influence sensor accuracy as external pressure can add internal resistance to the system, causing output no longer to be representative. Another cause for low reliability is a low linearity under 30° [14], meaning that the sensor should only be used to accurately measure higher degree angle values. Additionally, even more so than stretch sensors, bend sensors suffer from placement repeatability errors caused by displacement of the sensor through textile moving or warping. Lastly, bend sensors measuring capabilities are influenced by changes in humidity and temperature, both friction and ambient.

Table 1. Key advantages and disadvantages of using stretch, bend/flex, IMU based sensor systems in a maritime environment.

	Advantage	Disadvantage
Dielectric elastomer sensor (DES)	1) High linearity	1) Measurement errors caused by moving and warping garments
	2) Unobtrusive	
	3) Durable [15]	
Bend/flex sensor	1) one bidirectional sensor both measuring flexion and extension	1) External pressure sensitivity
	2) Unobtrusive	2) Measurement errors caused by moving and warping garments
		3) Low linearity under 30° [14]
		4) Humidity [14]
		5) Ambient and friction temperature variations [17]
Inertial measurement unit (IMU)	1) IMU chaining	1) Electromagnetic Interference
	2) Established technology for biomechanical motion tracking. = more mature	2) Obtrusive as they are rigid boxes, uncomfortable straps e.g. Wearnotch [7]
	3) High expected accuracy	3) Drift + need to re-adjust absolute position

Wearable inertial systems (IMU) are currently being used indoors and outdoors for biomechanical risk assessment [6, 8]. They have proven capabilities in this use case. They are lightweight, easy to use but sometimes reveal being affected by magnetic fields. This last aspect is important in a maritime environment especially for dockworkers who perform lashing and de-lashing of containers.

Wearables sensors as IMU sensors can be integrated in smart bands or garments that can be worn by the dockworkers. However existing mounting solutions reveal being uncomfortable. Indeed, these aspects should be considered for designing a device that should be used for monitoring dockworkers without affecting their performance. Overall, considering dielectric elastomer sensors, bend/flex sensors and inertial measurement unit, their body placement for monitoring specific gesture as wrist and hand gesture, we can conclude that IMU revealed easy to use and affordable device that can be integrated in dockworkers-system environment (Table 1). From these previous analysis, emerged the necessity to design a wearable smart glove for monitoring dockworkers.

Taking into account advantages and disadvantages of each sensor type (Table 1), it was concluded that contemporary IMU based motion monitoring devices carry an edge compared to DES and bend sensors. Compared to the other two sensor types they don't suffer as hard from placement repeatability issues and can be calibrated more easily. Additionally, IMU chaining allows for relatively easy expansion to a upper or full body measuring system. However, IMUs limitations need to be taken into close consideration since electromagnetic interferences could strongly affect its performance as well as drift over time.

3 Smart Glove Ideation

A biomechanical monitoring system must be designed to be unobtrusive, light, durable, reliable and safe. Workers overall health can be monitored, through physiological signals e.g. respiration, heartbeat. In order to maximize the effectiveness to prevent MSDs the biomechanical monitoring system will have to gather kinematic data of the hand, and with later expansion, lower back, neck and shoulder, which would need to be continuously monitored before, during and after shifts. Development of a full body biomechanical measurement tool could prove necessary as it would allow physiotherapists to see where and how an individual is compensating e.g. using their leg muscles to assist in lifting a load from the ground. A full system would provide feedback on body postures, ensuring correct usage of learned methods.

There are two key benefits the smart glove concept brings. However, measuring only hand and wrist kinematics can only afford limited insight.

One is increasing workers awareness of correct wrist posture, allowing to see for themselves what they are doing wrong and motivate them intrinsically to adopt correct techniques. The purpose of the smart gloves would primarily be as learning tools for new workers to adopt correct hand postures, while holding loads or while tensioning turn buckles, two tasks that induce MSDs.

Secondly, when the smart gloves were to be used during active shifts, data journals would augment organizations' capabilities to take preventative action e.g. ergotherapy, rest or additional training to prevent occurrence of long term RSI's.

3.1 Smart Glove Concept

Two design concepts are presented (Fig. 3). The first system (left) allows container lashers to detach their tokens, place them in their bracelets and halt monitoring of their hand in a quick and intuitive way. While being stored in their bracelet, the IMU token will recharge. Having the gloves and bracelets be a separate system allows for the gloves to be washed as IMUs cannot be placed in a washing machine.

The design trajectory is based around the main parts, bracelet, IMU token, glove and hand strap. Each part has different specifications assigned to it which need to be fulfilled. First, an estimation of the volume required for the bracelets' electronic components is required based on the required autonomy and battery life. Afterwards, low fidelity prototypes focusing on wearability, unobtrusiveness, adherence and intuitive user interaction will be made. With those, user scenarios and the proposed vision can be better simulated and adjusted. Simultaneous with the bracelet' design, development of the gloves token mount and the IMU token need to be performed with a focus on fulfillment of the specifications.

The gloves should be based on designs currently being used in container lashing and de-lashing. Ideally, gloves would be fully designed and produced by existing manufacturers in addition to their physical mounts for the IMU-token. Besides that, the mount on the glove should be as unobtrusive as possible, while being really adherent.

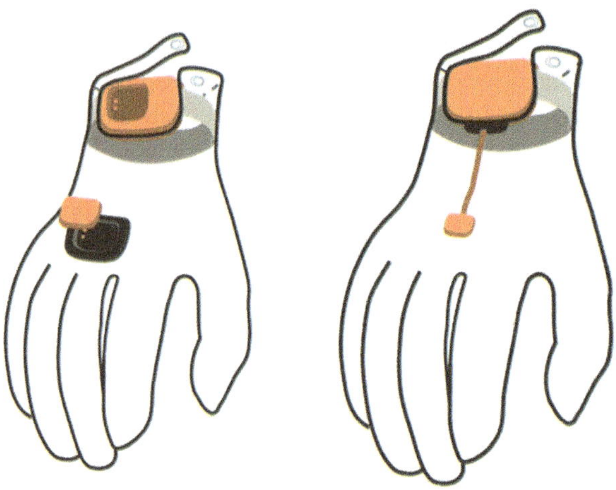

Fig. 3. Fully setup IMU-based hand motion monitoring devices.

The system would be trained to look for indicators that are cause for concern. One such indicator could be fatigue, which over an 8h shift, results in slowed reaction time, loss of awareness, reduced decision-making ability and poor judgment. To do this, flexion- extension and ulnar- radial deviation are monitored separately and divided into zones of 10° and 5° respectively. In flexion and extension, zones are larger because the range of motion is greater than for ulnar and radial deviation. Using this data physiotherapists would be able to analyze whether people, whose hands spend a large amount of time in higher zones, also suffer from more MSDs or RSIs. Lastly, when used during active shifts, IMU's internal accelerometers can also be used as fall detection. Before the design process can start, more research needs to be performed to identify specific indicators for wrist and hand MSDs.

In Fig. 4. the pipeline through which data will be acquired and handled is presented. Data is acquired e.g. angular rotation, acceleration, orientation, time through the proposed smart glove concept. This data combined with a biomechanical model of a hand allows us to derive joint angle values e.g. F/E, U/R and S/P. This data will be stored in a

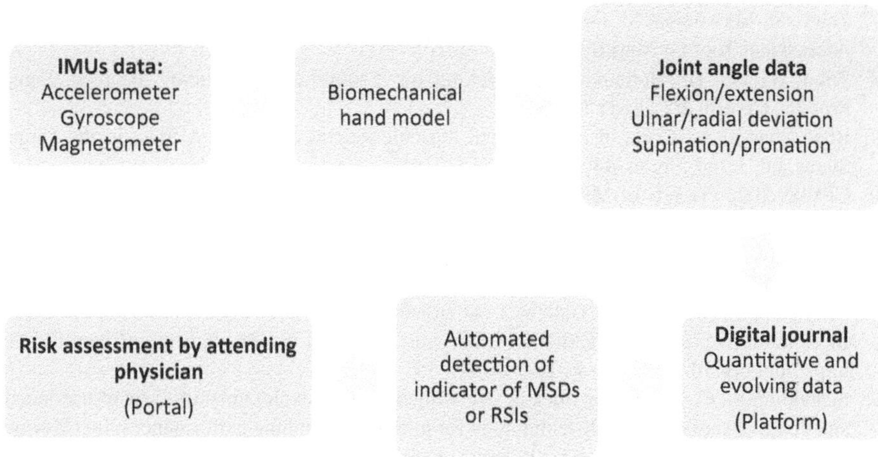

Fig. 4. Pipeline: data acquisition, processing and preliminary data points.

data journal specific to each worker (platform). Onwards the system will continuously look out for indicators. Through a portal attending physiotherapists are able to perform risk assessments and provide intervention strategies to improve their health.

4 Conclusion

In this paper, we presented a preliminary evaluation for designing a smart garment for monitoring dockworkers gestures during lashing and de-lashing task.

Three different sensors were studied and evaluated for ideating the smart glove concept presented in this paper. Future perspective will be addressed to developing a smart glove as tool for ergonomic risk assessment and training.

References

1. CommercialVesselsnz. Lashing Containers - Maritime New Zealand. [video]. <shorturl.at/adBI8> (2012). Accessed 11 Dec 2020
2. BC Maritime Employers Association. BCMEA Container Lashing Orientation Video -Lashing 101. [video] (2018). <shorturl.at/goX67>. Accessed 11 Dec 2020.
3. CDC—NIOSH Publications and Products. Ergonomic Guidelines for Manual Material Handling (2007-131). https://www.cdc.gov/niosh/docs/2007-131/default.html. Accessed 7 Mar 2020
4. Kingma, I., et al.: Lumbar loading during lifting: a comparative study of three measurement techniques. J. Electromyogr. Kinesiol. **11**(5), 337–345 (2001)
5. Marras, W., Parakkat, J., Chany, A., Yang, G., Burr, D., Lavender, S.: Spine loading as a function of lift frequency, exposure duration, and work experience. Clin. Biomech. **21**(4), 345–352 (2006)
6. Ranavolo, A., Draicchio, F., Varrecchia, T., Silvetti, A., Iavicoli, S.: Wearable monitoring devices for biomechanical risk assessment at work: current status and future challenges—a systematic review. Int. J. Environ. Res. Public Health **15**(9), 2001 (2018)

7. Zelck, S., Verwulgen, S., Denteneer, L., Scataglini, S.: Digital Human Modeling as a Risk Assessment Tool for Maritime Workers (2020)
8. Zhou, H., Hu, H.: Human motion tracking for rehabilitation—a survey. Biomed. Signal Process. Control **3**(1), 1–18 (2008)
9. Bhattacharya, A.: Costs of occupational musculoskeletal disorders (MSDs) in the United States. Int. J. Ind. Ergon. **44**(3), 448–454 (2014)
10. CPWR, 2020. Trends of Musculoskeletal Disorders and Interventions in the Construction Industry. CPWR, p.1. https://www.researchgate.net/publication/345180995_Quarterly_Trends_of_Musculoskeletal_Disorders_and_Interventions_in_the_Construction_Industry. Accessed 11 Dec 2020
11. Scataglini, S., Andreoni, G., Gallant, J.: Smart clothing design issues in military applications. In: Ahram, T.Z. (ed.) AHFE 2018. AISC, vol. 795, pp. 158–168. Springer, Cham (2019). https://doi.org/10.1007/978-3-319-94619-1_15
12. Scataglini, S., Cools, E., Neyrinck, J., Verwulgen, S.: An exploratory analysis of user needs and design issues of wearable technology for monitoring running performances. In: Cassenti, D.N., Scataglini, S., Rajulu, S.L., Wright, J.L. (eds.) AHFE 2020. AISC, vol. 1206, pp. 207–215. Springer, Cham (2021). https://doi.org/10.1007/978-3-030-51064-0_27
13. Scataglini, S., Danckaers, F., Huysmans, T., Sijbers, J., Andreoni, G.: Design smart clothing using digital human models. In: DHM and Posturography, pp.683–698 (2019)
14. Borghetti, M., Sardini, E., Serpelloni, M.: Evaluation of bend sensors for limb motion monitoring. In: 2014 IEEE International Symposium on Medical Measurements and Applications (MeMeA) (2014)
15. Huang, B., et al.: Wearable stretch sensors for motion measurement of the wrist joint based on dielectric elastomers. Sensors **17**(12), 2708 (2017)
16. Carnevale, A., et al.: Conductive textile element embedded in a wearable device for joint motion monitoring. In: 2020 IEEE International Symposium on Medical Measurements and Applications (MeMeA) (2020)
17. Simone, L., Kamper, D.: Tactilus Flex Sensor| Compression Force Sensing Resistor (fsr)|Force Sensing Resistors|Force Sensing Resistors|Tactilus|Surface Pressure Indicator|Mapping|Force Sensing and Profiling. Sensorprod.com (2021)

Stability Measurements of Overhead Tasks in Combination with Prismatic Glasses Using IMU Mocap System: Presenting an Experiment Protocol

Sander Zelck[1](✉), Stijn Verwulgen[1], Lenie Denteneer[2], Maite Dillen[1], Isabel Prieto García[1], Chiara Rousseau[1], Sep Sauter[1], and Sofia Scataglini[1]

[1] Department Product Development, Faculty of Design Sciences, University of Antwerp, Antwerp, Belgium
{Sander.Zelck,Stijn.Verwulgen,Sofia.Scataglini}@uantwerpen.be,
{Maite.Dillen,Isabel.Prieto.Garcia,Chiara.Rousseau, Sep.Sauter}@student.uantwerpen.be

[2] Department of Rehabilitation Sciences and Physiotherapy, Faculty of Medicine and Health Sciences, University of Antwerp, Antwerp, Belgium
Lenie.Denteneer@uantwerpen.be

Abstract. Although overhead work is closely related to musculoskeletal disorders, it is difficult to avoid the postures introduced by overhead work for heavy load professions. Prismatic glasses can help to avoid these disorders. As prismatic glasses obstruct the central view of the user, the sensory system linked with the posture control of the user can be disrupted. To understand if prismatic glasses propose an added value in an industrial setting, this paper reports the setup and evaluation of an experiment protocol that examines the effect of prismatic glasses on a person's stability while performing overhead work. Participants were asked to perform tasks representing different overhead work professions. Two experiments are composed: with and without wearing prismatic glasses. Inertial Measurement Units (IMU) mocap system, Xsens (MVN Awinda, Enschede, The Netherlands) are used to record the participants' center of mass (COM). During the experiments participants documented their perceived balance using the Activities-Specific Balance Confidence (ABC) scale. Observers used the Performance-oriented mobility assessment (POMA) test to document the balance of the participant. Future experiments need to focus on: representative and accurate population sample, the use of static non-computerized posturography in combination with the existing protocol and introduction/ comparison of posturography parameters using computerized posturography and body kinematics obtained by an IMU wearable motion capturing system.

Keywords: Prismatic glasses · Heavy load workers · Overhead workers · IMU wearable mocap system · Center of mass (COM) · Stability

J. L. Wright et al. (Eds.): AHFE 2021, LNNS 264, pp. 377–383, 2021.
https://doi.org/10.1007/978-3-030-79763-8_45

1 Introduction

1.1 The Impact of Work-Related Musculoskeletal Disorders Within Overhead Work

A common concern in working populations is work-related musculoskeletal disorders that can produce neck and shoulder pain. The risk for these disorders is higher for those with jobs related to overhead work [1–3]. Professions, where this type of labor is required, are for example: window cleaners, dock workers, warehouse workers, car mechanics, belayers and painters [4]. There is a tendency for the worker to over-extend the neck while performing overhead tasks. Therefore the neck might experiences monotonous stress [5]. This posture could be in long-term harmful for the neck and shoulders [6]. Studies show that performing overhead tasks with maximal neck extension have a higher risk of muscle strain from overhead work [6, 7]. A similarity in musculoskeletal disorders due to over extension of the neck was found in belayers when climbing.

1.2 The Use of Prismatic Glasses in Climbing Sports

A belayer is a person who controls the safety rope of a climber. Belayers use prismatic glasses to reduce neck pain caused by continuously looking up, as seen in Fig. 1 [4, 8]. High-quality mirror prisms reflect and refracts the light at an angle, projecting a sharp image of the climber. The glasses allow the belayer to continuously observe his climbing partner without having to change the neutral head position [4, 6, 7]. While wearing prismatic glasses, the user experiences the prismatic as an obstruction within the central view. Figure 2 illustrates that the peripheral view is not altered by the prism due to the small size of the prism. Therefore, users focus their attention between the different views.

Fig. 1. Belayer operating prismatic glasses [9]

A preliminary observation with dockworkers indicates an effectiveness in reducing hyper extension in the neck while performing overhead tasks. As the operational space on a container vessel is very limited, different kind of equipment can form an obstruction while performing overhead tasks. Losing one's balance on a ship can result in serious injuries and trauma. To understand the benefit of the prismatic glasses in a work environment, information about its effect on the user's stability must be acquired.

Fig. 2. Belayer experiences two view angles [10]

1.3 The Aim of the Study

To indicate added value in the use of prismatic glasses, thorough research of its effect on the users' posturography is needed. The purpose of this study is to set up and evaluate an experiment protocol to determine whether wearing prismatic glasses affects the stability of a worker while performing an overhead task.

2 Method

Participants executed the experiment by performing different tasks, simulating overhead work while the wearing the full wearable inertial mocap system (Xsens, Enschede, The Netherlands) [11]. A general floor plan (Fig. 3) is built to extrapolate to different locations. One video camera captures the different performed tasks. Next to the camera, one observer standing out of the frame will dictate the different tasks. After every task participants are asked to describe their perspective on balance changes using the Activities-Specific Balance Confidence (ABC) scale [12]. Simultaneously the observer notes their empiric results in the Performance-oriented mobility assessment (POMA) [13]. Posturography measures of the participants are recorded and exported in excel. Sway path, trajectory of the center of mass (COM) were used for evaluating the effect of wearing the prismatic glasses on balance control [14, 15].

The experiment is performed in three different setups. For each setup, a list of tasks is compiled to simulate the different overhead tasks that can occur in the previous mentioned professions. In the first setup participants perform general tasks: walking in a straight line for five meters (1), sitting on a chair (2), standing up and rotation 360° (3), walking in a straight line over obstacles (4), walking backwards in a straight line for five meters (5), read a text max. 3 m above them (6), walking while holding an obstacle (6), walking sideways while holding a stick (7), reaching with the stick a circle max. 3 m above the participant (8), standing on a static 360 degrees of freedom balance board also known as wobble board (BB) (9), stretching arms vertically and finally while standing on the balance board (10), bending the legs to touch the ground (11) and stepping back of the BB (12). This setup originates from the Timed Up & Go (TUG) test [16] and is altered in order to maximize the representation with overhead professions.

The second setup and last setup emphasizes more on the overhead tasks than on the TUG test. In the second setup, tthe observer requests the participants to grab a cord (13) and vertically pull to open curtains (14). The last setup requires participants to grab an object from the ground (15), walk up the stairs with the object (16), on top of the stairs the participants need to reach for a marking on the wall (17) and finally while walking down the stairs, placing the object in the designated area on the ground (18).

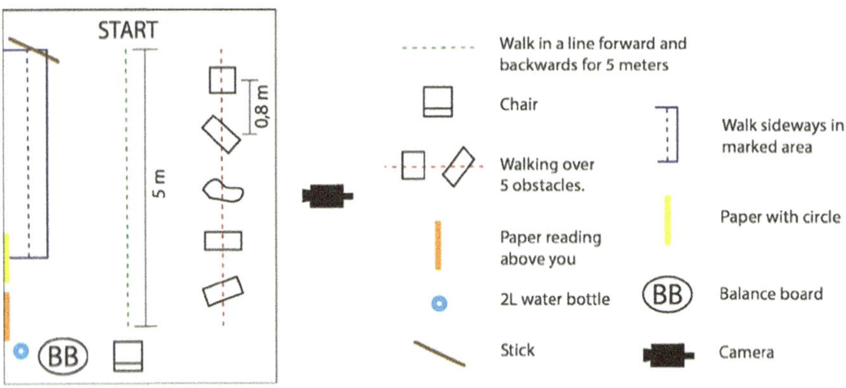

Fig. 3. Floor plan of the observation setups

The ABC scale indicates the level of confidence perceived by the participants in doing a task without losing balance or becoming unsteady [12]. Scores lower than 50 indicate a low level of functioning; scores above 50 but below 80 indicate a medium level, and those over 80 indicate a high level of functioning [16, 17]. The POMA [13] is an instrument to evaluate the mobility of participants in order to assess their risk of falling [13]. It is scored via an ordinal ranging from 0 to 1 or 2. A score of 50 to 62 indicates a normal risk of falling. Scores below 50 show an increased risk of falling. For this experiment, the existing POMA test is adapted so in order to correspond with the task simulating overhead work. These tasks are recorded by Xsens and exported into an Excel-file. The personal generated digital human models (DHM) are later analyzed to determine the relevant frames that correspond with the performed tasks. According to the frames that represent the different tasks, linear regression analysis of the coordinates from the center of mass (CoM) in SPSS is conducted.

3 Results and Discussion

Due to current developments in safety measures concerning Covid-19, experiments were performed in separated observation spaces with a limited amount of six participants. Difference of results between participants with and without experience with overhead tasks, experience with prismatic glasses and medical history is noticed.

For a more accurate sample, the group of participants should exist of heavy load workers with decent spread in anthropometric values. The use of static non-computerized posturography test as the Thumbs test [18] and Fukuda's test [18, 19] can be used as valid fundament to compare with alteration in posture rendered by the performed tasks.

A more detailed medical history of the participants needs to be considered in future experiments. Especially for tests with prismatic glasses, vision impairment is an important differentiator. As the acquired IMU mocap system [11] records segment and sensor velocity and acceleration, forthcoming studies need to integrate gait speed of participants [20].

In order to measure the gait speed in the experiment's participants should be required to familiarize with the tasks before recording [21]. This way performance is less affected by a participant being unfamiliar with the prismatic glasses.

Analyzing the prismatic glasses helps with understanding the effect of vision impairment on the stability of the wearer of prismatic glasses. Understanding the difference in usability of the prismatic glasses to regular glasses, can influence the optimalisation of the protocol [22].

In order to counter interference between recording session with the IMU [23], the session should be limited to a specific task. After the performed task, the observer will stop the recording and the participant will return the starting point. From this point the recording will start again and the participant will perform the following task. To further optimize data analysis, the protocol should include an evaluation of sensor coordinates before analyzing the recorded data. Comparing the recordings from the IMU [11] with computerized posturography plate, as the adaptation test (ADT) could improve the statistical data [18, 22, 24].

4 Conclusions

This paper presents constraints encountered while performing a preliminary study to estimate the impact of wearing of prismatic glasses. In order to measure the impact of the prismatic glasses on the stability of the user, the current experiment protocol needs to be adjusted using the validated posturography evaluation. Future work includes: 1) a verification of the assessment of the ABC and POMA questionnaire in comparison with validated posturography evaluation; 2) the introduction and the comparison of posturography parameters using a computerized posturography plate and the body's kinematics obtained by using the IMU wearable mocap system during the different tasks; and 3) a more representative and accurate study involving overhead workers.

Acknowledgement. This research was supported by Cepa, OCHA and the University of Antwerp. This paper was written as preliminary investigation in applied posturography of heavy load workers within a broader PhD study concerning the implementation of PSS in maritime transportation inducing injury prevention with its dockworkers.

References

1. Shin, S., Yoo, W.: Effects of two workstation positions for below-knee assembly work on upper extremity muscle activity. J. Phys. Ther. Sci. **27**, 5–6 (2015)
2. Tamene, A., Mulugeta, H., Ashenafi, T., Thygerson, S.M.: Musculoskeletal disorders and associated factors among vehicle repair workers in Hawassa City. South. Ethiopia. J. Environ Public Health **2020**, 1–11 (2020)
3. Van Der Molen, H.F., Foresti, C., Daams, J.G., Frings-, M.H.W., Kuijer, P.P.F.M.: Work-related risk factors for specific shoulder disorders: asystematic review and meta-analysis. Occup. Environ. Med. **74**, 745–755 (2017)
4. More Uses For Belay Glasses. http://www.streetarticles.com/climbing/more-uses-for-belay-glasses
5. Frost, P., et al.: Risk of shoulder tendinitis in relation to shoulder loads in monotonous repetitive work. Am. J. Ind. Med. **41**, 11–18 (2002)
6. Shin, S.J., Yoo, W.G.: Effects of overhead work involving different heights and distances on neck and shoulder muscle activity. Work **51**, 321–326 (2015)
7. Belayer's Neck – The Climbing Doctor. https://theclimbingdoctor.com/portfolio-items/belayers-neck/
8. Schmid, A., Eberhardt, U., Oberhoffer, R., Kern, C.: Cervical strains while belaying - the influence of different types of belaying on the mobility of the cervical strain. Sport **29**, 229–236 (2013)
9. Simond. https://www.simond.com/belayer-glasses-belayer-navy-blue-id_8495297
10. Y&Y Belay Glasses. http://www.rannveigaamodt.com/yy-belay-glasses-review/
11. MTw Awinda. https://www.xsens.com/products/mtw-awinda
12. Powell, L.E., Myers, A.M.: The activities-specific balance confidence (ABC) scale. J. Gerontol. **50A**, 28–34 (1995)
13. Mitchell, K.D., Newton, R.A.: Performance-oriented mobility assessment (POMA) balance score indicates need for assistive device. Disabil. Rehabil. Assist. Technol. **1**, 183–189 (2006)
14. Fuschillo, V.L., Bagalà, F., Chiari, L., Cappello, A.: Accelerometry-based prediction of movement dynamics for balance monitoring. Med. Biol. Eng. Comput. **50**, 925–936 (2012). https://doi.org/10.1007/s11517-012-0940-6
15. Xsens BASE. https://base.xsens.com/hc/en-us/community/posts/360009377919-How-does-the-MVN-system-estimates-the-center-of-mass-
16. Kear, B.M., Guck, T.P., McGaha, A.L.: Timed up and go (TUG) test: normative reference values for ages 20 to 59 years and relationships with physical and mental health risk factors. J. Prim. Care Community Health **8**, 9–13 (2017)
17. Nilsagård, Y., Carling, A., Forsberg, A.: Activities-specific balance confidence in people with multiple sclerosis. Multiple Scler. Int. **2012**, 1–8 (2012)
18. Scataglini, S.: Posturography. In: Scataglini, S., Paul, G. (eds.) DHM and Posturography, pp. 351–364. Academic Press, London (2019)
19. Fukuda, T.: The stepping test: two phases of the labyrinthine reflex. Acta Otolaryngol **50**, 95–108 (1959)
20. Yang, Y.R., Lee, Y.Y., Cheng, S.J., Lin, P.Y., Wang, R.Y.: Relationships between gait and dynamic balance in early Parkinson's disease. Gait Posture **27**, 611–615 (2008)

21. Rigby, B.R., Ray, C.T.: Measurement of gait and postural control in aging. In: Gatchel, R.J., Schultz, I.Z., Ray, C.T. (eds.) Handbook of Rehabilitation in Older Adults. HHWD, pp. 85–121. Springer, Cham (2018). https://doi.org/10.1007/978-3-030-03916-5_5
22. Olchowik, G., Czwalik, A., Kowalczyk, B.: The changes in postural stability of women in early old age. J. Nutr. Health Aging **24**(7), 739–744 (2020). https://doi.org/10.1007/s12603-020-1399-z
23. Pavei, G., Salis, F., Cereatti, A., Bergamini, E.: Body center of mass trajectory and mechanical energy using inertial sensors: a feasible stride? Gait Posture **80**, 199–205 (2020)
24. Fjeldstad, C., Pardo, G., Frederiksen, C., Bemben, D., Bemben, M.: Assessment of postural balance in multiple sclerosis. Int. J. MS Care **11**, 1–5 (2009)

Automatic Printed Fabric Defect Detection Based on Image Classification Using Modified VGG Network

Samit Chakraborty(✉), Marguerite Moore, and Lisa Parrillo-Chapman

Department of Textile and Apparel, Technology and Management,
North Carolina State University, NC 27695 Raleigh, USA
{schakr22,mmmoore4,llparril}@ncsu.edu

Abstract. Fabric defect detection is a critical operation in textile manufacturing. The inaccuracy in manual fabric inspection system may cause high fabric wastage. Therefore, deep learning algorithm-based automatic fabric defect detection (ADD) method is expected to gradually replace manual inspection method. However, the existing literature on ADD shows scarcity in the number of research conducted on printed fabric defect detection. A recent research on ADD of printed fabric showed that Virtual Geometric Group-16 (VGG16) network showed better performance in defect detection than that of a customized convolutional neural network (CNN) [1]. Our research proposes a modified structure of VGG16 and uses an extended printed fabric dataset to identify color spots and misprints on the printed fabric. The results show that proposed methodology enables the model to achieve higher accuracy (76%) than that of the accuracy (62.28%) achieved in the previous research [1]].

Keywords: Automatic defect detection (ADD) · Printed fabric · Convolutional neural network (CNN) · VGG16 · Defect detection accuracy

1 Introduction

During the textile manufacturing process, the defects during the printing process are common. Quality inspection and defect detection can decrease the number of occurrences of producing defective fabric and hence reduce fabric wastage. However, traditional inspection methods are time consuming and inefficient, which have the potential to cause disruption to the whole production line [2]. To increase the efficiency of defect detection technique, several research projects have been conducted over the last decade, which are primarily based on statistical, spectrum, model-based, and learning-based analysis [3–5]. Although, some of these techniques can achieve high accuracy, they may have poor generalizability. Deep learning algorithms such as convolutional neural network (CNN) has been widely studied since the advancement of computing hardware and learning based approaches. For instance, [6] proposed a CNN model for feature extraction and classification defects including slub, holes, oil spots, and knots. Researchers developed a real-time ADD system to identify defects in knit fabric such as needle mark and fabric holes [7]. The research conducted by [8] presented a convolutional Autoencoder network to detect woven fabric defects including oil spots, holes, fly yarn and

J. L. Wright et al. (Eds.): AHFE 2021, LNNS 264, pp. 384–393, 2021.
https://doi.org/10.1007/978-3-030-79763-8_46

perforated fabric surface. Researchers also used CNN for feature extraction and classification of fabric defects including end or pick missing, fabric holes and double yarns [9]. Additionally, researchers verified and reported that a deep CNN-based ADD offer superior performance compared traditional model-based and learning-based approaches [2, 10–12]. These research shows that CNN can provide outstanding performance in feature extraction and classification of fabric defects. However, previous research rarely explored printed fabric defect detection using CNN except Chakraborty et al. [1]. As a part of the state-of-the-art research approach this research explores further avenues of improving ADD model proposed by [1].

Existing studies related to computer vision and ADD recommends that the expansion of image dataset, adding more pooling layers and dropout layers to the network, applying batch normalization and utilizing proper optimizer can substantially increase accuracy of a CNN model [2, 12–16]. Researchers suggest expanding small dataset through image augmentation to help the model to learn various image features more closely, and achieve higher accuracy in image classification and defect detection [13, 14, 16]. Image augmentation is performed to manipulate images through rotation at various angles, flipping horizontally & vertically, moving upside down, adding random noise, cropping special areas and performing logarithmic correction of input images [3, 13, 17, 18]. Pooling is an important operation used in constructing CNN architecture. Max pooling is most commonly used to down sample the feature map extracted from a convoluted image. The image information collected from this operation is then transmitted to the next layer. Unlike max pooling, the global pooling reduces entire feature map it to a single value instead of reducing sampling patches of the feature map [19]. The addition of a global pooling layer before a fully connected layer in the transition from feature maps to an output prediction for the model gives the image features a global attention and reduces the computation of predicted values for the fully connected layer [2, 19]. The addition of dropout layers also helps to improve accuracy by dropping unwarranted connections from the network [13]. Batch normalization technique is used to train deep CNN by standardizing and distributing the inputs to a layer for each mini batch [20]. Thus, it enables a faster and more steady training of CNN [21]. With batch normalization it is possible to converge faster to the local minima and generalize output with less data [22]. Optimizers are also crucial for CNN implementation. Optimizers are algorithms, used to modify the weights and learning rates of a neural network, which helps to reach the local minima. Hence it assists to provide the most precise results by reducing the loss or error [14, 23]. Researchers reported adaptive moment estimation optimizer (Adam), Nesterov and Adam optimizer (Nadam) and Nesterov accelerated gradient (NAG) as the best optimizers compared to root mean square propagation (RMSProp); maximum adaptive moment estimation optimizer (AdaMax); and the adaptive gradient (AdaGrad) optimizers for image classification [14]. Hence, this paper considers these factors while addressing the following contributions.

1. Adding more images into the dataset by image augmentation method to increase the effectiveness of automatic fabric defect detection model in a complex rotary screen-printing environment.
2. Adding a global average pooling layer and dropout layers to the VGG16 network proposed [1], performs batch normalization on it and uses Adam, Nadam and NAG

optimizers to compare the performance of the generated models in terms of printed fabric defect detection accuracy.

2 Method

2.1 Database Development

Firstly, a library was developed with 400 random images of rotary screen-printed fabric, where the images were collected from eight knitwear-manufacturing industries in Bangladesh. The library consists of defective print fabric images with color spots and misprints as well as non-defective print fabric images. The images were resized to 512 × 512 pixels. Primarily, the entire dataset was divided into ratio of 50% (for training, n = 200), 25%(for validation, n = 100) and 25% (for testing, n = 100). The subsets were selected randomly. Table 1 shows distribution of images for training, validation and testing dataset.

Table 1. Image distribution across different dataset

Dataset	Image proportion for each dataset		
	Non-defective	Color spots	Misprints
Training	120	50	30
Validation	35	45	20
Testing	35	45	20

Then, image augmentation (by a factor of five) was performed to expand the training dataset 200 images to 1000 images, whereas [1] employed an image augmentation (by a factor of four) to expand the training dataset to 800 images. Image augmentation was carried out using four operations: image rotation, flipping upside down, adding random noise and logarithmic correction of images to generate additional 1000 image for expanding the training dataset. The images were labeled as: 0 for non-defective, 1 for color spot and 2 for misprint. Figure 1 presents print fabric images with (a) color spot and (b) misprint

(a) (b)

Fig. 1. Defective fabric images (a) Colorspot; (b) misprint

3 Model Development

3.1 Convolutional Neural Network (CNN)

The convolutional neural network (CNN) is an effective technique used for feature extraction and image classification [13, 24]. CNN architecture comprises of input, output and multiple hidden layers comprising of nodes, where the input layer contains image data. The hidden convolutional and pooling layers can be of different numbers and sizes [24, 25]. Moreover, the weight and bias are applied on neuron of a convolutional layer [13, 26]. The fully connected input layer receives the output from adjoining layers and flattens them into a single vector to represent a specific feature that belongs to a class. The output layer displays final probabilities or class scores or arbitrary real-valued numbers for image classification [13, 24, 25].

3.2 Architecture of the Modified VGG Network

To address the second contribution, a multi-layer convolution neural network (CNN) was employed to extract features and identify defective & non-defective fabric images. The hyper parameters selected to run model were learning rate, batch size, hidden layers, dropout probability, regularization and activation function. These were selected based on previous research and empirical [13, 15, 27–32]. A simple iteration of 20 epochs was employed sequentially on each hyperparameters to search for a value that gives the highest validation accuracy. The hyperparameters and their respective values were selected from the research by [1] (Table 2). Although [1] concluded that VGG16 model with Softplus activation function yielded highest validation accuracy, our research utilized all of the four activation functions (i.e., Relu, Tanh, Sigmoid, Softplus) to assess the defect detection accuracy.

Table 2. Hyperparameters and their values selected for model training

Hyperparameters	Tested range
Learning rate	0.0003
Batch size	16
Number of hidden neurons	256
Dropout probability	0.2
Regularization	0.001
Activation function	Relu, Tanh, Sigmoid, Softplus

In this research, global average pooling layer and a dropout layers was added before the fully connected layer to get more precise feature map as input to the next layer. Batch normalization was used after global average pooling, dropout and dense layers to improve the feature extraction performance (Fig. 2).

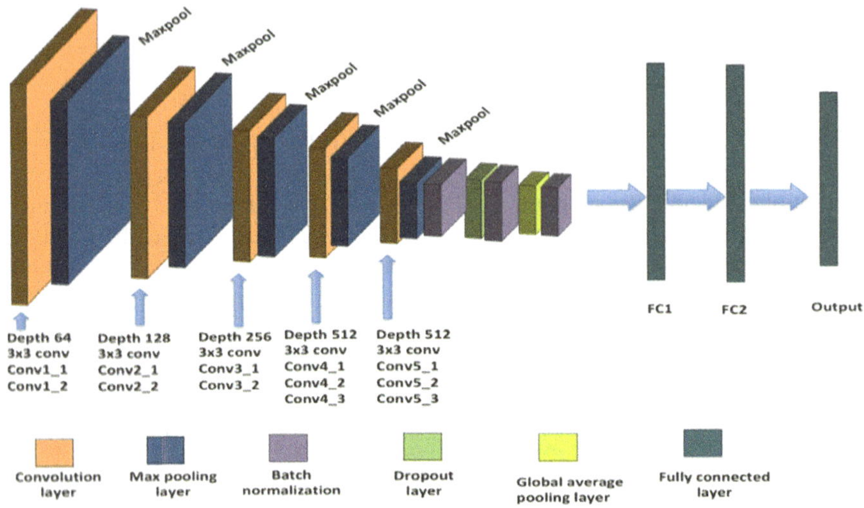

Fig. 2. Structure of modified VGG16 network

The study employed the Adam, Nadam and NAG optimizers for the loss function because of their popularity and effectiveness in image classification [14]. The model with best optimizer was selected based on the highest recall, precision and test accuracy. The evaluation metrics including recall, precision and test accuracy were based on true positive (TP), false positive (FP), true negative (TN) and false negative (FN) values, as shown in the following equations [5].

$$\text{Recall} = \frac{\text{TP}}{\text{TP} + \text{FN}} \tag{1}$$

$$\text{Precision} = \frac{\text{TP}}{\text{TP} + \text{FP}} \tag{2}$$

$$\text{Defect Detection Accuracy} = \frac{\text{TP}}{\text{TP} + \text{FP} + \text{TN} + \text{FN}} \tag{3}$$

4 Results and Discussion

After training the model with each activation function, the best-trained model was selected based on the highest validation accuracy. This operation was repeated three times by replacing the optimizer (i.e., Adam, Nadam, NAG) each time. Figure 3a show that for VGG16 model with Adam optimizer, ReLU outperform the other activation functions in terms of validation accuracy and training accuracy. The VGG16 model with ReLU achieved about 87% validation accuracy, which was higher than that of models developed with Softplus, Sigmoid and Tanh. Figure 3b shows the training and validation loss and accuracy plots found based on model trained with ReLU. The confusion matrix generated from the VGG16 with Adam optimizer indicates that model accuracy of 64% for detecting misprints and 77% for detecting colour spots (Fig. 4).

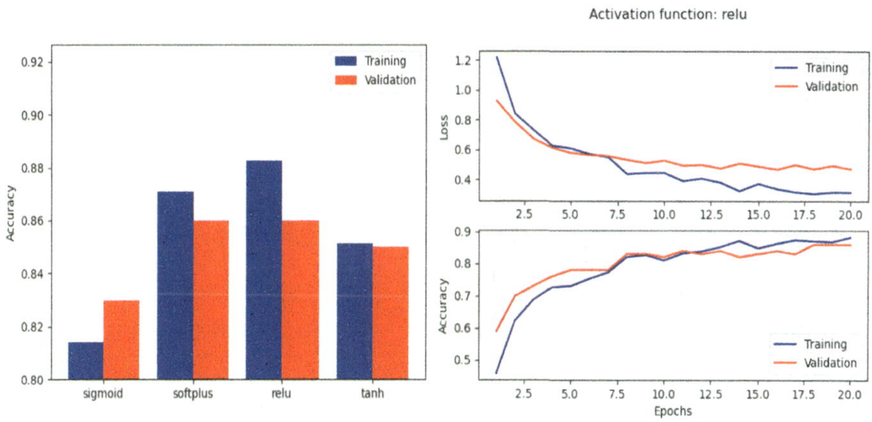

(a) (b)

Fig. 3. (a) Training and validation accuracy of various activation functions for VGG16 with Adam optimizer, (b) Plots of loss and accuracy vs. epochs for the best model (with ReLU)

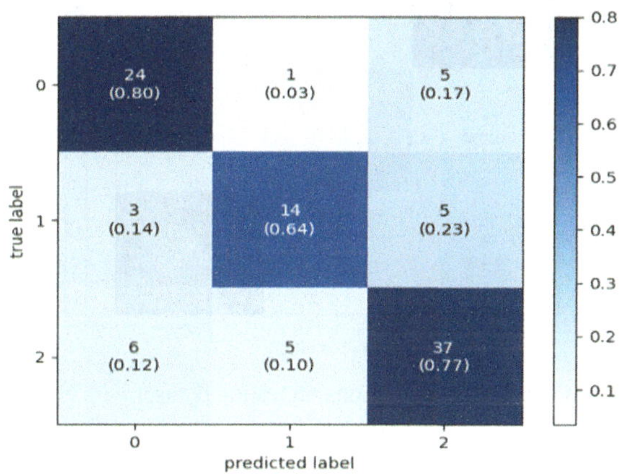

Fig. 4. Confusion matrix of predictions on testing dataset with Adam and ReLU

Similarly, for VGG16 with Nadam optimizer ReLU performs better than other activation functions in terms of validation accuracy and training accuracy (Fig. 5a). Figure 5b shows the training and validation loss, and accuracy plots found based on model trained with ReLU. The VGG16 model with ReLU achieved about 84% validation accuracy, which was higher than that of models developed with Softplus, Sigmoid and Tanh. The VGG16 model for Nadam optimizer with ReLU achieved 59% accuracy for detecting misprints and 75% for detecting colour spots (Fig. 6).

Lastly, for VGG16 with NAG optimizer ReLU outperform the other activation functions in terms of validation accuracy and training accuracy (Fig. 7a). Figure 7b shows the training and validation loss, and accuracy plots found based on model trained with ReLU. The VGG16 model with ReLU achieved about 83% validation accuracy, which

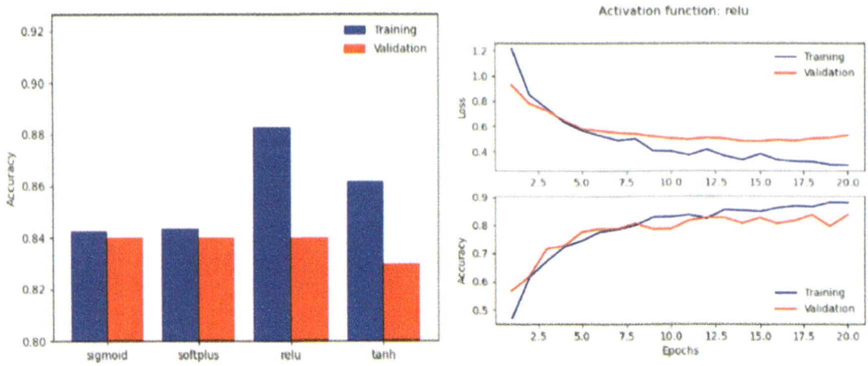

(a) (b)

Fig. 5. (a) Training and validation accuracy of various activation functions for VGG16 with Nadam optimizer, (b) Plots of loss and accuracy vs. epochs for the best model (with ReLU)

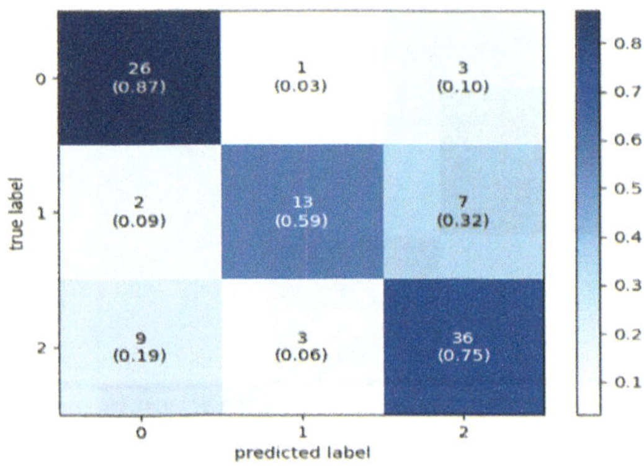

Fig. 6. Confusion matrix of predictions on testing dataset with Nadam and ReLU

was higher than that of models developed with Softplus, Sigmoid and Tanh. The VGG16 model with ReLU achieved 55% accuracy for detecting misprints and 71% for detecting colour spots (Fig. 8).

The VGG-16 models trained with Adam, Nadam and NAG optimizes were compared to determine the final model for automatic printed fabric defect detection (Table 3). The table indicated highest recall (0.75), precision (0.74) and test accuracy (76%) compared to both models with Nadam and Adam optimizers.

5 Conclusion

Automatic printed fabric printing is particularly challenging for manufacturers because it involves numerous colors and complicated designs. This research presents an extensive analysis of developing a modified VGG16 based CNN model combined with newly

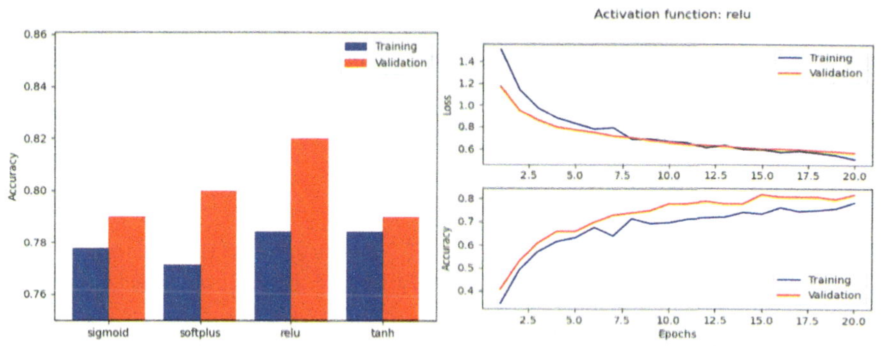

(a) (b)

Fig. 7. (a) Training and validation accuracy of various activation functions for VGG16 with NAG optimizer, (b) Plots of loss and accuracy vs. epochs for the best model (with ReLU)

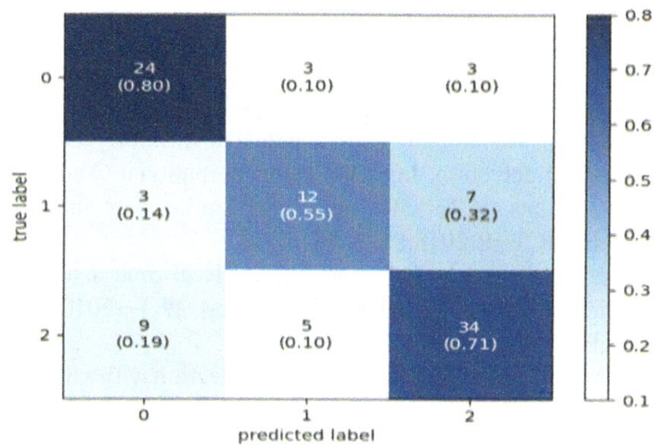

Fig. 8. Confusion matrix of predictions on testing dataset with NAG and ReLU

Table 3. Comparison among various network architectures

Network architectures	Precision	Recall	Test accuracy
Modified VGG16 with Adam	0.75	0.74	76%
Modified VGG16 with Nadam	0.75	0.73	75%
Modified VGG16 with NAG	0.70	0.70	70.56%
Conventional VGG16 with Adam [1]	0.65	0.63	62.28%

added dropout and global average pooling layers. Additionally, five data augmentation schemes, batch normalization operation and three effective optimizers are used to improve the performance of ADD of printed fabrics. The experimental results show that the optimized VGG16 model with Adam optimizer higher recall, precision and defect detection accuracy compared to conventional VGG16 model. The proposed methodology can significantly improve the image classification ability of a CNN. Future research can propose newly developed loss function and different values of hyperparameters to assess the improvement in model performance.

References

1. Chakraborty, S., Moore, M., Parrillo-Chapman, L.: Automatic Defect Detection of Print Fabric Using Convolutional Neural Network, arXiv:2101.00703, Jan 2021. Accessed 16 Jan 2021. http://arxiv.org/abs/2101.00703
2. Zhu, Z., Han, G., Jia, G., Shu, L.: Modified DenseNet for automatic fabric defect detection with edge computing for minimizing latency. IEEE Internet Things J. **7**(10), 9623–9636 (2020). https://doi.org/10.1109/JIOT.2020.2983050
3. Eldessouki, M.: Computer vision and its application in detecting fabric defects. In: Applications of Computer Vision in Fashion and Textiles, pp. 61–101. Elsevier (2018)
4. Hanbay, K., Talu, M.F., Özgüven, Ö.F.: Fabric defect detection systems and methods—a systematic literature review. Optik **127**(24), 11960–11973 (2016). https://doi.org/10.1016/j.ijleo.2016.09.110
5. Ngan, H.Y.T., Pang, G.K.H., Yung, N.H.C.: Automated fabric defect detection—a review. Image Vis. Comput. **29**(7), 442–458 (2011). https://doi.org/10.1016/j.imavis.2011.02.002
6. Bandara, P., Bandara, T., Ranatunga, T., Vimarshana, V., Sooriyaarachchi, T., Silva, C.D.: Automated fabric defect detection. In: 2018 18th International Conference on Advances in ICT for Emerging Regions (ICTer), Colombo, Sri Lanka, September 2018, pp. 119–125. https://doi.org/10.1109/ICTER.2018.8615491.
7. Hanbay, K., Talu, M.F., Özgüven, Ö.F., Öztürk, D.: Real-time detection of knitting fabric defects using shearlet transform, Tekstil ve Konfeksiyon, **29**(1) (2019). Art. no. 1. https://doi.org/10.32710/tekstilvekonfeksiyon.482888
8. Mei, S., Wang, Y., Wen, G.: Automatic fabric defect detection with a multi-scale convolutional denoising autoencoder network model. Sensors **18**(4), 1064 (2018). https://doi.org/10.3390/s18041064
9. . Guan, M., Zhong, Z., Rui, Y.: Automatic defect segmentation for plain woven fabric images. In: 2019 International Conference on Communications, Information System and Computer Engineering (CISCE), Haikou, China, July 2019, pp. 465–468. https://doi.org/10.1109/CISCE.2019.00108.
10. Calik, R.C., Demirci, M.F.: Cifar-10 image classification with convolutional neural networks for embedded systems. In: 2018 IEEE/ACS 15th International Conference on Computer Systems and Applications (AICCSA), Aqaba, October 2018, pp. 1–2. https://doi.org/10.1109/AICCSA.2018.8612873
11. Czimmermann, T., et al.: Visual-based defect detection and classification approaches for industrial applications—a survey. Sensors **20**(5), 1459 (2020). https://doi.org/10.3390/s20051459
12. Li, Y., Huang, H., Xie, Q., Yao, L., Chen, Q.: Research on a surface defect detection algorithm based on mobilenet-SSD. Appl. Sci. **8**(9), 1678 (2018). https://doi.org/10.3390/app8091678
13. Goodfellow, I., Bengio, Y., Courville, A.: Deep Learning. The MIT Press, Cambridge (2016)
14. Kandel, I., Castelli, M., Popovič, A.: Comparative study of first order optimizers for image classification using convolutional neural networks on histopathology images. J. Imaging **6**(9), 92 (2020). https://doi.org/10.3390/jimaging6090092
15. Luo, P., Wang, X., Shao, W., Peng, Z.: Towards Understanding Regularization in Batch Normalization. arXiv:1809.00846, April 2019. Accessed 1 Nov 2020. http://arxiv.org/abs/1809.00846

16. Szeliski, R.: Computer vision: Algorithms and applications. Springer, London (2011)
17. Ballard, D.H., Brown, C.M.: Computer Vision. Prentice-Hall, Englewood Cliffs (1982)
18. Kumar, G., Bhatia, P.K.: A detailed review of feature extraction in image processing systems. In: 2014 Fourth International Conference on Advanced Computing and Communication Technologies, Rohtak, India, February 2014, pp. 5–12. https://doi.org/10.1109/ACCT.2014.74
19. Brownlee, J.: A Gentle Introduction to Pooling Layers for Convolutional Neural Networks. Machine learning mastery (2019). https://machinelearningmastery.com/pooling-layers-for-convolutional-neural-networks/. Accessed 12 Mar 2021
20. Brownlee, J.: A Gentle Introduction to Batch Normalization for Deep Neural Networks. Machine learning mastery (2019). https://machinelearningmastery.com/batch-normalization-for-training-of-deep-neural-networks/. Accessed 12 Mar 2021
21. Santurkar, S., Tsipras, D., Ilyas, A., Madry, A.: How Does Batch Normalization Help Optimization? arXiv:1805.11604, April 2019. Accessed 12 Mar 2021. http://arxiv.org/abs/1805.11604
22. Uppal, S.: Curse of Batch Normalization. Towards data science (2020). https://towardsdatascience.com/curse-of-batch-normalization-8e6dd20bc304. Accessed 12 Mar 2021
23. Doshi, S.: Various Optimization Algorithms For Training Neural Network. Towards data science, 12 March 2021. https://towardsdatascience.com/optimizers-for-training-neural-network-59450d71caf6
24. Gao, C., Zhou, J., Wong, W.K., Gao, T.: Woven fabric defect detection based on convolutional neural network for binary classification. In: Wong, W.K. (ed.) AITA 2018. AISC, vol. 849, pp. 307–313. Springer, Cham (2019). https://doi.org/10.1007/978-3-319-99695-0_37
25. Missinglink, Convolutional Neural Networks Fully Connected Layers in Convolutional Neural Networks: The Complete Guide, missinglink.ai 2020. https://missinglink.ai/guides/convolutional-neural-networks/fully-connected-layers-convolutional-neural-networks-complete-guide/
26. LeCun, Y.: LeNet-5, convolutional neural networks. yann.lecun.com (2020). http://yann.lecun.com/exdb/lenet/. Accessed 31 Oct 2020
27. Albawi, S., Mohammed, T.A., Al-Zawi, S.: Understanding of a convolutional neural network. In: 2017 International Conference on Engineering and Technology (ICET), Antalya, August 2017, pp. 1–6. https://doi.org/10.1109/ICEngTechnol.2017.8308186.
28. Basirat, M., Roth, P.M.: The Quest for the Golden Activation Function. arXiv:1808.00783, August 2018. Accessed 1 Nov 2020. http://arxiv.org/abs/1808.00783
29. Devarakonda, A., Naumov, M., Garland, M.: AdaBatch: Adaptive Batch Sizes for Training Deep Neural Networks. arXiv:1712.02029, February 2018. http://arxiv.org/abs/1712.02029. Accessed 1 Nov 2020
30. Ide, H., Kurita, T.: Improvement of learning for CNN with ReLU activation by sparse regularization. In: 2017 International Joint Conference on Neural Networks (IJCNN), Anchorage, AK, USA, May 2017, pp. 2684–2691.https://doi.org/10.1109/IJCNN.2017.7966185
31. Park, S., Kwak, N.: Analysis on the dropout effect in convolutional neural networks. In: Lai, S.-H., Lepetit, V., Nishino, Ko., Sato, Y. (eds.) ACCV 2016. LNCS, vol. 10112, pp. 189–204. Springer, Cham (2017). https://doi.org/10.1007/978-3-319-54184-6_12
32. Takase, T., Oyama, S., Kurihara, M.: Effective neural network training with adaptive learning rate based on training loss. Neural Netw. **101**, 68–78 (2018). https://doi.org/10.1016/j.neunet.2018.01.016

Author Index

J. L. Wright et al. (Eds.): AHFE 2021, LNNS 264, pp. 395–397, 2021.
https://doi.org/10.1007/978-3-030-79763-8

The manufacturer's authorised representative in the EU is Springer Nature Customer Service Centre GmbH, Europaplatz 3, 69115 Heidelberg, Germany. If you have any concerns regarding our products, please contact ProductSafety@springernature.com

Printed and bound by CPI Group (UK) Ltd, Croydon, CR0 4YY
30/09/2026
02225292-0007